Alcohol Fuels in Advanced Combustion Strategy Mode Diesel Engines

Alcohol Fuels in Advanced Combustion Strategy Mode Diesel Engines investigates the utilization of alcohol-based fuels in diesel engines with specific emphasis on advanced combustion modes, including homogeneous charge compression ignition (HCCI), premixed charge compression ignition (PCCI), partially premixed compression ignition (PPCI), and dual fuel reactivity-controlled compression ignition (RCCI).

Exploring the application of wavelet power spectrum analysis, the book demonstrates its use as a technique for evaluating the stability of combustion in a diesel engine operating in low-temperature combustion (LTC) mode. It identifies the most effective injection timing when using alcohol-diesel blended fuels across different engine loads to achieve the low-temperature PPC strategy.

The book will interest industry professionals in automotive and fuels engineering working with alcohol fuels, advanced combustion mode diesel engines, engine instrumentation, and environmental challenges.

Alcohol Fuels in Advanced Combustion Strategy Mode Diesel Engines

Edited by

Sarbjot Singh Sandhu, Prem Kumar,
and Mandeep Singh

CRC Press
Taylor & Francis Group
Boca Raton London New York

CRC Press is an imprint of the
Taylor & Francis Group, an **informa** business

Designed cover image: Getty Images

First edition published 2026
by CRC Press
2385 NW Executive Center Drive, Suite 320, Boca Raton FL 33431

and by CRC Press
4 Park Square, Milton Park, Abingdon, Oxon, OX14 4RN

CRC Press is an imprint of Taylor & Francis Group, LLC

ISBN: 978-1-032-77288-2 (hbk)
ISBN: 978-1-032-77289-9 (pbk)
ISBN: 978-1-003-48227-7 (ebk)

DOI: 10.1201/9781003482277

Typeset in Times
by KnowledgeWorks Global Ltd.

This book is dedicated to ***my mentors and colleagues****, whose guidance, collaboration, and insights have played a crucial role in shaping this work. Their dedication to advancing knowledge in internal combustion engines and sustainable energy solutions has been invaluable.*

It is also dedicated to ***the researchers, engineers, and students*** *committed to exploring alternative fuels and advanced combustion strategies, striving toward a cleaner and more sustainable future. May this book serve as a valuable resource in pursuing innovation and excellence.*

With deep gratitude and appreciation,

Dr Sarbjot Singh Sandhu

Contents

Preface

The rapid evolution of internal combustion (IC) engines, driven by environmental concerns and the global push toward cleaner energy sources, has led to significant interest in alternative fuels. Alcohol fuels have emerged as promising candidates due to their cleaner combustion characteristics and renewable nature. This book, *Alcohol Fuels in Advanced Combustion Strategy Mode Diesel Engines*, aims to provide an in-depth analysis of the role of alcohol fuels in advanced combustion modes such as reactivity-controlled compression ignition (RCCI), partially premixed compression ignition (PPCI), and homogeneous charge compression ignition (HCCI). Through rigorous experimental investigations, theoretical analyses, and case studies, this book presents a comprehensive overview of the feasibility, benefits, and challenges of integrating alcohol fuels into next-generation IC engines.

This book is intended for researchers, engineers, graduate students, and professionals in engine technology, alternative fuels, and sustainable transportation. Addressing key aspects of fuel properties, combustion behavior, emissions, and performance metrics offers valuable insights into the transition toward a cleaner and more efficient energy future.

Acknowledgments

The editors express their sincere gratitude to the numerous authors worldwide whose timely submissions of high-quality work and their responsiveness to revisions have been instrumental in creating this book. This book comprehensively examines how fuel properties influence combustion within internal combustion engines. It delves into analyzing biofuels (including alcohol fuels and biodiesel) and biogases (such as natural gas and hydrogen), providing key insights into their effects on engine performance and emissions. Several chapters showcase recent research findings, emphasizing current trends in fuel use within the transportation sector, while others present state-of-the-art reviews of clean fuel applications.

Furthermore, this book explores cutting-edge technologies for understanding combustion stability in diesel engines, addressing critical challenges and innovative approaches to improve both engine efficiency and emissions control.

Designed for professionals, researchers, and postgraduate students in fuels, IC engines, engine instrumentation, and environmental research, this book aims to be a valuable resource for advancing knowledge and inspiring further investigation into sustainable fuel technologies and advanced combustion strategies.

Finally, we extend our deepest appreciation to everyone who contributed to this book's realization, including our colleagues, institutions, and research collaborators, for their unwavering support.

Editor Biographies

Sarbjot Singh Sandhu is Associate Professor in the Department of Mechanical Engineering at Dr B R Ambedkar National Institute of Technology, Jalandhar. He graduated in Mechanical Engineering from Guru Nanak Dev Engineering College, Ludhiana and obtained his Ph.D. from Indian Institute of Technology (IIT), Delhi in the area of alternative fuels for IC engines. He has over 20 years of academic experience. Sandhu's research focuses on engine emission control, fuel-efficient engines, alternative fuels for combustion engines, and advanced combustion strategies in CI engines. He has, to his credit, over 40 peer-reviewed published research papers and has co-authored two book chapters. He has supervised 14 M.Tech. projects and 5 doctoral theses and is a tutor of 4 doctoral students. He has successfully completed two external research projects sponsored by the Department of Science and Technology, New Delhi, and the Bureau of Indian Standards, Delhi.

Prem Kumar is Assistant Professor in the Department of Mechanical Engineering at Government Engineering College, Samastipur, under the Science, Technology, and Technical Education Department. He graduated in Mechanical Engineering from Guru Nanak Dev Engineering College, Ludhiana. He obtained his M.Tech. from the Indian Institute of Technology (IIT), Roorkee, in the area of alternative fuels for IC engines. His research focuses on alternative fuels, advanced combustion strategies, and IC engine performance optimization. He has published 12 peer-reviewed research papers, authored a book chapter, and published a utility patent. Mr. Prem Kumar has actively contributed to organizing national and international conferences and has presented his research at several international conferences. In addition to his academic and research contributions, he is engaged in research projects on sustainable energy solutions and continues to explore innovative advancements in Mechanical Engineering and combustion technologies.

Mandeep Singh is a distinguished mechanical engineer, researcher, and academician with extensive experience in alternative fuels, internal combustion engines, and industrial consultancy. Holding a PhD in Mechanical Engineering from Dr B R Ambedkar National Institute of Technology, Jalandhar, he has contributed significantly to sustainable energy solutions, particularly in biodiesel applications. With over a decade of experience spanning academia, industry, and government consultancy, he has worked as Assistant Professor at Lovely Professional University and currently serves as Technical Consultant at the Ministry of Heavy Industries, India. Dr Singh has an impressive research portfolio, featuring multiple publications with renowned publishers such as Elsevier, Springer, and Taylor & Francis. Additionally, he has published a patent on renewable energy-powered poles designed to detect noise pollution.

Editor Biographies

[illegible]

Contributors

Vijayanta Ahuja
DGM, National Automotive Board
India

Vinutha Ashokan
Ministry of Heavy Industries
India

Prabakaran B.
Department of Automobile Engineering
Hindustan Institute of Technology and Science
India

Abhishek Barua
Department of Automobile Engineering
Parala Maharaja Engineering College
India

Akash Deep
Lovely Professional University
India

Gaurav Dwivedi
Department of Mechanical Engineering
Maulana Azad National Institute of Technology
India

Nataraj G.
School of Mechanical Engineering
Vellore Institute of Technology
India

N. V. V. Jyothi
Department of Chemistry
Sri Venkateswara University
India

Sumit Kanchan
Department of Mechanical Engineering
Lovely Professional University
India

K. Keerthi
Department of Chemistry
Sri Venkateswara University
India

Sujeet Kesharvani
Energy Centre
Maulana Azad National Institute of Technology
India

K. Siva Kumar
Department of Chemistry
S.V. Arts Degree & P.G. College
India

Prem Kumar
Department of Mechanical Engineering
Government Engineering College
Samastipur, Bihar, India

Kanchan Kumari
Department of Mechanical Engineering
Parala Maharaja Engineering College
India

E. A. Lohith
Department of Chemistry
Sri Venkateswara University
India

Ananda Murugan M.
School of Mechanical Engineering
Vellore Institute of Technology
India

Dipak Kumar Mandal
Department of Mechanical Engineering
Government Engineering College
India

Tikendra Nath
Energy Centre
Maulana Azad National Institute of Technology
India

Suyasha Pandey
Energy Centre
Maulana Azad National Institute of Technology
India

Swastik Pradhan
School of Mechanical Engineering
Lovely Professional University
India

Manisha Priyadarshini
Department of Materials Engineering
Indian Institute of Science
India

Luiz Alberto Oliveira Rocha
Universidade Federal do Rio Grande do Sul
Brazil

Josimar Souza Rosa
Universidade do Vale do Rio dos Sinos
Brazil

Sarbjot Singh Sandhu
Department of Mechanical Engineering
Dr B R Ambedkar National Institute of Technology
India

Ankit Sharma
Department of Mechanical and Aerospace Engineering
Case Western Reserve University
USA

Mandeep Singh
Ministry of Heavy Industries
India

Satsimran Singh
Department of Mechanical Engineering
Dr. B R Ambedkar National Institute of Technology
India

Giovani Dambros Telli
Universidade de Caxias do Sul
Brazil

Sada Venkateswarlu
Department of Chemistry
Gachon University
Republic of Korea

Prashant Baredar Verma
Energy Centre
Maulana Azad National Institute of Technology
India

Paulo Roberto Wander
Universidade do Vale do Rio dos Sinoa
Brazil

1 Introduction to Application of Alcoholic Fuels in Combustion Engine

Kanchan Kumari, Abhishek Barua,
Swastik Pradhan, and Manisha Priyadarshini

1.1 INTRODUCTION

Alcoholic fuels, commonly referred to as biofuels, are organic compounds derived from renewable resources such as biomass. The most widely used alcoholic fuels include methanol (CH_3OH) and ethanol (C_2H_5OH), though other alcohols like butanol and propanol are also considered. These fuels are gaining popularity due to their potential to reduce greenhouse gas (GHG) emissions, decrease reliance on fossil fuels, and promote energy security. These fuels are characterized by high octane numbers, making them suitable for spark-ignition engines. Additionally, they burn cleaner than gasoline, producing fewer emissions of carbon monoxide, unburned hydrocarbons (HCs), and particulate matter (PM) [1].

The use of alcohol as a fuel dates back to the 19th century. Ethanol was used in internal combustion engines as early as 1826 and gained attention during the late 19th and early 20th centuries due to its availability and clean-burning properties. However, with the discovery and mass production of petroleum, ethanol's use as a fuel declined. The oil crises of the 1970s renewed interest in alternative fuels. Countries like Brazil initiated large-scale ethanol production programs, leading to the development of the Proálcool program in 1975, which significantly expanded ethanol use in transportation. In the United States, the Clean Air Act Amendments of 1990 and the Energy Policy Act of 1992 encouraged the use of cleaner fuels, leading to the increased adoption of ethanol-gasoline blends. The Renewable Fuel Standard (RFS) program, established in 2005 and expanded in 2007, mandated the blending of renewable fuels with gasoline, further boosting ethanol production. Methanol's history as a fuel is intertwined with its use as a chemical feedstock. Interest in methanol as a fuel peaked during the 1980s and 1990s, with research and demonstration projects exploring methanol's potential in transportation. However, challenges related to toxicity, infrastructure, and the rise of ethanol limited its widespread adoption. Butanol's potential as a fuel has been recognized, more recently, with advancements in biotechnology enabling more efficient production methods. Research and development efforts continue to explore its feasibility and benefits as a biofuel [2].

DOI: 10.1201/9781003482277-1

The primary objective of this chapter is to provide a comprehensive introduction to the application of alcoholic fuels in combustion engines. This includes understanding their chemical properties, production methods, historical context, and the current state of their usage in the transportation sector. This chapter explores the chemical properties and combustion characteristics of alcoholic fuels, specifically methanol, ethanol, and butanol. It examines the historical development and evolution of these fuels, highlighting significant milestones and technological advancements. The chapter evaluates the benefits and challenges associated with their use in combustion engines, including environmental, economic, and technical aspects. It also analyzes the impact of policies and regulations on their adoption and development. The chapter also explores future trends and potential innovations in the production and utilization of alcoholic fuels. It aims to inform readers about the potential of alcoholic fuels to contribute to a more sustainable and energy-secure future by providing a comprehensive understanding of these aspects.

1.2 TYPES OF ALCOHOLIC FUELS

Alcoholic fuels are a class of biofuels that are derived from alcohols. They are increasingly considered as alternative fuels for internal combustion engines due to their potential for reducing GHG emissions and enhancing energy security. The main types of alcoholic fuels are methanol, ethanol, and butanol. Each alcoholic fuel has unique advantages and challenges. Methanol and ethanol are more established with broader production methods, while butanol offers higher energy content and potential for improved sustainability. The choice of fuel depends on factors such as feedstock availability, production technology, and specific engine requirements [3].

1.2.1 Methanol

Methanol (CH_3OH) is the simplest alcohol, consisting of one carbon atom, four hydrogen atoms, and one hydroxyl group (–OH). It is a colorless, volatile liquid with a distinctive odor. The key chemical properties of methanol are included in Table 1.1.

TABLE 1.1
Different Properties of Ethanol [4]

Molecular weight	32.04 g/mol
Boiling point	64.7°C
Melting point	–97.6°C
Density	0.7918 g/cm^3 at 20°C
Energy content	15.6 MJ/L (lower than gasoline, which has approximately 32 MJ/L)
Solubility	Completely miscible with water and many organic solvents
Octane number	High, typically around 108, which improves anti-knock properties in engines

Methanol can be produced through various methods [4]:

- *Natural Gas Reforming* – The most common method involves steam reforming of natural gas (methane) to produce syngas (a mixture of hydrogen, carbon monoxide, and carbon dioxide), which is then catalytically converted into methanol.
- *Coal Gasification* – Coal can be gasified to produce syngas, which is subsequently converted to methanol.
- *Biomass Gasification* – Biomass materials like wood, agricultural residues, and waste can be gasified to produce syngas for methanol production.
- *CO_2 Hydrogenation* – Carbon dioxide can be hydrogenated using renewable hydrogen to produce methanol, contributing to carbon recycling.

1.2.2 Ethanol

Ethanol (C_2H_5OH) is a two-carbon alcohol with a hydroxyl group. It is widely used as a fuel and fuel additive. Key chemical properties of ethanol are included in Table 1.2.

Ethanol can be produced via several methods [5]:

- *Fermentation* – The most common method involves the fermentation of sugars derived from crops like corn, sugarcane, and other biomass materials using yeast. The fermentation process converts sugars into ethanol and carbon dioxide.
- *Chemical Synthesis* – Ethanol can also be synthesized from ethylene via the hydration process, where ethylene reacts with water in the presence of a catalyst.
- *Cellulosic Ethanol* – Advanced technologies allow the production of ethanol from lignocellulosic biomass (such as wood, grasses, and agricultural residues) through pretreatment, enzymatic hydrolysis, and fermentation.

TABLE 1.2
Different Properties of Methanol [5]

Molecular weight	46.07 g/mol
Melting point	−114.1°C
Boiling point	78.37°C
Density	0.789 g/cm³ at 20°C
Energy content	21.1 MJ/L (lower than gasoline but higher than methanol)
Solubility	Completely miscible with water and many organic solvents
Octane number	Typically around 108, which provides good anti-knock properties

1.2.3 Butanol

Butanol (C_4H_9OH) is a four-carbon alcohol with several isomers, the most common being n-butanol. It is considered a superior biofuel due to its favorable properties. Key chemical properties of butanol are included in Table 1.3.

Butanol can be produced through several methods [6]:

- *Fermentation* – Traditional Acetone-Butanol-Ethanol (ABE) fermentation uses Clostridium bacteria to ferment sugars from biomass into a mixture of acetone, butanol, and ethanol.
- *Chemical Synthesis* – Butanol can be synthesized through the hydroformylation of propylene followed by hydrogenation.
- *Advanced Fermentation* – Genetic engineering of microbes is being explored to improve yields and selectivity in butanol production from various biomass sources.

1.2.4 Propanol

Propanol (C_3H_7OH), also known as propyl alcohol, exists in two isomeric forms: 1-propanol and 2-propanol (isopropanol). Both isomers have potential as alternative fuels, although 1-propanol is more commonly considered for fuel applications due to its higher energy content and favorable combustion properties (Table 1.4).

Propanol can be produced through two different methods [7]:

- Fermentation:
  - *Biological Process*: Produced by fermenting sugars and other organic materials using specific microorganisms, such as genetically modified bacteria or yeast.
  - *Feedstocks*: Can be derived from various biomass sources, including agricultural residues, energy crops, and waste materials.
- Chemical Synthesis:
  - *Hydroformylation*: This process involves the hydroformylation of ethylene to produce propanal, which is then hydrogenated to form propanol.
  - *Oxidation of Propylene*: Propylene can be oxidized to form propanol.

TABLE 1.3
Different Properties of Butanol [6]

Molecular weight	74.12 g/mol
Boiling point	117.7°C
Melting point	–89.8°C
Density	0.81 g/cm^3 at 20°C
Energy content	29.2 MJ/L (closer to gasoline than ethanol or methanol)
Solubility	Moderately soluble in water, fully miscible with organic solvents
Octane number	Typically around 96 for n-butanol, providing good anti-knock properties

TABLE 1.4
Different Properties of Propanol [7]

Property	1-Propanol	2-Propanol
Molecular formula	C_3H_7OH	C_3H_7OH
Molecular weight	60.10 g/mol	60.10 g/mol
Boiling point	97.2°C (207°F)	82.6°C (180.7°F)
Melting point	−126.5°C	−89.5°C
Density	0.803 g/cm^3 at 20°C	0.785 g/cm^3 at 20°C
Solubility	Completely miscible with water and many organic solvents	Completely miscible with water and many organic solvents
Energy density	24.4 MJ/L	24.4 MJ/L
Octane rating	108	108

1.2.5 Comparison of Alcoholic Fuels

- *Energy Content and Efficiency* – Butanol has the highest energy content among the three, making it more comparable to gasoline. Methanol and ethanol have lower energy contents, requiring larger volumes for the same energy output.
- *Octane Number and Combustion Properties* – All four alcohols have high octane numbers, beneficial for anti-knock properties. Methanol and ethanol have better vaporization characteristics due to lower boiling points.
- *Production Methods and Sustainability* – Methanol and ethanol can be produced from a wide range of feedstocks, including natural gas, coal, and biomass. Ethanol production is more mature, with well-established fermentation processes. Butanol production is advancing with potential for higher sustainability through genetic engineering and improved fermentation techniques.
- *Environmental Impact* – Alcoholic fuels generally produce lower emissions of particulates, NOx, and other pollutants compared to gasoline. Methanol and ethanol can contribute to carbon recycling when produced from renewable sources.

1.3 COMBUSTION CHARACTERISTICS OF ALCOHOLIC FUELS

Alcoholic fuels, such as methanol and ethanol, are considered alternative fuels for internal combustion engines. Their combustion characteristics (see Figure 1.1) differ significantly from traditional HC fuels like gasoline and diesel. The use of alcoholic fuels in internal combustion engines presents a range of benefits and challenges. Their higher flame speed and lower CO_2 and PM emissions make them attractive from an efficiency and environmental standpoint. However, their lower energy density and higher NOx emissions require careful consideration and management. Understanding these combustion characteristics is crucial for optimizing engine performance and achieving a balance between efficiency, power, and emissions [1, 8].

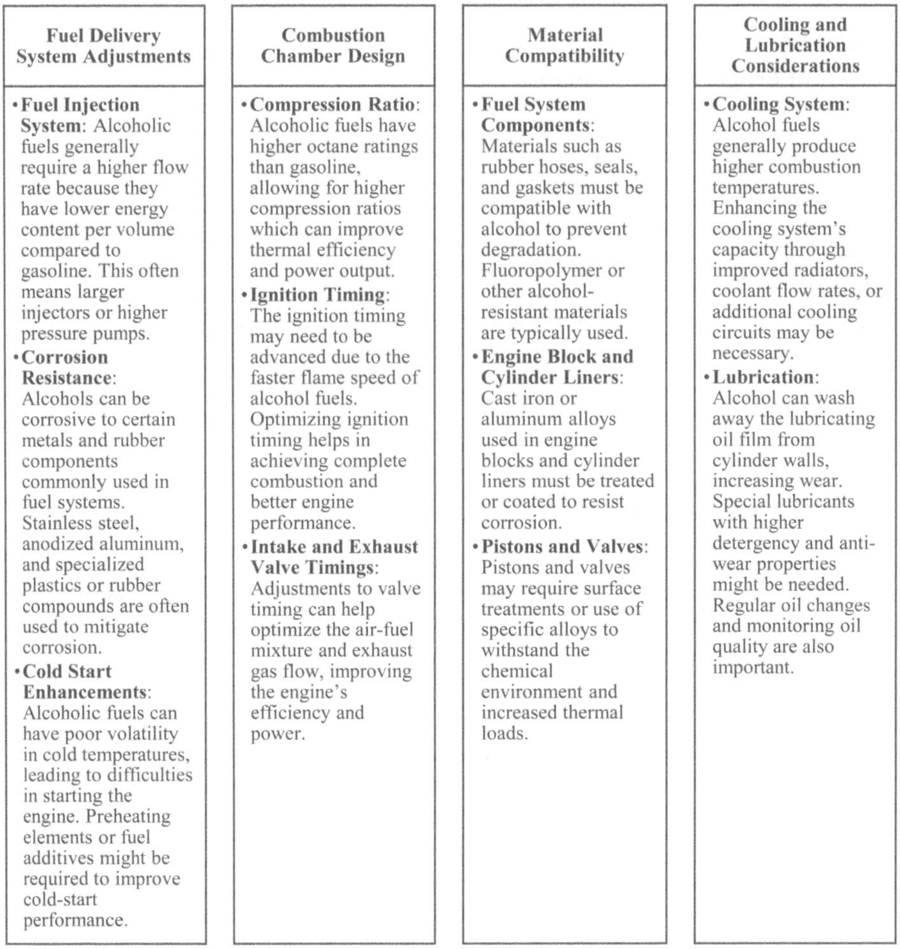

FIGURE 1.1 Different engine modifications.

Alcoholic fuels primarily contain carbon, hydrogen, and oxygen atoms. Methanol (CH_3OH) and ethanol (C_2H_5OH) undergo combustion in the presence of oxygen to produce carbon dioxide (CO_2) and water (H_2O). The balanced chemical reactions for their combustion are the following [9]:

$$CH_3OH + 1.5O_2 \rightarrow CO_2 + 2H_2O \tag{1.1}$$

$$C_2H_5OH + 3O_2 \rightarrow 2CO_2 + 3H_2O \tag{1.2}$$

Alcoholic fuels have lower energy content compared to gasoline, with a lower heating value (LHV) of 27 MJ/kg compared to 44 MJ/kg. This results in more fuel volume needed for the same energy output. Alcoholic fuels generally have higher laminar flame speeds, leading to faster and more efficient combustion. They are less prone to pre-ignition and knocking in high-compression engines due to their higher autoignition temperature of 365°C. Ignition delay is shorter in alcoholic fuels compared to diesel fuel due to their higher volatility and faster evaporation rates. The flame propagation speed in alcoholic fuels is higher due to their oxygen content and molecular structure, leading to more complete combustion and higher engine efficiency. Heat release rate affects engine performance and efficiency, with alcoholic fuels exhibiting a rapid heat release rate due to their high oxygen content and low carbon-to-hydrogen ratio. Their lower energy density means they produce less energy per unit of energy, but their higher flame speed and heat release rate partially compensate for this by improving engine thermal efficiency. Emissions profile of alcoholic fuels includes lower carbon dioxide (CO_2) per unit of energy compared to gasoline, lower carbon monoxide (CO), increased nitrogen oxides (NOx), fewer PM emissions than diesel, and lower unburned HC. However, incomplete combustion can lead to higher HC emissions at lower temperatures.

1.4 ENGINE MODIFICATIONS FOR ALCOHOLIC FUELS

Alcoholic fuels like ethanol and methanol have unique chemical properties compared to conventional gasoline and diesel fuels, necessitating engine modifications to optimize performance and reliability. These modifications include changes to the fuel delivery system, combustion chamber design, material compatibility, and cooling and lubrication systems. These changes enable engines to use alcoholic fuels, such as higher octane ratings and cleaner combustion, while mitigating potential drawbacks like corrosion and increased thermal stress. As technology advances, these modifications continue to evolve, making alcoholic fuels a viable alternative to conventional fuels in internal combustion engines [9–13].

1.5 PERFORMANCE AND EFFICIENCY

In summary, the application of alcoholic fuels in combustion engines presents both challenges and opportunities in terms of performance and efficiency. While alcoholic fuels generally result in higher fuel consumption due to their lower energy density, their high octane rating, better combustion characteristics, and potential for higher thermal efficiency can enhance overall engine performance. Comparative analysis of performance metrics, thermal efficiency, power output, and fuel consumption is essential to fully understand and optimize the use of alcoholic fuels in modern combustion engines [8].

1.5.1 Thermal Efficiency

Thermal efficiency is a critical parameter in assessing the performance of combustion engines using alcoholic fuels. It is defined as the ratio of the useful work obtained from the engine to the total energy input from the fuel. Ethanol and methanol have higher octane ratings than gasoline, allowing for higher compression ratios and improved thermal efficiency. They also have a higher latent heat of vaporization, allowing better cooling of the intake charge and increasing air-fuel mixture density. Alcoholic fuels have faster flame speeds and complete combustion, enhancing thermal efficiency but also having lower energy content per unit volume, which must be considered in overall efficiency analysis.

1.5.2 Power Output

Power output is a measure of the engine's ability to generate mechanical work. The power output of an engine running on alcoholic fuels can be influenced by several factors [14]. Alcoholic fuels have lower energy content per unit volume compared to gasoline, but their high octane rating allows engines to operate at higher compression ratios. They require different air-fuel mixtures for optimal combustion, with ethanol having a 9:1 stoichiometric air-fuel ratio compared to 14.7:1 for gasoline. This can increase power output under certain conditions. Alcoholic fuels can be beneficial in turbocharged or supercharged engines due to their cooling effect, allowing higher boost pressures and increased power output without knocking.

1.5.3 Fuel Consumption

Fuel consumption is a critical metric for assessing the economic and environmental viability of alcoholic fuels in combustion engines [8]. Alcoholic fuels have lower energy content per unit volume, requiring more volume for the same energy output compared to gasoline. However, optimized engines can increase thermal efficiency and power output, reducing fuel consumption. Flex-fuel vehicles, such as E85, which combines ethanol and gasoline, often experience variations in fuel consumption depending on the fuel blend. Optimizing engine parameters for specific fuel blends can help achieve the best balance between performance and fuel economy, reducing the relative increase in fuel consumption.

1.5.4 Comparative Analysis of Performance Metrics

When evaluating the performance of alcoholic fuels in combustion engines, it is crucial to compare various performance metrics against traditional fossil fuels such as gasoline and diesel. Key performance metrics include the following:

- *Power Output*: Measures the engine's ability to convert fuel energy into mechanical work. Alcoholic fuels like ethanol and methanol generally have a higher octane rating, allowing for higher compression ratios and potentially greater power output.
- *Thermal Efficiency*: Indicates the efficiency with which the engine converts fuel energy into useful work. Alcoholic fuels can achieve higher thermal efficiency due to their higher latent heat of vaporization and better knock resistance.
- *Fuel Consumption*: Assesses the amount of fuel required to produce a specific amount of energy or travel a certain distance. While alcoholic fuels

often result in higher fuel consumption due to lower energy density compared to gasoline, their efficiency improvements can offset this disadvantage in some applications.

1.5.5 Comparison of Alcoholic Fuels with Diesel Fuel and Engines

Alcoholic fuels, primarily ethanol and methanol, are being explored as sustainable alternatives to traditional diesel fuel in the automotive sector. Diesel fuel is preferred due to its high energy density and efficiency in compression-ignition (CI) engines. However, concerns over GHG emissions, air pollutants, and dependence on fossil fuels have led to interest in alternative fuels. Alcoholic fuels, derived from biomass or synthetic processes, offer environmental and economic benefits but present unique challenges. Experimental research from countries like India, Brazil, China, and the United States has shown the potential of ethanol and methanol as alternative fuels in CI engines. Success depends on continued technological innovation, supportive government policies, and substantial infrastructure investments [15, 16].

1.5.5.1 Properties of Alcoholic Fuels vs. Diesel

Understanding the fundamental differences between alcoholic fuels and diesel is crucial for assessing their compatibility and performance in diesel engines (see Table 1.5). Diesel fuels have higher energy density, longer driving ranges, and better fuel economy compared to alcoholic fuels. Diesel's high cetane number facilitates efficient autoignition in CI engines, while ethanol and methanol have low cetane numbers, making ignition more challenging and potentially leading to incomplete combustion. Alcoholic fuels have wider flammability limits, leading to more stable and complete combustion under varying conditions, reducing emissions of pollutants like PM and HCs. However, alcoholic fuels have higher latent heat of vaporization, absorbing more heat during vaporization, which may lead to combustion difficulties in cold-start conditions. Diesel's higher viscosity and lubricating properties protect

TABLE 1.5
Different Properties of Fuels [17, 18]

Property	Diesel Fuel	Ethanol (C_2H_5OH)	Methanol (CH_3OH)
Energy density	~45.5 MJ/kg	~30 MJ/kg	~19.7 MJ/kg
Cetane number	~51–55	~8	~12
Viscosity	Higher viscosity (~2–4 cP)	Lower viscosity (~1.2 cP)	Lower viscosity (~0.6 cP)
Boiling point	180–360°C	78.37°C	64.7°C
Vapor pressure	Lower vapor pressure	Higher vapor pressure	Higher vapor pressure
Corrosiveness	Low corrosiveness	High corrosiveness	High corrosiveness
Lubricity	Good lubricating properties	Poor lubricity	Poor lubricity
Emissions profile	Higher PM and NOx, lower CO_2	Lower PM and CO, higher NOx	Lower PM and CO, higher NOx

engine components, while alcoholic fuels' lower viscosity and poor lubricity can lead to increased engine wear and require modifications to fuel delivery systems. Alcoholic fuels generally produce fewer PM and carbon monoxide emissions but may result in higher nitrogen oxides (NOx) emissions.

1.5.5.2 Engine Compatibility and Necessary Modifications

Adapting diesel engines to utilize alcoholic fuels involves addressing several technical challenges due to the inherent differences in fuel properties. Engine designers can improve ignition characteristics by incorporating cetane improvers or modifying injection timing and pressure. Pilot injection strategies can initiate combustion with a small amount of diesel. Fuel delivery systems must be corrosion resistant and use materials like stainless steel or specialized polymers. Higher injection pressures may be necessary for efficient combustion. Lubricity additives can mitigate the poor lubricating properties of alcoholic fuels, while fuel pump modifications can prevent issues related to fuel flow and pressure. Advanced engine control systems can optimize combustion parameters dynamically, accommodating different combustion characteristics. Enhanced cooling systems may be necessary to manage the exothermic nature of alcohol combustion and prevent overheating.

1.5.5.3 Performance Metrics

The performance of diesel engines using ethanol and methanol fuels is evaluated by assessing power output, fuel efficiency, and emission profiles. Diesel engines with ethanol and methanol produce slightly less power compared to pure diesel, with lower thermal efficiency due to different combustion properties. Efficient injection timing and pressure can mitigate efficiency losses. Ethanol and methanol engines have slightly lower fuel efficiency due to lower energy density, which may offset environmental benefits unless efficiency improvements are made. Both ethanol and methanol significantly reduce PM emissions, with methanol achieving the highest reduction. Proper engine tuning can reduce incomplete combustion, while selective catalytic reduction (SCR) can address higher combustion temperatures and increased NOx emissions (see Figure 1.2).

1.5.5.4 Findings from Experimental Research in Different Countries

Numerous studies worldwide have explored the feasibility and performance of alcoholic fuels in diesel engines. Below are key findings from selected countries (Table 1.6).

1.5.5.5 Challenges Identified in Research

Experimental studies across different regions have consistently highlighted several challenges in adopting alcoholic fuels for diesel engines:

- *Low Cetane Number*: Difficulties in achieving reliable ignition and stable combustion, leading to misfires and incomplete combustion.
- *Fuel Economy*: The lower energy density of alcoholic fuels results in increased fuel consumption, potentially negating some environmental benefits unless offset by higher efficiency or alternative energy savings.

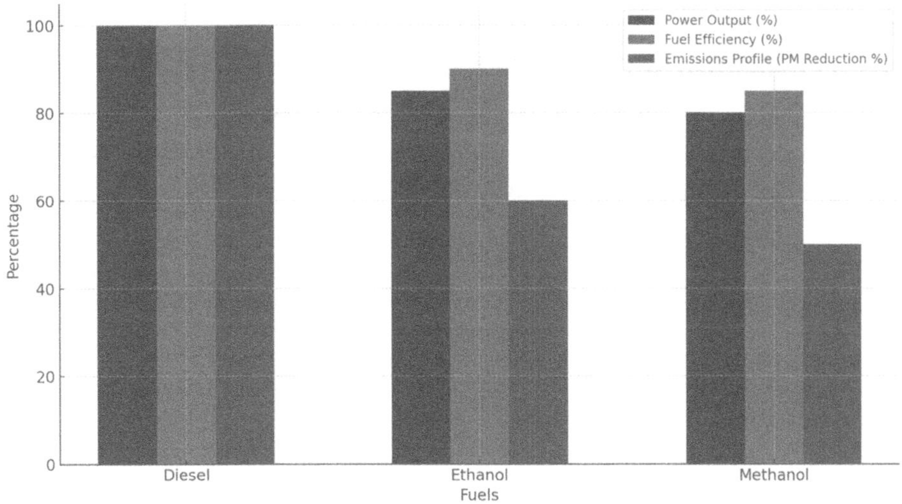

FIGURE 1.2 Performance and efficiency evaluation of diesel, ethanol, and methanol fuels in diesel engines [16].

TABLE 1.6
Studies Conducted by Different Institute across the World [15–18]

Research Institution	Study Focus	Findings
Technische Universität Dresden (TU Dresden), Germany	Dual-fuel systems combining diesel and ethanol.	• *Emissions*: Significant reduction in PM and CO emissions. • *NOx Emissions*: Notable increase, requiring additional after-treatment. • *Engine Performance*: Slight decrease in power output, manageable with optimized injection strategies.
U.S. Department of Energy (DOE), United States	Biodiesel and ethanol blends in CI engines.	• *Emissions*: Improved emissions profiles with reduced PM and CO. • *Fuel Stability*: Issues with phase separation and fuel stability over time. • *Engine Wear*: Increased wear observed due to poor lubricity, mitigated with additives. • *Economic Viability*: Higher costs associated with fuel production and engine modifications.
Beijing Institute of Technology, China	Methanol-diesel blends in heavy-duty engines.	• *Emissions*: Reduced PM and CO emissions, slight increase in NOx. • *Engine Modifications*: Required significant alterations to fuel injection systems. • *Performance*: Acceptable power output with optimized engine tuning. • *Adoption*: Limited to pilot regions due to infrastructure constraints.

(Continued)

TABLE 1.6 *(Continued)*
Studies Conducted by Different Institute across the World [15–18]

Research Institution	Study Focus	Findings
Brazilian Agricultural Research Corporation (Embrapa), Brazil	Ethanol and biodiesel dual-fuel applications.	• *Emissions*: Marked reduction in GHG emissions and PM. • *Fuel Availability*: Leveraged existing ethanol infrastructure from the Proálcool program. • *Engine Performance*: Maintained acceptable performance levels with minor efficiency losses. • *Scalability*: High potential for scaling due to established biofuel industry.
Indian Institute of Technology (IIT) Delhi, India	Ethanol-diesel blends for urban transport.	• *Emissions*: Significant reductions in urban air pollutants. • *Fuel Blending*: Stability issues with higher ethanol concentrations (above E10). • *Economic Factors*: Potential cost savings with domestic ethanol production from agricultural waste. • *Adoption Barriers*: Limited infrastructure and higher initial modification costs.
Universidad Politécnica de Madrid, Spain	Methanol as a supplementary fuel in CI engines.	• *Emissions*: Reduction in PM and CO, increase in NOx emissions. • *Fuel System Modifications*: Enhanced injection systems are required for methanol compatibility. • *Performance*: Improved combustion efficiency with optimized injection timing. • *Future Prospects*: Potential integration with renewable methanol production pathways.
Toyota Central R&D Labs, Japan	Advanced diesel engines with ethanol supplementation.	• *Emissions*: Lowered GHG emissions and improved fuel economy in hybrid systems. • *Engine Control*: Complex engine management systems are necessary for real-time fuel blending. • *Consumer Acceptance*: Positive response in markets with strong environmental regulations. • *Technological Innovations*: Development of flexible-fuel engines capable of handling variable ethanol blends.

- *Engine Wear and Corrosion*: Alcoholic fuels' corrosive nature and poor lubricity can lead to accelerated engine wear and component degradation, necessitating more frequent maintenance and part replacements.
- *Infrastructure Limitations*: Limited availability of alcoholic fuels, especially methanol, hampers widespread adoption. Retrofitting existing fuel stations and distribution networks is capital-intensive.
- *Economic Viability*: Higher production costs for alcoholic fuels and the expenses associated with engine modifications can deter investment and consumer adoption without substantial subsidies or incentives.

- *Fuel Stability and Storage*: Alcoholic fuels can absorb moisture and are prone to phase separation, complicating storage and transportation logistics.

1.5.5.6 Adaptation Strategies for Alcoholic Fuels in Diesel Engines

- *Dual-Fuel Systems*: Dual-fuel operation is a common method for using alcoholic fuels in diesel engines. Alcohol is injected into the intake manifold, while diesel is used as a pilot fuel. This strategy uses alcohol's clean combustion benefits and diesel's high cetane number for ignition stability. Systems can switch between diesel and alcoholic fuels based on engine load and operating conditions.
- *Engine Modifications*: Modifications such as higher compression ratios, optimized fuel injectors, and turbocharging have been employed to improve the combustion characteristics of alcoholic fuels in CI engines. Electronic control units (ECUs) play a crucial role in adjusting fuel injection timing and quantity based on the fuel mixture.
- *Fuel Formulation and Additives*: Incorporating additives to increase the cetane number of alcoholic fuels, improving ignition quality. Adding lubricants to mitigate wear and tear on engine components. Formulating fuels with stabilizers to prevent phase separation and moisture absorption.
- *Advanced Combustion Techniques*: Homogeneous Charge Compression Ignition (HCCI) to achieve more uniform combustion, enhancing efficiency and reducing emissions. Utilizing engines with adjustable compression ratios to optimize combustion for different fuel blends.
- *Policy and Incentive Programs*: Providing financial incentives for the production, distribution, and adoption of alcoholic fuels. Establishing standards for fuel quality and engine compatibility to ensure safe and efficient use of alcoholic fuels. Allocating funds to support ongoing research into overcoming technical barriers and improving fuel formulations.
- *Infrastructure Development*: Investing in the expansion and modification of fuel distribution systems to accommodate alcoholic fuels. Developing refueling infrastructure equipped to handle the unique properties of alcoholic fuels, including corrosion-resistant dispensers and storage tanks.

1.6 ENVIRONMENTAL IMPACT OF ALCOHOLIC FUELS COMPARED TO DIESEL FUEL AND DIESEL ENGINES

The use of alcoholic fuels in combustion engines is gaining attention due to its potential to reduce environmental impact compared to traditional fossil fuels. The application of alcoholic fuels in combustion engines presents a promising pathway to reducing environmental impacts, especially in terms of GHG emissions and air quality improvement. However, the overall benefits depend on the entire lifecycle of the fuels, from production to end-use. A thorough understanding and optimization of each stage in the lifecycle are essential to maximize the environmental advantages of alcoholic fuels [19].

1.6.1 Environmental Concerns with Diesel Engines

Alcoholic fuels, derived from renewable sources, offer environmental benefits but also present new challenges. Comparative analysis of their environmental impacts is crucial for understanding their potential in diesel engines. Diesel fuel, derived from fossil sources, is widely used due to its high energy density, thermal efficiency, and economic feasibility, but its substantial environmental costs contribute to environmental issues.

- *GHG Emissions*: Diesel engines produce a significant amount of carbon dioxide (CO_2), contributing to global climate change.
- *Air Pollutants*: Diesel combustion emits harmful pollutants, such as PM, nitrogen oxides (NOx), sulfur oxides (SOx), and carbon monoxide (CO), contributing to air quality deterioration and health risks.
- *Energy Sustainability*: Diesel is a non-renewable resource, and its continued use accelerates the depletion of fossil reserves.

1.6.2 Greenhouse Gas (GHG) Emissions

Alcoholic fuels, particularly when derived from biomass or renewable sources, have the potential to lower overall GHG emissions when compared to diesel. The extent of the reduction depends on the production method, land use, and lifecycle analysis of the fuels.

a. *Diesel Fuel*: Diesel engines are known for high CO_2 emissions due to the complete combustion of HC chains in the fuel. Diesel combustion in internal combustion engines releases approximately 2.68 kg of CO_2 per liter of diesel burned. The overall carbon footprint of diesel extends beyond combustion, including emissions during extraction, refining, and transportation processes.
b. *Alcoholic Fuels*: Ethanol: Typically derived from renewable sources like sugarcane, corn, or cellulosic biomass, ethanol has a lower lifecycle CO_2 emission footprint than diesel. When burned, ethanol releases 1.5–2 kg of CO_2 per liter, which is significantly lower than diesel. However, its carbon neutrality depends on the sustainability of the production process. Ethanol production through agricultural processes is often considered carbon-neutral since plants absorb CO_2 during photosynthesis. However, emissions from fertilizers, land-use changes, and farming equipment offset some of the CO_2 absorbed. Methanol: Produced from natural gas, coal, or renewable sources (biomass, CO_2 capture), methanol's lifecycle CO_2 emissions can vary widely. Methanol from biomass can be carbon-neutral, while methanol from fossil fuels emits more GHG than ethanol but still less than diesel. Methanol produces 1.37 kg of CO_2 per liter when combusted. However, the key to reducing methanol's carbon footprint lies in its production method—biomethanol from renewable sources or CO_2 capture can significantly reduce its environmental impact.

1.6.3 Air Pollutants

- *PM*: Diesel engines produce significant amounts of PM, contributing to respiratory diseases and cardiovascular problems. Diesel engines without particulate filters emit 10–100 times more PM than gasoline or alcoholic fuels. Alcoholic fuels like ethanol and methanol result in near-zero PM emissions due to their combustion process, reducing carbon-based particles.
- *Nitrogen Oxides (NOx)*: Diesel engines produce high combustion temperatures, leading to NOx emissions. Alcoholic fuels, like ethanol, generally lower NOx emissions due to their higher oxygen content.
- *Carbon Monoxide (CO)*: Diesel engines emit relatively low levels of CO due to high combustion temperatures. Alcoholic fuels, like ethanol, produce lower CO emissions due to their higher oxygen content.
- *Sulfur Oxides (SOx)*: Diesel contains sulfur, leading to acid rain. Alcoholic fuels, like ethanol and methanol, are sulfur-free, reducing SOx emissions in diesel engines.

1.6.4 Environmental Impact of Fuel Production

a. *Diesel Fuel Production*:
  - Diesel production involves crude oil extraction, refining, and transportation, which have significant environmental impacts. Oil extraction can lead to habitat destruction, oil spills, and water contamination.
  - The refining process is energy-intensive, emitting large amounts of GHGs and pollutants. Additionally, the transportation of diesel from refineries to distribution centers further contributes to the carbon footprint.

b. Alcoholic Fuel Production
  - *Ethanol Production*: Ethanol is primarily produced via fermentation of crops like corn, sugarcane, and other biomass. The environmental impacts of ethanol production vary based on the feedstock and farming practices used. Large-scale ethanol production from crops like corn and sugarcane can lead to deforestation, loss of biodiversity, and competition with food production. Land-use changes associated with crop-based ethanol can release stored carbon from soil and vegetation, negating some of its environmental benefits. Ethanol production is water-intensive, particularly in areas where irrigation is required. Fertilizer and pesticide use in farming can lead to water contamination, affecting aquatic ecosystems.
  - *Methanol Production*: Methanol can be produced from a variety of sources, including natural gas, coal, and biomass. Methanol from biomass or renewable electricity has a significantly lower environmental impact compared to fossil-based methanol production. Methanol production from CO_2 capture or renewable electricity (e.g., from wind or solar) offers a pathway to net-zero emissions, making it a promising fuel from an environmental perspective.

1.6.5 End-of-Life and Waste Considerations

- *Diesel*: Diesel engines produce oil waste, chemical residues, and exhaust after-treatment waste (from catalysts, filters), all of which require proper disposal. Improper handling can lead to soil and water contamination. Diesel spills during transportation and storage can have long-lasting effects on soil and water bodies, harming ecosystems and groundwater supplies.
- *Alcoholic Fuels*: Alcoholic fuels, particularly ethanol and methanol, are more biodegradable than diesel. In case of leaks or spills, these fuels break down more rapidly in the environment, reducing the risk of long-term contamination. The combustion of alcoholic fuels does not generate the same hazardous by-products as diesel. Furthermore, their production processes (especially from biomass) generate less waste compared to fossil-based diesel refining.

1.7 ECONOMIC AND LOGISTICAL CONSIDERATIONS

a. Adoption of Alcoholic Fuels in Combustion Engines
    - The integration of alcoholic fuels into the fuel landscape requires strategic management of economic and logistical factors.
    - Production and distribution costs include raw materials, processing, waste management, transport, storage, and blending.
    - Infrastructure requirements include refineries, production plants, feedstock supply chains, pipelines, transport vehicles, fuel stations, and end-user infrastructure [10,11].

b. Alcoholic Fuel Policy Framework Challenges in Adoption
    - Challenges remain in infrastructure, feedstock sustainability, and consumer acceptance.
    - Successful examples from Brazil, the United States, and China demonstrate the potential of alcoholic fuels in reducing emissions and enhancing energy security [12].

c. Ethanol and Methanol Adoption
    - Ethanol and methanol adoption face significant barriers, including limited distribution infrastructure and corrosive properties.
    - The energy sector must adapt equipment to handle larger volumes of these fuels and retrofit gasoline pipelines and refueling stations.

d. Regional Strategies
    - Brazil leads the world with the highest adoption rate of ethanol, while the United States has an adoption rate of about 10% ethanol in its national gasoline pool.
    - Ethanol adoption varies by country, with Sweden and France leading in higher ethanol blends.

1.8 CASE STUDIES AND REAL-WORLD APPLICATIONS

The application of alcoholic fuels in combustion engines has been explored through various case studies and real-world applications. The integration of alcoholic fuels into combustion engines presents both opportunities and challenges. Success stories from Brazil, the United States, and the marine industry demonstrate the potential for significant environmental and economic benefits. However, challenges such as material compatibility, energy density, cold-start issues, and infrastructure development need to be addressed. Comparative studies highlight the advantages in emissions and performance, while also noting the trade-offs in fuel economy and cost. As technology and infrastructure continue to evolve, the application of alcoholic fuels is likely to become more widespread, contributing to a more sustainable future in transportation.

1.8.1 Case Studies and Their Success Stories

i. *Brazil's Ethanol Program*: Brazil's Proálcool program, launched in the 1970s, aimed to reduce the country's dependence on oil imports by promoting the use of ethanol as a fuel. Ethanol, derived from sugarcane, has since become a major fuel source for Brazilian vehicles. Flex-fuel vehicles, which can run on ethanol, gasoline, or a combination of both, dominate the market. This program has significantly reduced Brazil's GHG emissions and has provided a model for other countries [20–22].
ii. *Ethanol-Blended Fuels in the United States*: In the United States, the RFS mandates the blending of renewable fuels like ethanol into gasoline. E10 (10% ethanol, 90% gasoline) and E85 (85% ethanol, 15% gasoline) are common blends. This initiative has supported rural economies, reduced oil imports, and contributed to lower GHG emissions [23].
iii. *Methanol as a Marine Fuel*: Methanol is being used as an alternative marine fuel, providing a cleaner option compared to heavy fuel oil. The Stena Germanica, a ferry operating between Sweden and Germany, successfully converted to methanol fuel, reducing sulfur oxide (SOx), nitrogen oxide (NOx), and PM emissions [23].

1.9 FUTURE PROSPECTS OF ALCOHOLIC FUELS

Alcoholic fuels in combustion engines offer promising future prospects due to their renewable nature and lower carbon emissions. Technological innovations, integration with hybrid and electric technologies, and ongoing research and development efforts are key areas for future advancements. These advancements not only enhance engine performance and efficiency but also contribute to a more sustainable and environmentally friendly transportation sector. Advanced engine design, fuel injection systems, catalyst and exhaust treatment, and hybrid vehicle integration are key areas for optimizing the use of alcoholic fuels [24]. Modern engines can be specifically engineered or adapted to maximize fuel efficiency and reduce emissions. New catalyst technologies and exhaust treatment systems are being developed to handle the specific challenges posed by alcoholic fuels, such as advanced catalytic converters and particulate filters. Integration with hybrid and electric technologies presents opportunities for leveraging the environmental benefits of alcoholic fuels while maintaining the efficiency and extended range provided by hybrid systems. Alcoholic fuels can serve as range extenders for electric vehicles (EVs), addressing range anxiety issues associated with EVs and making them more appealing to a broader audience [25]. Dual-fuel systems that use a combination of alcoholic fuels and conventional fuels can dynamically adjust the fuel mix based on driving conditions, optimizing performance and emissions. Research and development directions focus on understanding the combustion characteristics of alcoholic fuels, material compatibility and durability, renewable production methods, emission control technologies, and integration with digital technologies. These advancements aim to optimize fuel delivery, combustion processes, and emission control systems in real time, leading to smarter and cleaner engines [26].

1.10 CONCLUSION

Alcoholic fuels, primarily ethanol and methanol, have gained significant interest as an alternative fuel for internal combustion engines due to their potential to reduce GHG emissions, improve energy security, and provide economic benefits. These fuels possess unique chemical and physical properties, such as higher octane ratings compared to gasoline, which allows for higher compression ratios and greater thermal efficiency. Alcoholic fuels can significantly reduce CO_2 emissions and other pollutants when compared to conventional fossil fuels, helping mitigate the effects of climate change and reduce the automotive industry's carbon footprint. Engines optimized for alcoholic fuels can achieve higher performance and efficiency levels, with a higher heat of vaporization leading to improved cooling of the intake charge and better engine knock resistance. Economic and energy security can be enhanced by using domestic feedstocks for alcohol production, reducing dependence on imported oil, and producing alcoholic fuels from renewable resources. However, the widespread adoption of alcoholic fuels faces several challenges, including infrastructure development, engine compatibility, and consumer acceptance. There are also concerns about the energy balance and environmental impact of producing alcoholic fuels, particularly from non-renewable resources. The implications for the automotive industry include technological advancements in engine technology, fuel delivery systems, and material compatibility, as well as regulatory and policy support from governments. Government policies and regulations play a crucial role in promoting the adoption of alcoholic fuels, with incentives for renewable fuel production, emissions reduction targets, and support for infrastructure development. Market dynamics will be influenced by the introduction of alcoholic fuels into the market, requiring automotive manufacturers to adapt their product lines to include vehicles capable of running on alcoholic fuels while educating consumers on the benefits and usage of these fuels. Alcoholic fuels offer a promising alternative to traditional fossil fuels for internal combustion engines, but realizing this potential requires concerted efforts in research, policy support, and consumer education. With the right strategies in place, alcoholic fuels can play a significant role in creating a sustainable and efficient transportation system.

REFERENCES

1. Göktaş, M., Balki, M. K., Sayin, C., & Canakci, M. (2021). An evaluation of the use of alcohol fuels in SI engines in terms of performance, emission and combustion characteristics: A review. *Fuel*, *286*, 119425. https://doi.org/10.1016/j.fuel.2020.119425
2. Awad, O. I., Mamat, R., Ali, O. M., Sidik, N., Yusaf, T., Kadirgama, K., & Kettner, M. (2018). Alcohol and ether as alternative fuels in spark ignition engine: A review. *Renewable and Sustainable Energy Reviews*, *82*, 2586–2605. https://doi.org/10.1016/j.rser.2017.09.074

3. Bae, C., & Kim, J. (2017). Alternative fuels for internal combustion engines. *Proceedings of the Combustion Institute, 36*(3), 3389–3413. https://doi.org/10.1016/j.proci.2016.09.009
4. Verhelst, S., Turner, J. W., Sileghem, L., & Vancoillie, J. (2019). Methanol as a fuel for internal combustion engines. *Progress in Energy and Combustion Science, 70*, 43–88. https://doi.org/10.1016/j.pecs.2018.10.001
5. Balat, M., & Balat, H. (2009). Recent trends in global production and utilization of bio-ethanol fuel. *Applied Energy, 86*(11), 2273–2282. https://doi.org/10.1016/j.apenergy.2009.03.015
6. Harvey, B. G., & Meylemans, H. A. (2010). The role of butanol in the development of sustainable fuel technologies. *Journal of Chemical Technology & Biotechnology, 86*(1), 2–9. https://doi.org/10.1002/jctb.2540
7. Qian, Y., Guo, J., Zhang, Y., Tao, W., & Lu, X. (2018). Combustion and emission behavior of n-propanol as partially alternative fuel in a direct injection spark ignition engine. *Applied Thermal Engineering, 144*, 126–136. https://doi.org/10.1016/j.applthermaleng.2018.08.044
8. Balki, M. K., Sayin, C., & Canakci, M. (2014). The effect of different alcohol fuels on the performance, emission and combustion characteristics of a gasoline engine. *Fuel, 115*, 901–906. https://doi.org/10.1016/j.fuel.2012.09.020
9. Agarwal, A. K. (2007). Biofuels (alcohols and biodiesel) applications as fuels for internal combustion engines. *Progress in Energy and Combustion Science, 33*(3), 233–271. https://doi.org/10.1016/j.pecs.2006.08.003
10. Comandini, A., Nativel, D., & Chaumeix, N. (2021). Laminar flame speeds and ignition delay times of Gasoline/Air and Gasoline/Alcohol/Air mixtures: The effects of heavy alcohol compared to light alcohol. *Energy & Fuels, 35*(18), 14913–14923. https://doi.org/10.1021/acs.energyfuels.1c01708
11. Tutak, W., Lukács, K., Szwaja, S., & Bereczky, Á (2015). Alcohol–diesel fuel combustion in the compression ignition engine. *Fuel, 154*, 196–206. https://doi.org/10.1016/j.fuel.2015.03.071
12. Stansfield, P. A., Bisordi, A., OudeNijeweme, D., Williams, J., Gold, M., & Ali, R. (2012). The performance of a modern vehicle on a variety of alcohol-gasoline fuel blends. *SAE International Journal of Fuels and Lubricants, 5*(2). https://doi.org/10.4271/2012-01-1272
13. Surisetty, V. R., Dalai, A. K., & Kozinski, J. (2011). Alcohols as alternative fuels: An overview. *Applied Catalysis a General*. https://doi.org/10.1016/j.apcata.2011.07.021
14. Turner, J. W., Lewis, A. G., Akehurst, S., Brace, C. J., Verhelst, S., Vancoillie, J., Sileghem, L., Leach, F., & Edwards, P. P. (2018). Alcohol fuels for spark-ignition engines: Performance, efficiency and emission effects at mid to high blend rates for binary mixtures and pure components. *Proceedings of the Institution of Mechanical Engineers Part D Journal of Automobile Engineering, 232*(1), 36–56. https://doi.org/10.1177/0954407017752832
15. Campos-Fernández, J., Arnal, J. M., Gómez, J., & Dorado, M. P. (2012). A comparison of performance of higher alcohols/diesel fuel blends in a diesel engine. *Applied Energy, 95*, 267–275. https://doi.org/10.1016/j.apenergy.2012.02.051
16. Geng, P., Cao, E., Tan, Q., & Wei, L. (2017). Effects of alternative fuels on the combustion characteristics and emission products from diesel engines: A review. *Renewable and Sustainable Energy Reviews, 71*, 523–534. https://doi.org/10.1016/j.rser.2016.12.080
17. Çelebi, Y., & Aydın, H. (2019). An overview on the light alcohol fuels in diesel engines. *Fuel, 236*, 890–911. https://doi.org/10.1016/j.fuel.2018.08.138
18. Ning, L., Duan, Q., Chen, Z., Kou, H., Liu, B., Yang, B., & Zeng, K. (2020). A comparative study on the combustion and emissions of a non-road common rail diesel engine fueled with primary alcohol fuels (methanol, ethanol, and n-butanol)/diesel dual fuel. *Fuel, 266*, 117034. https://doi.org/10.1016/j.fuel.2020.117034

19. Erdiwansyah, N., Mamat, R., Sani, M., Sudhakar, K., Kadarohman, A., & Sardjono, R. (2019). An overview of higher alcohol and biodiesel as alternative fuels in engines. *Energy Reports*, *5*, 467–479. https://doi.org/10.1016/j.egyr.2019.04.009
20. Walker, G. M. (2011). 125th anniversary review: Fuel alcohol: Current production and future challenges. *Journal of the Institute of Brewing*, *117*(1), 3–22. https://doi.org/10.1002/j.2050-0416.2011.tb00438.x
21. Alckmin-Governor, G., & Goldemberg-Secretary, J. (2004). Assessment of greenhouse gas emissions in the production and use of fuel ethanol in Brazil. *Government of the State of São Paulo, São Paulo.*
22. Gaffney, J. S., & Marley, N. A. (2009). The impacts of combustion emissions on air quality and climate – From coal to biofuels and beyond. *Atmospheric Environment*, *43*(1), 23–36. https://doi.org/10.1016/j.atmosenv.2008.09.016
23. Abrar, I., Arora, T., & Khandelwal, R. (2023). Bioalcohols as an alternative fuel for transportation: Cradle to grave analysis. *Fuel Processing Technology*, *242*, 107646. https://doi.org/10.1016/j.fuproc.2022.107646
24. Belincanta, J., Alchorne, J. A., & Da Silva, M. T. (2016). The Brazilian experience with ethanol fuel: Aspects of production, use, quality and distribution logistics. *Brazilian Journal of Chemical Engineering*, *33*(4), 1091–1102. https://doi.org/10.1590/0104-6632.20160334s20150088
25. McGill, R. N.. (1989). *Alcohol and blends as motor fuels* (No. ORNL/FTR-3490). Oak Ridge National Lab.(ORNL), Oak Ridge, TN (United States).
26. Yun, Y. (2020). Alcohol Fuels: Current status and future direction. In *IntechOpen eBooks*. https://doi.org/10.5772/intechopen.89788

2 Advancements in Low-Temperature Combustion

Exploring the Efficacy of Alcohol Fuels

Suyasha Pandey, Sujeet Kesharvani, Tikendra Nath Verma, Prashant Baredar, Gaurav Dwivedi, and Ankit Sharma

2.1 INTRODUCTION

Since the inception of the diesel engine, it has played a pivotal role in various sectors such as public transit, hefty machinery, generation of power, agribusiness, and industry, owing to its remarkable attributes of high production power, durability, and thermal and fuel efficiency (Sayed et al., 2021; Zhang, Li, et al., 2021).

Nonetheless, the diesel engine is not without its drawbacks, notably the emission of high levels of soot and nitrogen oxides (NOx) (Singh et al., 2020), posing significant environmental and health concerns. Addressing these challenges has become imperative, driven by the rapid growth of global oil demand (Chen et al., 2019) due to an exponential growth of the global population and increasingly stringent emission regulations, also the outbreak of the COVID-19 virus in 2019 further intensified the focus on energy issues across many nations (Li et al., 2023). However, reliance on traditional non-renewable energy sources has presented significant challenges, including environmental degradation, greenhouse gas emissions, dwindling fossil fuel reserves, and fluctuating fuel prices (Mahdi et al., 2018).

Moreover, the unequal distribution of fuel reserves across various regions aggravates the problems which in turn hinders economic development and makes the global energy prospect look less optimistic (Dey et al., 2021; Yang et al., 2020). Figure 2.1 illustrates the diverse array of fuel consumption to match up cardinal demands of energy over the past five decades and projects future requirements up to 2040 (Nanthagopal et al., 2020).

Because of the rapid increase in global energy consumption, there is a corresponding rise in carbon emissions, exacerbating issues (Zhang, Ye, et al., 2021). Compression ignition (CI) engines, renowned for their fuel efficiency, reliability, and adaptability, have become ubiquitous in various sectors, including transportation,

DOI: 10.1201/9781003482277-2

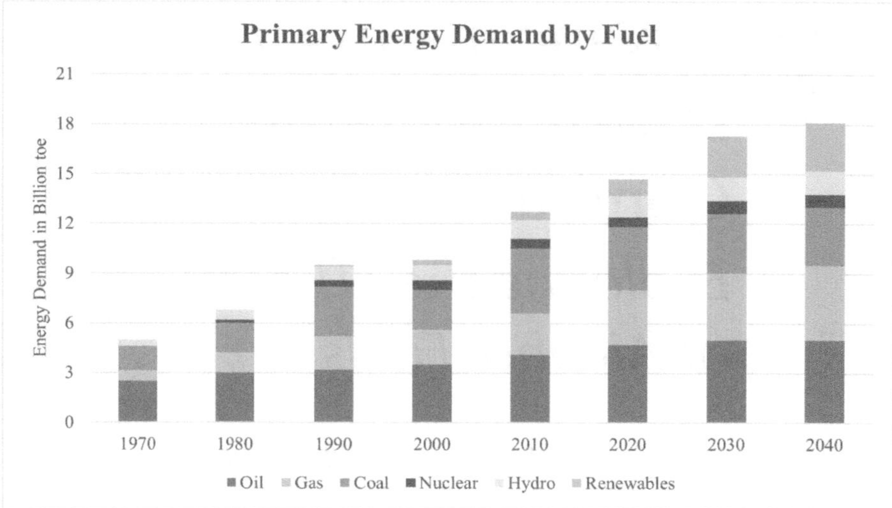

FIGURE 2.1 The rise in demand for cardinal energy. (Nanthagopal et al., 2020.)

agriculture, marine, and power generation (Duan et al., 2017). Among various sectors that consume energy, transportation stands out as a significant contributor to both energy usage and the discharge of carbon dioxide (CO_2) primarily due to the widespread reliance on fossil fuels. It is estimated that the travel industry utilizes almost 20% of global cardinal energy and is accountable for about 23% of CO_2 discharge (Cai & Zhao, 2022).

Moreover, speaking of the urgency to transition to cleaner alternatives is influenced by several key factors, one of them being high energy consumption. Since the transportation sector is one of the largest consumers of energy globally, it includes all modes of transport such as cars, trucks, airplanes, and ships. The sheer volume of energy consumed by these vehicles contributes substantially to overall global energy demand. Second, beyond the discharge of CO_2, transportation sector also emits other pollutants such as nitrogen oxides (NOx) and particulate matter (PM), which have detrimental effects on air quality and public health. The need to reduce these emissions is another driving force behind the transition to cleaner alternatives. Also, the finite nature of fossil fuel resources poses a sustainability issue. On top of that, technological advancements, such as the development of low-temperature combustion (LTC) and the use of alternative fuels like alcohols (methanol, ethanol, n-butanol, and isopropyl alcohol), present viable solutions to reduce the environmental footprint of transportation. These fuels can operate efficiently under LTC conditions, leading to lower NOx emissions, reduced soot formation, and improved overall combustion efficiency thus underscoring the urgency to transition to cleaner alternatives and foster sustainable energy practices.

Currently, internal combustion (IC) engines powered by crude oil-based fuels preside over the transit sector because of their high thermal capacity and versatile province of power. According to data from the World Resources Institute (WRI)

(Cai, Zhao, et al., 2021; Ye et al., 2023; Zameer et al., 2021), traditional energy sources accounted for over 30% of global carbon dioxide (CO_2) emissions between 2003 and 2012 (Nathaniel & Adeleye, 2021), exacerbating the urgency to mitigate greenhouse gas emissions.

However, in response to climate change, global warming, and the need for energy security, many countries aim to achieve "carbon neutrality" by 2050–2070 (Cai, Becker, et al., 2021; Zhang, Li, et al., 2022) necessitating a transition from petroleum-based fuels to cleaner alternatives in the transportation sector. International agreements like the Paris Agreement have been established to curtail greenhouse gas emissions (Gallagher et al., 2019). Nations such as China have pledged commitments, aiming to peak carbon emissions by 2030, aligning with the objectives of the Paris Agreement. The pursuit of sustainable development goals has led to concerted efforts to reduce emissions and explore renewable energy alternatives (Hu et al., 2023).

The global focus on environmental preservation has spurred research into biofuels as promising alternatives (Jiaqiang et al., 2018). Initiatives led by the European Union like the Biofuels Research Advisory Council and the Energy Independence and Security Act in the United States underscore the growing momentum toward increasing biofuel usage (Osman et al., 2021). Biofuels, derived from biomass notably oils derived from vegetables and alcohols, offer the possibility to abate exhaust emissions and supplement hydrocarbon (HC) deposits (Gautam & Kumar, 2020).

However, challenges persist with biofuels, particularly regarding their impact on engine performance and emissions (Viswanathan & Thomai, 2021). While biodiesel shows promise in reducing HC, carbon monoxide (CO), and PM emissions, concerns arise regarding increased nitrogen oxide (NOx) emissions and engine performance degradation (Cai, Becker, et al., 2021).

In recent years, extensive research efforts, both domestically and internationally, have focused on exploring unsullied and inexhaustible alternative combustibles and fuel additives for diesel engines (Chen et al., 2020). Additionally, the adoption of after-treatment systems such as diesel oxidation catalysts (DOC) (Ghadikolaei et al., 2020), diesel particulate filters (DPF) (Fayyazbakhsh & Pirouzfar, 2017), and selective catalytic reduction (SCR) (Kumar et al., 2022) has proven effective in reducing pollutant emissions, albeit with increased costs and complexities.

The quest for clean fuels to replace petrochemical counterparts has led to the investigation of various alternatives, including natural gas, alcohols, ethers such as dimethyl ether (DME) (Liu et al., 2019) and polymethoxydimethyl ether (PODE) (Gao et al., 2021), and biodiesel. Alcohol fuels, in particular, have garnered attention due to their high oxygen content, low carbon footprint, and renewable nature. Among alcohol fuels, methanol, ethanol, and n-butanol (Nanthagopal et al., 2020) have emerged as promising candidates for diesel engine applications (Kar et al., 2018).

Methanol, with its wide availability, ease of production, and favorable combustion characteristics, presents a compelling alternate fuel option for CI engines (Han et al., 2016). Notably, its high-octane number and built-in oxygen content contribute to reduced soot emissions when blended with diesel. Extensive research, as demonstrated by studies carried out by Dou et al. (2017), Xu et al. (2022), and Wei et al. (2015), has elucidated the impact of methanol substitution proportion (MSP) on engine emissions and combustion characteristics.

Similarly, ethanol, renowned for its clean-burning properties and oxygenated nature, offers significant potential as a diesel engine fuel. Studies by Wei et al. (2022), Shadidi et al. (2021), and Feng et al. (2012) have highlighted the efficacy of ethanol-diesel blends in reducing discharge of HC and CO, albeit with varying impacts on discharge of NOx and combustion characteristics.

Blending alcohols with diesel fuel is acknowledged as a strategy to enhance both the efficiency of fuel and combustion along with augmenting gross engine power. Nonetheless, head-on administration of alcohols in diesel engines poses challenges, necessitating partial replacement of conventional diesel while ensuring stable miscibility. Alcohols, characterized by their hydroxyl group (OH), serve as oxygenated additives, facilitating soot oxidation and reducing PM emissions. Despite the advantages offered by methanol and ethanol, challenges persist, including increased fuel consumption and cold-start difficulties, their inferior physical and chemical properties, including lower lubricity. Additionally, the inherent polarity differences between methanol, ethanol, and diesel pose challenges in achieving optimal fuel blending. They also have lower auto-ignition temperature, cetane number (CN), and calorific value, contributing to the diminished performance of engine and increased discharge of CO and HC.

While earlier research primarily focused on lower alcohols, recent years have seen increased interest in higher alcohols due to advancements in production processes. In the last few years, n-butanol has garnered considerable recognition as an alternative fuel or additive for diesel engines. With its higher viscosity, heat value, and superior miscibility with mineral diesel when juxtaposed against methyl and ethyl alcohols, n-butanol presents a promising solution to reduce particulate emissions but has little to no amount of impact on CN or NOx discharge. Research efforts, as exemplified by studies conducted by Jamrozik et al. (2021) and Li et al. (2019), have provided valuable insights into the discharge and ignition attributes of n-butanol-diesel blends.

Other than n-butanol among the long chain alcohols, pentanol, propanol, heptanol, hexanol, and octanol have also been successfully mixed with mineral diesel without encountering blending issues. These blends, when partially added to diesel propellant, could be utilized in contemporary power engines without requiring alterations. However, the incorporation of higher alcohols directly impacts engine functioning, combustion, and discharge attributes.

Research work administered by Zhao et al. (2022a,b) and others have demonstrated the result of higher distilled HC blends on ignition efficiency and emissions in diesel engines. For example, blends of hexanol and pentanol with diesel propellant have shown promising end results at excessive loads regarding the ignition phase, albeit with slight penalties in fuel burning efficiency and NOx discharge. Similarly, the addition of propanol and butanol in varying proportions has led to changes in combustion delay, rate of heat release, and emissions profile.

Ternary blends incorporating butanol and isopropanol with diesel fuel have also been investigated to enhance the rate of detonation of diesel-vegetable oil blends, meliorating various attributes of fuel and providing phase stability. Heptanol, with its longer carbon chain, presents a promising alternative owing to its higher CN and energy density. However, challenges such as cold-start operations remain. Research into ternary alcohol blends in diesel engines remains limited, highlighting the need for further exploration of suitable alternative fuel blends, particularly those involving higher alcohols.

Long chain alcohols, including both straight chain and aromatic variants, due to advancements in production processes, offer short-term disintegration procedures, reduced level of power consumption during production, elevated energy density, prominent CNs, ameliorated mixing stability, and reduced level of subsuming of water (humidity) content compared to lower alcohols. These attributes make them promising contender fuels for CI engines and advanced combustion modes, including homogeneous charge CI (HCCI) and reactivity-controlled CI (RCCI). Furthermore, the existence of alcohols in diesel fuel influences fuel injection characteristics, affecting combustion behavior. Comparative studies on isopropyl alcohol/diesel and ethyl alcohol/diesel blends have revealed differences in injection timing and combustion characteristics at various engine speeds and operating conditions (Geo et al., 2019; Kumar & Saravanan, 2016).

Additionally, alcohol fuels, such as ethanol and methanol, offer several advantages over traditional HC fuels when it comes to combustion such as alcohol fuels produce less soot PM during the combustive reaction process than conventional HC fuels like petrol or mineral diesel. This happens because alcohols have oxygen in their chemical makeup, facilitating more complete combustion. Soot formation is mitigated because oxygen allows the combustible to burn more cleanly, culminating in fewer unburned HCs that can form soot particles. Propellants derived from alcohol tend to have a higher-octane rating than petrol, enabling engines to work more proficiently at leaner air-to-fuel ratios without experiencing problems like knocking. This is especially essential for engines endeavoring for better fuel efficiency and lowered emissions. Moreover, the presence of oxygen in their atomic structure means that there is no need for excess air ensuring a more complete combustive reaction, and also they have lower carbon content in their chemical makeup which results in curbing down the value of harmful pollutants. In summary, alcohol fuels mitigate challenges associated with standard combustion methods by promoting cleaner combustion, reducing soot formation, extending lean combustibility limits, and lowering emissions of harmful pollutants.

This chapter focuses extensively on encapsulating the outcomes of long chain linear alcohols (e.g., 1-hexanol, 1-pentanol, 1-octanol, 1-heptanol) and aromatic alcohols (e.g., benzyl alcohol, phenethyl alcohol) on performance of engine, combustion, and emissions under copious operation modes (Zhou et al., 2014). These include binary or ternary mixtures with diesel, biodiesel, waste oil, and lower alcohols, together with their implementation in bi-fuel engines and advanced combustion modes like RCCI and Premixed Charge CI (PCCI). Additionally, comparisons between long carbon chain alcohols and other alcohols and ether fuels will be provided, along with directions for future investigative work in higher alcohol fuels in IC engines. This chapter aims to provide insights into the potential of higher alcohols as alternative fuels and guide the optimization of engine designs fueled by these fuels. It also sheds light on the cyclic combustion process and injection attributes of engines powered with quaternary blends, underscoring the importance of continued research in this area.

2.2 ELEMENTAL TRAITS OF HIGHER ALCOHOLS AND PRODUCTION METHODS

2.2.1 Elemental Traits

The physicochemical properties of long chain carbon alcohols are depicted in Table 2.1. As the carbon chain in the alcohol molecule increases, the kinematic viscosity, density, CN, low heat value, boiling point, and flash point all exhibit gradual increments. With the exception of aromatic alcohols, the density of other long chain carbon alcohols remains comparable to that of mineral diesel. Although, the viscosity of C6 to C8 alcohols significantly surpasses that of diesel, enhancing lubrication performance while potentially impacting spray characteristics negatively. In terms of heat value, long chain carbon alcohols closely resemble conventional fuels, with 1-octanol exhibiting a relatively high heat value even surpassing the gasoline and the diesel. Consequently, the energy density of long carbon chain alcohols surpasses that of short-chain alcohols.

Speaking of CN, it is a measure of a fuel's ignition quality, specifically its readiness to auto-ignite when subjected to the high pressures and temperatures present in a CI engine. It is named after cetane (n-hexadecane), which has excellent ignition properties and is assigned a CN of 100. The CN is determined by comparing the combustion characteristics of the fuel to a standard mixture of cetane and heptamethylnonane, with the latter having poor ignition qualities and a CN of 15. CN is crucial for determining fuel spontaneous combustion in CI engines. While diesel typically ranges between 45 and 55 in CN, higher alcohols exhibit lower CN values than diesel.

Moreover, CN is of very high significance in CI engines:

a. *Shorter Ignition Delay (ID)*: A higher CN indicates a shorter ID, meaning the fuel ignites more readily upon injection into the combustion chamber. This leads to smoother engine operation and reduced noise.
b. *Combustion Efficiency*: Fuels with higher CNs tend to ignite more uniformly and completely, enhancing combustion efficiency. This results in better fuel economy and lower emissions of unburned HCs.
c. *Engine Performance*: High-cetane fuels can improve cold-starting performance and reduce the likelihood of misfires in CI engines. They ensure consistent and reliable engine operation across a range of temperatures and operating conditions.
d. *Emission Characteristics*: Higher cetane fuels typically produce lower levels of nitrogen oxides (NOx) and PM. This is because the more complete and quicker combustion process reduces the formation of these pollutants.
e. *Fuel Injection Timing*: The CN affects the timing of fuel injection in CI engines. Manufacturers often design engines with specific CN requirements to optimize performance. Using a fuel with a CN outside, this range can lead to suboptimal engine performance and increased wear.
f. *Fuel Standards*: Diesel fuel standards often specify a minimum CN to ensure that the fuel meets the ignition quality requirements of most CI engines. For instance, in the United States, diesel fuel typically must have a CN of at least 40, while in Europe, the minimum is often 51.

TABLE 2.1
Various Physiochemical Characteristics of Alcohols Having Four or More Carbon Atoms

Fuels	Diesel	1-Butanol	1-Pentanol	1-Hexanol	1-Heptanol	1-Octanol	Benzyl Alcohol
Chemical formula	–	C_4H_9OH	$C_5H_{11}OH$	$C_6H_{13}OH$	$C_7H_{15}OH$	$C_8H_{17}OH$	C_7H_7O
Atomic composition	–						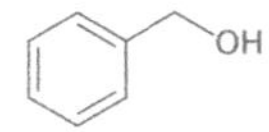
Molar mass (kg/kmol)	189–212	74.12	88.1482	102.177	116.2	130.2279	108.14
wt.% of carbon	86.13	64.82	68.13	70.52	72.16	73.72	77.78
wt.% of hydrogen	13.87	13.49	13.61	13.70	13.71	13.82	7.41
wt.% of oxygen	0	21.59	18.15	15.70	14.13	12.29	14.81
Density (kg/m^3)	820–840	809.8	814.6	820	821.9	827	1041.9
Self-ignition temperature (°C)	–	343	320	290	275	253	436
Kinematic viscosity (mm^2/s)	1.3–4.71	2.2–2.7	3.47–5	4.59–5.32	5.75–5.97	7.59	5.47–5.6
Boiling point (°C)	163–357	117.7	138	157	174.9–175.8	195	204.9–205.3
Flash point (°C)	52–96	21–37	32–43	59–63	70–71	81	93
Low heat value (MJ/kg)	42.8	33	34.65	39	39.92	52.94	34.65
Heat of vaporization (kJ/kg)	275–365	582	508.6	605.3	575	545	466.8
Stoichiometric ratio of air fuel mixture	14.8	11.27	11.84	12.26	12.52	12.82	10.95
Cetane No.	45–55	17	20	23	29.5	39	29
Water solubility (g/L)	–	77	22	5.9	1.67	0.54	42.9

g. *Engine Design*: Engine manufacturers design CI engines to operate efficiently with fuels within a certain CN range. Using fuels with CNs significantly lower or higher than this range can result in performance issues and potential engine damage.

In summary, the CN is a vital indicator of a fuel's suitability for use in CI engines, directly impacting ID, combustion efficiency, engine performance, and emissions. High-cetane fuels enhance overall engine operation, making them preferred for achieving optimal performance and compliance with environmental regulations.

Thus, blending higher alcohols with highly inflammable alternate propellants (for instance, ethers) may facilitate their complete replacement of diesel oil. However, as the number of carbon atoms in higher alcohols increases, unconstrained ignition characteristics improve while performance concerning anti-knock element diminishes, rendering long carbon chain alcohols more apt for CI engines compared to SI engines. Another essential characteristic, being the latent heat of vaporization, influences in-cylinder mixture formation and combustion. Although higher alcohols exhibit slightly higher evaporation latent heat than diesel, it wasn't anticipated that they would cause significant cold-start issues. Flash and Boiling points of C4 to C6 alcohols are marginally lower when juxtaposed to diesel, while C7 to C8 alcohols demonstrate similar boiling points. Aromatic alcohols have higher boiling and flash points compared to linear alcohols. Moreover, long carbon chain alcohols display superior blending stability with gasoline/diesel than short-chain alcohols and lower hygroscopicity. These conducive physicochemical characteristics position long carbon chain alcohols as desirable replacement for short-chain alcohols.

2.2.2 Production Methods

Long carbon chain alcohols, derived from inexhaustible raw materials, represent attractive second- and third-generation biofuels (No, 2020). Recent advancements in production processes have facilitated the mass production of higher alcohols from engineered microbial biosynthetic pathways and lignocellulosic biomass (Christensen et al., 2011). Additionally, progress has been made in processing long chain alcohols via syngas which is in turn produced from waste generated by industrial plants (Sharma & Duraisamy, 2019). Pentanol, for instance, could be processed from fermentation of sugar (Cann & Liao, 2010) or lignocellulosic biomass gasification, followed by catalytic conversion. Organic pathways, for instance, intrinsic bacterial fermentation by the adoption of skillfully maneuvered microbes, are also typical paths for pentanol production. 1-hexanol, synthesized by ethylene or ethanol oligomerization, can be synthesized from innumerable sustainable raw materials such as lignocellulosic biomass derived from agricultural residue of straw and corn strove, algae, waste oils and fats, sugar and starch, and also via other vegetable oils such as palm oil and soyabean oil. Various renewable and environmentally friendly production pathways for 1-hexanol are as follows:

1. Microbial Fermentation of *Escherichia coli* or *Saccharomyces cerevisiae*: It can be used to convert glucose into 1-hexanol through metabolic pathways tailored for long chain alcohol production.

2. Chemical synthesis from biomass-derived intermediates such as Hydroformylation of Biomass-derived ethylene can produce butyraldehyde, which can then be converted into 1-hexanol through aldol condensation followed by hydrogenation. Furthermore, alcohols can be dehydrated to form alkenes, which can then be hydrated to produce longer-chain alcohols like 1-hexanol.
3. *Biocatalysis*: Enzymatic processes can convert biomass-derived fatty acids or their derivatives into alcohols. Lipases and other enzymes are used to catalyze the reactions under mild conditions, which is beneficial for sustainable production.
4. *Guerbet Reaction*: This process involves the condensation of ethanol (obtained from fermentation of biomass) to produce higher alcohols, including 1-hexanol. The reaction involves the formation of intermediates such as butanol, which can further react to form 1-hexanol. (Bergthorson & Thomson, 2015).

Economic production processes, including syngas fermentation using Clostridium carboxidivorans (Phillips et al., 2015), have been reported for n-hexanol production. Similarly, by means of ethylene oligomerization or biosynthesis using engineered microorganisms like *E. coli* and Clostridium, the production of octanol could be easily done (Xia et al., 2015). A multifunctional catalyst system has been utilized to obtain dioctyl ether and octanol from lignocellulosic biomass (Julis & Leitner, 2012). A novel pathway designed specifically for n-octanol biosynthesis in *E. coli* has shown promise. Aromatic alcohols, unlike straight-chain higher alcohols, can be straight-forwardly processed by lignin depolymerization, offering a shorter decomposition process with less energy consumption (Zhou et al., 2014).

2.3 UTILIZATION OF HIGHER ALCOHOLS AS ADVANCED PROPELLANTS IN INTERNAL COMBUSTION ENGINES

Presently, higher alcohols find application in IC engines primarily through blending with mineral diesel, waste oil, biodiesel, and gasoline. Engine bench tests are commonly employed to evaluate various performance parameters and discharge attributes resulting from the addition of long carbon chain alcohols. Among others, these parameters encompass brake thermal efficiency (BTE), in-cylinder pressure, heat release rate (HRR), Exhaust Gas Temperature (EGT), ID, brake-specific energy consumption (BSEC), combustion phasing, Brake Specific Fuel Consumption (BSFC), Pressure rise rate, and discharge of gases of order of nitrogen oxides (NOx), carbon dioxide (CO_2) HC, carbon monoxide (CO), PM, soot, and more.

2.3.1 Low-Temperature Combustion in Pentanol

Among higher alcohols, n-Pentanol stands out as one of the most promising due to its five-carbon straight-chain structure (Pinzi et al., 2021). Although its low CN restricts its use as a pure fuel, research has extensively explored its application as an alternate substitute in CI engines. For instance, studies by Campos-Fernández et al. (2012, 2013) have investigated the short-term performance of direct-injection CI engines fueled with 1-pentanol/diesel blends, revealing no difficulty in engine start-up even at 25% 1-pentanol concentration. Further researches carried out by eminent researchers like Wei et al. (2014) and Zhao et al. (2022) delved into the efficacy of the addition of 1-pentanol on engine performance, ignition, and emission characteristics, highlighting aspects such as rate of heat release, ID, and amount of concentration of pollutant. Similarly, investigations by Atmanli and Yilmaz (2018) evaluated the impact of pentan-1-ol blending on engine parameters like BSFC, BTE, and discharge of exhaust gases like HC, CO, and NOx.

Further studies have explored strategies to augment the characteristics concerning performance and emissions of pentan-1-ol/diesel blends. These include the addition of CN improvers like 2-ethylhexyl nitrate (EHN) to mitigate issues related to ID and emissions. Additionally, studies by Wang et al. (2021) examined the likely outcomes of di-n-butyl ether (DBE) addition to n-pentyl alcohol/diesel blends, revealing improvements in BTE, ID, and reduced emissions of smoke, CO, and NOx.

Research has also focused on optimizing engine operation conditions, for instance, injection strategies and, exhaust gas recirculation (EGR) rates to achieve better performance and emissions trade-offs. Kumar and Saravanan et al. (2015) and Santhosh and Radheshyam et al. (2020) explored the impact of medium EGR rates on reducing smoke and NOx emissions in pentan-1-ol/diesel mixtures. Similarly, studies by Şen et al. (2019) investigated the effect of injection timing on combustion and exhaust characteristics.

Moreover, investigations into the emissions of non-compliant pollutants, for instance, carbonyl compounds and polycyclic aromatic hydrocarbons (PAHs), have provided insights into the environmental implications of pentan-1-ol/diesel amalgamations. Investigative analyses by Yilmaz and Atmanli (2017) quantified the reduction in PAH emissions with n-pentanol addition, while Arias et al. (2022) analyzed the impact of pentan-1-ol blending on carbonyl emissions.

Other than mineral diesel, n-pentanol has also been mixed with waste oil and biodiesel to explore its potency as an alternate combustible. Studies by Zhu et al. (2016) and Yilmaz et al. (2022) scrutinized the combustion and emissions properties of n-pentanol/biodiesel blends, revealing improvements in BTE and particle emissions compared to pure biodiesel and mineral diesel. Moreover, the ternary blend of pentan-1-ol/mineral diesel/biodiesel has been studied, showing enhanced combustion and emission characteristics compared to binary blends.

Furthermore, recent research has explored the utilization of n-pentanol in spark ignition (SI) engines, demonstrating its potential as a substitutive measure for gasoline propellant. Studies by Qian et al. (2019) and Yaman and Yesilyurt (2021) evaluated the combustive reaction and emissions traits of n-pentanol/gasoline blends, highlighting improvements in the rate of heat release, in-cylinder pressure, and emanations reduction contradistinguished to pure gasoline. Additionally, studies by Kothare et al. (2023) proposed ternary mixtures of ethyl alcohol-petrol-pentan-1-ol as a means to attain optimized operation and emanation characteristics in SI engines.

In summary, the application of higher alcohols, particularly n-pentanol, in IC engines shows promise for optimizing performance, combustion efficiency, and mitigating emissions across a range of operating conditions. Continued research in this area aims to optimize fuel blends and engine operation strategies to maximize the benefits of higher alcohol utilization in IC engines.

2.3.2 Low-Temperature Combustive Reaction in Hexanol

Hexanol is a common additive to diesel propellant for CI engines, often utilized in various blends. An analysis carried out by Sundar Raj and Saravanan et al. (2011) delved into the ramification of hexanol blending on a singular-cylinder diesel engine's effectiveness, combustion, and emanations. Their findings indicated that as the hexanol content increased, so did the rate of rise of pressure and peak pressure. This resulted in an overall improvement in the hybridized fuel's performance contradistinguished to pure diesel, manifesting notably in reduced smoke levels but higher nitrogen oxides (NOx) emissions.

Duraisamy et al. (2020) observed a slight increment in BTE and a reduction in total rate of consumption when blending 10%–20% hexanol with diesel. Furthermore, as the hexanol mixing ratio increased, CO emissions decreased while HC emissions escalated. The impact of n-hexanol addition on emissions, as noted by De Poures et al. (2017), was load-dependent. While the discharge of NOx gas diminished at reduced to moderate loads with growing hexan-1-ol content, they rose at higher loads. Additionally, CO emissions remained steady at reduced to moderate loads but dwindled at higher loads, whereas carbon-based emissions advanced across all load conditions, and PM/Plume declined universally.

Santhosh and Kumar (2020a,b) explored the effects of blending hexan-1-ol (10%–40%) with diesel in a CRDI engine accompanied by EGR. They observed reductions in in-cylinder pressure and average gas thermal energy with the n-hexanol/diesel mixture, alongside registered amelioration in net rate of heat release (HRR). Moreover, NOx emissions decreased significantly after blending with n-hexanol, although HC and CO emissions experienced slight increases. Despite the benefits of EGR in further alleviating NOx emissions, its effectiveness diminished with a 40% n-hexanol blend.

Subsequent investigative works conducted by Nour et al. (2021) and Santhosh and Kumar (2021) investigated higher concentrations of n-hexanol (up to 50%) in diesel blends, confirming improved premixed combustion and reduced diffusion combustion. Moreover, advanced injection timings were found to enhance engine efficiency, particularly with 50% n-hexanol blends. Other investigations explored the ramifications of hexanol homogenization on soot particle characteristics, performance of RCCI engine, and dual-fuel applications with hydrogen.

Additionally, studies have explored hexanol blending in conjunction with biodiesel, waste oil/mineral diesel, and biodiesel/diesel blends, highlighting improvements in fuel performance, engine combustion, and emissions characteristics. The optimal blending ratios and conditions vary depending on the specific application and engine type, suggesting a promising potential for hexanol-based alternative fuels in CI engines.

2.3.3 Low-Temperature Combustion in Heptanol

In this study, research on heptanol in IC engines is summarized, focusing mainly on CI engines. Nour et al. (2022) investigated the combustive reaction characteristics of heptan-1-ol/mineral diesel blends, noting a prolonged ignition latency and increased BSEC. Despite a registered decrease in BTE, the blends improved the smoke-NOx balance, particularly evident with 50% heptan-1-ol, which curtailed the discharge of Plume and NOx by 70% and 25%, respectively. Comparable insights were reported by El-Seesy et al. (2019) who discerned increased BSFC with n-heptanol addition but noted substantial minimizations in smoke and NOx emanations.

Further research by El-Seesy et al. (2019, 2020) investigated the outcomes of integrating graphene oxide (GO) nanoparticles to blends of butan-1-ol, octan-1-ol, and heptan-1-ol with diesel. The addition of GO increased BTE, in-cylinder peak pressure, and rate of heat release (HRR) in the meantime reducing HC, CO, and smoke emissions. On the other hand, NOx emissions escalated, suggesting that a 20% composition ratio of long carbon chain alcohols and GO with diesel could accomplish substantially reduced HC, CO, and smoke emissions while maintaining prime heat efficacy.

Heptan-1-ol has the potential to also be added to biodiesel, as demonstrated by El-Seesy et al. (2020) who observed reductions in smoke and NOx emissions when blending n-heptanol with biodiesel. The recommended blending ratio was found to be 20%, consistent with findings for blends with diesel. Additionally, the mixing of heptan-1-ol along with methyl oleate biodiesel led to a reduction in the discharge of smoke and NOx, albeit with a sparse increase in BSFC.

In addition to CI engines, Yaman et al. (2021) explored the usage of petrol-heptan-1-ol and gasohol-hexan-1-ol blends in SI engines. They observed increased consumption of combustibles and decreased thermal effectiveness with higher alcohol blending ratios. While NOx emissions rose, discharge of HC and CO decreased when analyzed against unadulterated gasoline, indicating potential for hexanol and heptanol as alternate propellants for SI engines under certain conditions.

2.3.4 Low-Temperature Combustion in Octanol

Octanol, characterized by its higher CN of 39, has emerged as an encouraging substitute for CI engines, owing to recent advancements in production processes. This section provides an in-depth overview of research conducted on octanol in CI engines, encompassing a wide array of studies.

Initial investigations focused on the blending of octan-1-ol with diesel propellant. McCormick et al. (1997) (Effect of Several Oxygenates on Regulated Emissions from Heavy-Duty Diesel Engines, 1997) delved into its effects on discharge of nitrogen oxides and PM, revealing notable reductions in PM levels (10%–15%) at oxygen levels of 1–2 wt.%, along with marginal decreases in NOx under specific conditions. Of particular interest was n-octanol's ability to attain a beneficial balance between soot and NOx emissions across various engine load ranges, maintaining high indicated efficiency while significantly mitigating PM emissions, especially at rated power.

Further exploration by Deep et al. (2014) scrutinized the impact of varied composition ratios of octan-1-ol (ranging from 10% to 40%) on the performance of engine and emanations. While BTE and BSFC exhibited minimal changes across the blending ratios, there were slight reductions in discharge of nitrogen oxide gas with rising content of n-octanol, in contrast to previous findings which reported a decrease in emanations of CO and HC.

The role of injection timing and EGR rate in shaping the efficacy and emanations of octanol-diesel blends was underscored in subsequent studies. Pan et al. (2019) highlighted the superior BTE of octan-1-ol/mineral diesel blends compared to diesel at low EGR rates, coupled with simultaneous reductions in emanations of nitrogen oxide fumes and PMs. Gopal et al. (2018) identified specific combinations of octan-1-ol blends, EGR rates, and injection timing synchronization parameters that minimized Plume, BSFC, and NOx.

Moreover, Rajesh Kumar et al. (2016) shed light on the potentiality of octan-1-ol to replace diesel by up to 30% under EGR conditions, demonstrating improvements in both emissions and performance.

Investigations into the combustive reaction characteristics of octyl alcohol isomer 2-ethylhexanol blended with diesel unveiled promising results in a poly-cylinder light-duty automotive engine, with reductions in soot and CO emissions relative to diesel. Additionally, blending n-octanol with biodiesel showcased improvements in BTE and reductions in NOx emissions, highlighting its potential for optimizing biodiesel engine performance.

Comparative studies between n-octanol and other alcohol or ether fuels unveiled disparities in rate of heat release, ignition latency, peak pressure, and emissions. While n-octanol-diesel blends showcased subsided values of CO and smoke discharged when contradistinguished to other higher alcohol-diesel blends, they also displayed elevated maximum heat release rates and lower HC emissions.

Furthermore, investigations into the ignition and emanation traits of di-n-butyl ether (DNBE) and octan-1-ol emphasized differences in ignition latency times and emission profiles, with DNBE demonstrating significant reductions in soot emissions compared to n-octanol across various engine loads.

In conclusion, research on octanol in CI engines underscores its potential as a sustainable fuel, offering improvements in combustion performance and emissions reduction compared to conventional diesel fuel. These findings underscore the importance of continued exploration and optimization of octanol-based fuels for sustainable energy solutions in the automotive sector.

2.3.5 Low-Temperature Combustion in Aromatic Alcohol

Aromatic alcohols, in comparison to their higher linear counterparts, have received less attention in research but have shown promising results. The primary investigations concerning aromatic alcohols in IC engines are deliberated here in detail.

In a study orchestrated by Mathan Raj et al. (2018) and Raj and Manoj (2021) on a one-cylinder CI engine, the impact of benzyl alcohol addiction (at 5% and 10% ratios) to Simarouba vegetable oil-diesel blends (B20, B40) was explored. Findings indicated that as the proportion of benzyl alcohol increased, the pre-ignition period extended, combustion characteristics considerably augmented, and emissions of HCs, carbon monoxide (CO), and smoke lessened. Despite proportionately high nitrogen oxide emissions, benzyl alcohol exhibited encouraging enhancements in performance metrics for instance, registered advancement in efficiency or combustion parameters. Furthermore, benzyl alcohol displayed a superior exothermic reaction rate compared to pentan-1-ol, suggesting that even diminutive supplements of Hydroxymethyl benzene could ameliorate the suboptimal combustion behavior of biodiesel-diesel blend.

Zhou et al. (2012) investigated the outcomes of Hydroxymethyl benzene and benzyl carbinol mixing with diesel fuel on the functioning and discharge characteristics of heavy-duty diesel engines with 2.59 wt.% of oxygen. They registered improved and balanced dissipation of soot-NOx and fuel consumption rate when contradistinguished to conventional diesel. Additionally, they examined the impact of Hydroxymethyl benzene, benzyl carbinol, and anisole blending with diesel under EGR conditions, noting that phenylethyl alcohol exhibited the utmost diesel particulate emission under non-EGR constraints. However, benzyl alcohol demonstrated the highest indicated efficiency under both EGR constraints and non-EGR, with benzyl carbinol exhibiting nethermost.

Comparative studies on SI engines have been conducted amid minor and major linear alcohols and aromatic alcohols. Edwin Geo et al. (2019) investigated the ramifications of blending ethyl alcohol and hydroxymethyl benzene with petrol (at 10% and 20% proportions) on combustive reaction, functioning, and exhaust gas discharge characteristics. They observed a significant rise in BTE and reductions in CO_2, NOx, HC, and CO emissions following homogenization of ethyl alcohol and hydroxymethyl benzene with petrol. Emphasizing functioning, emissions, and combustive reaction elements, they concluded that the blend of hydroxymethyl benzene with gasoline was ideal.

Similarly, Jesu Godwin et al. (2019) compared the impacts of blending isobutanol, n-butanol, benzyl alcohol (at 10% ratios), and n-pentanol with petrol on the performance and emanation parameters of SI engines. They found that blended fuels exhibited significantly higher performance than non-blended gasoline, with hydroxymethyl benzene-gasoline blended propellant providing greater power generation capacity at full load. Additionally, all blends demonstrated lower emissions of CO_2, HC, and CO contradistinguished to non-blended gasoline. Considering both functioning and discharge characteristics, they determined that the hydroxymethyl benzene-petrol blend outperformed other blends.

In summary, aromatic alcohols prove to be advantageous and immense potential when used with both gasoline and diesel as additives in both CI and SI engines (Table 2.2).

2.4 CONCLUSIONS AND PATH FOR FUTURE INVESTIGATION

2.4.1 Conclusions

Long carbon chain alcohols represent a promising class of bio propellants, characterized by renewability and carbon neutrality, typically comprising four or more carbon atoms. However, existing literature primarily focuses on butanol, with

TABLE 2.2
Prominent Researches Carried Out in Field of Higher Alcohols

Combustibles	Specifications	Result	Authors
ABE (acetone/butanol/ethanol)	The jet fuel blend consists of 10% by weight ethanol, 30% by weight acetone, and 60% by weight 1-butanol. The catalyst used is an ionic liquid composed of [bmim]Cl-2AlCl3.	The ABE conversion rate reached 89% while the yield of jet fuel was of the order of 71.5%.	Zhang, He, et al. (2023)
Bio-alcohols and bio-ketones	The blend comprises butanol, pentanol, cyclopentanol, and cyclopentanone	The blended propellant exhibited a decreased cetane number, culminating in protracted ignition latency and an extended phase of premixed combustive reaction. This coupled with higher average peak combustion temperature contributed to a reduction of 91% in particle concentration (NOx, Hydrocarbons, and particulate matter).	Doustdar et al. (2024)
Heptyl alcohol (C7) and octyl alcohol (C8)	The evaluation of RCCI (reactivity-controlled compression ignition) combustion was carried out using a modified mono-cylinder direct injection (DI) engine, contrasting it with the standard ignition approach fueled via diesel B7 combustible. The engine loads were varied based on 0%–60% of the maximal engine torque, with intermittent periods of 20% while maintaining an invariable engine velocity of 2400 rpm. Additionally, diverse proportioned blends of low reactivity fuel (LRF) were employed, ranging from 0% to 60% at intervals of 15%.	Under any engine loads, the most notable increase of approximately 40% was noted in cylinder pressure at 0.45 Rp and under 60% load while utilizing octyl alcohol. Meanwhile, the peak rate of heat release (HRR), around 111%, was recorded at 0.60Rp and under 60% load with heptyl alcohol. Significant reductions in NO emissions (decreased by 58%) were attained at 20% load and 0.60Rp when using heptyl alcohol. Conversely, the greatest reduction in smoke opaqueness, reaching 92%, was assessed at 0.60Rp and at 60% engine load when employing octyl alcohol. Comparatively, heptanol RCCI proved to be more productive contrary to octyl alcohol in alleviating discharge of NOx whereas the octyl alcohol/diesel RCCI demonstrated lower unburned hydrocarbon (uHC) emissions.	Okcu et al. (2022)

(Continued)

TABLE 2.2 *(Continued)*
Prominent Researches Carried Out in Field of Higher Alcohols

Combustibles	Specifications	Result	Authors
Diesel blended with methanol, ethanol, n-butanol in different proportions	The engine is a Four-cylinder, turbocharged system with water cooling, operating at a velocity of 2000 rpm.	The combustible blend consisting of 10% methyl alcohol + 10% butyl alcohol + 80% diesel, (referred to as D80M10B10) demonstrated significant advantages in lengthening ignition latency, enhancing cylinder pressure, and curtailing the period of combustive reaction. In addition, it effectively mitigated soot formation and lowered emissions of NOx, HC, and CO.	Zhang, Tian, et al. (2022)
Mélanges of heptyl alcohol/ butyl alcohol/isopropyl alcohol and diesel.	Two fuel blends were employed, namely D85HIB15 (comprising 5% heptyl alcohol, 5% isopropyl alcohol, 5% butyl alcohol and 85% diesel) and D70HIB30 (containing 10% butyl alcohol, 10% heptyl alcohol, 10% isopropyl alcohol and 70% diesel)	Lower cyclic variations be it in injection pressure, or cylinder pressure were noted when utilizing the mélanges vis-à-vis diesel fuel.	Sanh et al. (2023)
Butanol and pentanol	Usage of long carbon chain alcohols in CI engines	Low value of CN of long carbon chain alcohols contributes to prolonged ignition latency leading to more complete combustion and improved fuel efficiency. Additionally, the reduced value of viscosity of the long carbon chain alcohols elevates the spray properties facilitating better atomization and mixing with air within the combustion chamber. Reduced soot and NOx emissions	Zhang, Gao, et al. (2023)
Hydrotreated vegetable oil (HVO)	In dual-fuel compression-ignition engine, blends of methanol and ethanol were combined in different proportions with hydrotreated vegetable oil (HVO) or diesel	The combination of ethanol and HVO achieved an impressive 84% substitution ratio, leading to a remarkable diminution in discharge of NOx gas.	Domínguez et al. (2024)

Isopropanol/gasoline dual-fuel	Bi-fuel system	In this research, to address the issue of engine weakening, a coupling approach was employed by integrating the composite injection engine with oxyhydrogen dispensing mechanism, which comprises gasoline port injection (GPI), oxyhydrogen negative pressure inhalation (ONPI), and isopropanol direct injection (IDI). ONPI effectively enhances the engine's tolerance to isopropyl alcohol, and the ideal IDI rate progressively increases concurrently with the rise of ONPI value. Moreover, the integration of IDI and ONPI values also ameliorates the engine's combustive reaction characteristics. This integration results in improvements in Indicated Mean Effective Pressure (IMEP) and torque by approximately 8.77 % and 12.37 %, respectively. Furthermore, there's a diminution in discharge of both NOx and CO showcasing the environmental benefits of this approach.	Shi et al. (2024)
Methyl alcohol, Ethyl alcohol, and butyl alcohol.	In the common rail bi-fuel engine, a detailed examination was conducted on the implementation of R-OH fuel port injection in conjunction with diesel direct injection. The engine was run at a constant velocity of 2500 rpm, with an IMEP of 0.75 MPa.	After experimenting with various proportions of R-OH (alcohol-based fuel), it was observed that the blend of CH3OH with diesel showcased the lowest emissions of CO, NOX, and soot. However, it was noted that the emission of total hydrocarbon (THC) was highest in this particular blend. Furthermore, when utilizing alcohol-blended fuel, there was a reduction in BTE and ringing intensity (RI).	Ning et al. (2020)

(Continued)

TABLE 2.2 *(Continued)*
Prominent Researches Carried Out in Field of Higher Alcohols

Combustibles	Specifications	Result	Authors
Alcohol Biofuels	C1–C4 alcohols	They lack negative temperature coefficient (NTC) behavior which typically aids in controlling combustion stability. Additionally, these blends possess an exceptionally high cooling potential, resulting in complete combustion due to proportional air-to-fuel ratio.	Gainey and Lawler (2021)
	C5 + alcohols	They have nascent NTC behavior and low cooling potentials.	
Biodiesel, EGR, and diethyl ether.	In this study, EGR was deployed at a rate of 10% alongside the addition of DEE to biodiesel at varying proportions of 2.5%, 5%, and 7.5%.	Reductions in NOx emissions by up to 70% are accompanied by a 3% reduction in torque.	Ergen et al. (2020)
Elevated quantity of EGR/ sustainable alcohol propellants.	Split injection technique was utilized in diesel engines to attain semi premixed ignition.	Diminution in combustion temperature leads to low soot and NOx emissions	Kuppili et al. (2023)
Methyl alcohol, Ethyl alcohol, Propyl alcohol, and Butyl alcohol.	Bi-fuel mechanism HCCI engine	Among all the blends, methanol has the highest cooling potential which aided in pruning the NOx emissions.	Gawale et al. (2022)

limited attention given to long carbon chain alcohols containing four or beyond atoms of carbon. Analyses on alcohols with more than 4 carbon atoms have predominantly addressed the ramification of pentyl alcohol on CI engines, along with considerations of long carbon chain alcohols' effects on PM properties, basic combustion behavior, and CI engine emissions. Currently, there's a notable gap in comprehensive reviews addressing the overall consequences of alcohols having 4 or more carbon atoms (encompassing both straight chain and aromatic alcohols) on the functioning of engine, combustive reaction, and emanation characteristics across various application scenarios.

This review provides a comprehensive overview of investigation advancement on long carbon straight chain alcohols such as heptan-1-ol, hexan-1-ol, pentan-1-ol, octan-1-ol, and aromatic alcohols like phenyl methanol and benzyl carbinol as sustainable combustibles for IC engines. Key topics covered include traits and generation methods of long carbon chain alcohols, their effects when blended with mineral diesel, bio propellant, waste/residual oil, and lower alcohols in traditional CI engines, as well as their application in advanced combustion modes like RCCI and PCCI. Evaluation parameters encompass power, torque, fuel consumption, efficiency, combustion characteristics, and emissions profiles, including HC, NOx, CO_2, CO, Plume, carbon black, PAH, PM, and carbonyl compounds.

Results indicate that long carbon chain alcohols can be soundly integrated into CI engines as mixed-fuels or dual-fuels without significant modifications. Most studies focus on blending ratios of 10%–40% for linear alcohols and 5%–20% for aromatic alcohols. Optimal blending ratios typically range around 20% for CI and RCCI engines, and 10% for PCCI engines. Elevated alcohol levels show promise in diminishing PM emissions, enhancing the trade-off between NOx and PM while maintaining efficiency comparable to diesel. However, they may increase fuel consumption and their impact on NOx, HC, and CO emissions varies depending on engine operating conditions. Furthermore, blending long carbon chain alcohols with biodiesel, waste/residual oil, or lower alcohols improves combustion characteristics and engine performance.

In SI engines, substitution ratios of higher alcohols to gasoline usually remain around 20%. Notably, n-pentanol enhances efficiency and emanations, while hexenyl alcohol, heptenyl alcohol, and benzyl alcohol demonstrate reductions in discharge of exhaust gases benefits. Overall, higher alcohols exhibit promising suitability and potential as sustainable fuels for both conventional IC engines and novel combustive reaction modes.

2.4.2 Path for Future Investigation

Future research should focus on increasing the percentage of long carbon chain alcohols beyond 50% or up to and including 100%, necessitating long-term durability testing. Additionally, exploration of C4+ alcohols as partial substitutes for gasoline in SI engines is warranted. Investigation into higher alcohol-lower alcohol polyblends could enhance mixing stability and combustion characteristics, while also considering effects on unregulated emissions and emission treatment systems.

Further exploration of higher alcohols in new combustion concept engines like RCCI, PCCI, and HCCI, coupled with optimization of EGR rates, injection strategies, compression ratios, and hydrogen addition, is recommended. Understanding the low temperature reactivity and combustion behavior of higher alcohols is crucial for future engine applications. Advanced testing techniques and simulation methods should be employed to elucidate combustion reaction pathways and pollutant generation mechanisms.

ACKNOWLEDGMENT

The authors would like to express their deepest gratitude to Energy Centre, MANIT, Bhopal for their unwavering support and MPCST for the financial support provided for this research project entitled "Enhancement of cold flow properties of waste cooking biodiesel and diesel" under the file number A/RD/RP-2/345.

REFERENCES

Arias, S., Botero, M. L., Molina, F., & Agudelo, J. R. (2022). Pentanol/diesel fuel blends: Assessment of inhalation cancer risk and ozone formation potential from carbonyl emissions emitted by an automotive diesel engine. *Fuel*, *321*, 124054.

Atmanli, A., & Yilmaz, N. (2018). A comparative analysis of n-butanol/diesel and 1-pentanol/diesel blends in a compression ignition engine. *Fuel*, *234*, 161–169.

Bergthorson, J. M., & Thomson, M. J. (2015). A review of the combustion and emissions properties of advanced transportation biofuels and their impact on existing and future engines. *Renewable and Sustainable Energy Reviews*, *42*, 1393–1417.

Cai, T., Becker, S. M., Cao, F., Wang, B., Tang, A., Fu, J., … & Zhao, D. (2021). NOx emission performance assessment on a perforated plate-implemented premixed ammonia-oxygen micro-combustion system. *Chemical Engineering Journal*, *417*, 128033.

Cai, T., & Zhao, D. (2022). Enhancing and assessing ammonia-air combustion performance by blending with dimethyl ether. *Renewable and Sustainable Energy Reviews*, *156*, 112003.

Cai, T., Zhao, D., Sun, Y., Ni, S., Li, W., & Guan, D. (2021). Evaluation of NOx characteristics in a CO_2 emissions-free micro-power system by implementing a perforated plate. *Renewable and Sustainable Energy Reviews*, 145, 111150.

Campos-Fernández, J., Arnal, J. M., Gómez, J., & Dorado, M. P. (2012). A comparison of performance of higher alcohols/diesel fuel blends in a diesel engine. *Applied Energy*, *95*, 267–275.

Campos-Fernandez, J., Arnal, J. M., Gomez, J., Lacalle, N., & Dorado, M. P. (2013). Performance tests of a diesel engine fueled with pentanol/diesel fuel blends. *Fuel*, *107*, 866–872.

Cann, A. F., & Liao, J. C. (2010). Pentanol isomer synthesis in engineered microorganisms. *Applied Microbiology and Biotechnology*, *85*, 893–899.

Chen, C., Yao, A., Yao, C., Wang, H., Liu, M., &, & Li, ZZ. (2020). Selective catalytic reduction of nitrogen oxides with methanol over the (cobalt-molybdenum)/alumina dual catalysts under the diesel methanol dual fuel exhaust conditions. *Chemical Engineering Science*, *211*, 115320.

Chen, H., He, L., Chen, J., Yuan, B., Huang, T., & Cui, Q. (2019). Impacts of clean energy substitution for polluting fossil-fuels in terminal energy consumption on the economy and environment in China. *Sustainability*, *11*(22), 6419.

Christensen, E., Yanowitz, J., Ratcliff, M., & McCormick, R. L. (2011). Renewable oxygenate blending effects on gasoline properties. *Energy & Fuels, 25*(10), 4723–4733.

Deep, A., Kumar, N., Karnwal, A., Gupta, D., Vibhanshu, V., Sharma, A., & Patel, J. S. (2014). *Assessment of the performance and emission characteristics of 1-octanol/diesel fuel blends in a water cooled compression ignition engine* (No. 2014-01-2830). SAE Technical Paper.

De Poures, M. V., Sathiyagnanam, A. P., Rana, D., Kumar, B. R., & Saravanan, S. (2017). 1-Hexanol as a sustainable biofuel in DI diesel engines and its effect on combustion and emissions under the influence of injection timing and exhaust gas recirculation (EGR). *Applied Thermal Engineering, 113*, 1505–1513.

Dey, S., Reang, N. M., Das, P. K., & Deb, M. (2021). A comprehensive study on prospects of economy, environment, and efficiency of palm oil biodiesel as a renewable fuel. *Journal of Cleaner Production, 286*, 124981.

Domínguez, V. M., Hernández, J. J., Ramos, Á, Giménez, B., & Rodríguez-Fernández, J. (2024). Exploring the effect of methanol and ethanol on the overall performance and substitution window of a dual-fuel compression-ignition engine fueled with HVO. *Fuel, 359*, 130529.

Dou, Z., Yao, C., Wei, H., Wang, B., Liu, M., Chen, C., ... & Shi, J. (2017). Experimental study of the effect of engine parameters on ultrafine particle in diesel/methanol dual fuel engine. *Fuel, 192*, 45–52.

Doustdar, O., Zeraati-Rezaei, S., Herreros, J. M., Martos, F. J., Tsolakis, A., & Wyszynski, M. L. (2024). The significance of low carbon bio-alcohols and bio-ketones fuels for clean propulsion systems. *Fuel, 361*, 130641.

Duan, X., Fu, J., Zhang, Z., Liu, J., Zhao, D., & Zhu, G. (2017). Experimental study on the energy flow of a gasoline-powered vehicle under the NEDC of cold starting. *Applied Thermal Engineering, 115*, 1173–1186.

Duraisamy, S., Kasirajan, L., Krishnamoorthy, S., Kadasari, R., & Tiruchengode, R. C. (2020). Performance test and emission characteristics of diesel fuel blended with n-hexanol. *Thermal Science, 24*(1 Part B), 557–564.

El-Seesy, A., Nour, I., Attia, M., He, A. M., & Hassan, Z. (2020). Investigation the effect of adding graphene oxide into diesel/higher alcohols blends on a diesel engine performance. *International Journal of Green Energy, 17*(3), 233–253.

El-Seesy, A. I., Kayatas, Z., Hawi, M., Kosaka, H., & He, Z. (2020). Combustion and emission characteristics of a rapid compression-expansion machine operated with n-heptanol-methyl oleate biodiesel blends. *Renewable Energy, 147*, 2064–2076.

El-Seesy, A. I., Kosaka, H., Hassan, H., & Sato, S. (2019). Combustion and emission characteristics of a common rail diesel engine and RCEM fueled by n-heptanol-diesel blends and carbon nanomaterial additives. *Energy Conversion and Management, 196*, 370–394.

Fayyazbakhsh, A., & Pirouzfar, V. (2017). Comprehensive overview on diesel additives to reduce emissions, enhance fuel properties and improve engine performance. *Renewable and Sustainable Energy Reviews, 74*, 891–901.

Feng, X., Huo, M., Lee, C. F., & Liu, H. (2012). The effects of EGR and injection timing on the engine combustion and emission performances fueled by butanol-diesel blends. *SAE International Journal of Engines, 5*(3), 794–811.

Gainey, B., & Lawler, B. (2021). The role of alcohol biofuels in advanced combustion: An analysis. *Fuel, 283*, 118915.

Gallagher, K. S., Zhang, F., Orvis, R., Rissman, J., & Liu, Q. (2019). Assessing the policy gaps for achieving China's climate targets in the Paris Agreement. *Nature Communications, 10*(1), 1256.

Gao, Z., Wu, S., Luo, J., Zhang, B., Zhang, H., & Xiao, R. (2021). Optimize the co-solvent for methanol in diesel with group of oxygen-containing reagents: Molecular structure and intermolecular forces analysis. *Fuel Processing Technology, 222*, 106980.

Gautam, R., & Kumar, S. (2020). Performance and combustion analysis of diesel and tallow biodiesel in CI engine. *Energy Reports, 6*, 2785–2793.

Gawale, G. R., Srinivasulu, G. N., & Lingayat, A. (2022). Experimental investigation on diesel ignited dual fuel homogeneous charge compression ignition engine for the selection of suitable alcohol-based fuel. *Journal of Cleaner Production, 380*, 135048.

Geo, V. E., Godwin, D. J., Thiyagarajan, S., Saravanan, C. G., & Aloui, F. (2019). Effect of higher and lower order alcohol blending with gasoline on performance, emission and combustion characteristics of SI engine. *Fuel, 256*, 115806.

Ghadikolaei, M. A., Yung, K. F., Cheung, C. S., Ho, S. S. H., & Wong, P. K. (2020). Non-polar organic compounds, volatility and oxidation reactivity of particulate matter emitted from diesel engine fueled with ternary fuels in blended and fumigation modes. *Chemosphere, 249*, 126086.

Godwin, D. J., Geo, V. E., Thiyagarajan, S., Martin, M. L. J., Maiyalagan, T., Saravanan, C. G., & Aloui, F. (2019). Effect of hydroxyl (OH) group position in alcohol on performance, emission and combustion characteristics of SI engine. *Energy Conversion and Management, 189*, 195–201.

Gopal, K., Sathiyagnanam, A. P., Kumar, B. R., Saravanan, S., Rana, D., & Sethuramasamyraja, B. (2018). Prediction of emissions and performance of a diesel engine fueled with n-octanol/diesel blends using response surface methodology. *Journal of Cleaner Production, 184*, 423–439.

Han, B., Yang, Y., Xu, Y., Etim, U. J., Qiao, K., Xu, B., & Yan, Z. (2016). A review of the direct oxidation of methane to methanol. *Chinese Journal of Catalysis, 37*(8), 1206–1215.

Hu, W., Jiaqiang, E., Leng, E., Zhang, F., Chen, J., & Ma, Y. (2023). Investigation on harvesting characteristics of convective wind energy from vehicle driving on multi-lane highway. *Energy, 263*, 126062.

Jamrozik, A., Tutak, W., & Grab-Rogaliński, K. (2021). Combustion stability, performance and emission characteristics of a CI engine fueled with diesel/n-butanol blends. *Energies, 14*(10), 2817.

Jiaqiang, E., Zhang, Z., Chen, J., Pham, M., Zhao, X., Peng, Q., … & Yin, Z. (2018). Performance and emission evaluation of a marine diesel engine fueled by water biodiesel-diesel emulsion blends with a fuel additive of a cerium oxide nanoparticle. *Energy Conversion and Management, 169*, 194–205.

Julis, J., & Leitner, W. (2012). Synthesis of 1-octanol and 1, 1-dioctyl ether from biomass-derived platform chemicals. *Angewandte Chemie International Edition: A Journal of the Gesellschaft Deutscher Chemiker, 51*(34), 8615–8619.

Kar, S., Kothandaraman, J., Goeppert, A., & Prakash, G. S. (2018). Advances in catalytic homogeneous hydrogenation of carbon dioxide to methanol. *Journal of CO2 Utilization, 23*, 212–218.

Kothare, C. B., Kongre, S., Malwe, P., Sharma, K., Qasem, N. A., Ağbulut, Ü., … & Panchal, H. (2023). Performance improvement and CO and HC emission reduction of variable compression ratio spark-ignition engine using n-pentanol as a fuel additive. *Alexandria Engineering Journal, 74*, 107–119.

Kumar, B. R., & Saravanan, S. (2016). Use of higher alcohol biofuels in diesel engines: A review. *Renewable and Sustainable Energy Reviews, 60*, 84–115.

Kumar, B. R., Saravanan, S., Rana, D., Anish, V., & Nagendran, A. (2016). Effect of a sustainable biofuel–n-octanol–on the combustion, performance and emissions of a DI diesel engine under naturally aspirated and exhaust gas recirculation (EGR) modes. *Energy Conversion and Management, 118*, 275–286.

Kumar, D., Sonawane, U., Chandra, K., & Agarwal, A. K. (2022). Experimental investigations of methanol fumigation via port fuel injection in preheated intake air in a single cylinder dual-fuel diesel engine. *Fuel, 324*, 124340.

Kuppili, Y. S., Kumar, P., & Sandhu, S. S. (2023). A critical review of the impacts of split injection strategies on the performance, combustion and emissions characteristics of a diesel engine fueled with low carbon alcohol fuels. *Materials Today: Proceedings, 72*, 2928–2932.

Li, Y., Tang, W., Chen, Y., Liu, J., & Chia-fon, F. L. (2019). Potential of acetone-butanol-ethanol (ABE) as a biofuel. *Fuel, 242*, 673–686.

Li, Y., Wen, C., Luo, Z., & Jin, L. (2023). Dynamic characteristics analysis of a dual-rotor system with bolted-disk joint. *Proceedings of the Institution of Mechanical Engineers, Part C: Journal of Mechanical Engineering Science, 237*(3), 534–548.

Liu, H., Wang, X., Wu, Y., Zhang, X., Jin, C., & Zheng, Z. (2019). Effect of diesel/PODE/ethanol blends on combustion and emissions of a heavy duty diesel engine. *Fuel, 257*, 116064.

Mahdi, F. P., Vasant, P., Kallimani, V., Watada, J., Fai, P. Y. S., & Abdullah-Al-Wadud, M. (2018). A holistic review on optimization strategies for combined economic emission dispatch problem. *Renewable and Sustainable Energy Reviews, 81*, 3006–3020.

McCormick, R. L., Ross, J. D., & Graboski, M. S. (1997). Effect of several oxygenates on regulated emissions from heavy-duty diesel engines. *Environmental Science & Technology, 31*(4), 1144–1150.

Nanthagopal, K., Kishna, R. S., Atabani, A. E., Ala'a, H., Kumar, G., & Ashok, B. (2020). A compressive review on the effects of alcohols and nanoparticles as an oxygenated enhancer in compression ignition engine. *Energy Conversion and Management, 203*, 112244.

Nathaniel, S. P., & Adeleye, N. (2021). Environmental preservation amidst carbon emissions, energy consumption, and urbanization in selected African countries: Implication for sustainability. *Journal of Cleaner Production, 285*, 125409.

Ning, L., Duan, Q., Chen, Z., Kou, H., Liu, B., Yang, B., & Zeng, K. (2020). A comparative study on the combustion and emissions of a non-road common rail diesel engine fueled with primary alcohol fuels (methanol, ethanol, and n-butanol)/diesel dual fuel. *Fuel, 266*, 117034.

No, S. Y. (2020). *Application of liquid biofuels to internal combustion engines*. Springer Nature.

Nour, M., Nada, S., & Li, X. (2022). Experimental study on the combustion performance of a stationary CIDI engine fueled with 1-heptanol-diesel mixtures. *Fuel, 312*, 122902.

Nour, M., Sun, Z., El-Seesy, A. I., & Li, X. (2021). Experimental evaluation of the performance and emissions of a direct-injection compression-ignition engine fueled with n-hexanol–diesel blends. *Fuel, 302*, 121144.

Okcu, M., Fırat, M., Varol, Y., Altun, S., Kamışlı, F., & Atila, O. (2022). Combustion of high carbon (C7-C8) alcohol fuels in a reactivity controlled compression ignition (RCCI) engine as low reactivity fuels and ANN approach to predict RCCI emissions. *Fuel, 319*, 123735.

Osman, A. I., Qasim, U., Jamil, F., Ala'a, H., Jrai, A. A., Al-Riyami, M., ... & Rooney, D. W. (2021). Bioethanol and biodiesel: Bibliometric mapping, policies and future needs. *Renewable and Sustainable Energy Reviews, 152*, 111677.

Pan, M., Zheng, Z., Huang, R., Zhou, X., Huang, H., Pan, J., & Chen, Z. (2019). Reduction in PM and NOX of a diesel engine integrated with n-octanol fuel addition and exhaust gas recirculation. *Energy, 187*, 115946.

Phillips, J. R., Atiyeh, H. K., Tanner, R. S., Torres, J. R., Saxena, J., Wilkins, M. R., & Huhnke, R. L. (2015). Butanol and hexanol production in clostridium carboxidivorans syngas fermentation: Medium development and culture techniques. *Bioresource Technology, 190*, 114–121.

Pinzi, S., Redel-Macías, M. D., Carmona-Cabello, M., Cubero, A., Herreros, J. M., & Dorado, M. P. (2021). Influence of 1-butanol and 1-pentanol addition to diesel fuel on exhaust and noise emissions under stationary and transient conditions. *Fuel, 301*, 121046.

Qian, Y., Chen, F., Zhang, Y., Tao, W., Han, D., & Lu, X. (2019). Combustion and regulated/unregulated emissions of a direct injection spark ignition engine fueled with C3-C5 alcohol/gasoline surrogate blends. *Energy, 174*, 779–791.

Raj, V. M., & Manoj, A. (2021, April). Improvement in performance and combustion of CRDI engine with simarouba blended fuel as alternative fuel. In *IOP Conference Series: Materials Science and Engineering* (Vol. 1130, No. 1, p. 012043). IOP Publishing.

Raj, V. M., Subramanian, L. G., Thiyagarajan, S., & Geo, V. E. (2018). Effects of minor addition of aliphatic (1-pentanol) and aromatic (benzyl alcohol) alcohols in Simarouba Glauca-diesel blend fuelled CI engine. *Fuel, 234*, 934–943.

Santhosh, K., & Kumar, G. N. (2020a). Effect of 1-pentanol addition and EGR on the combustion, performance and emission characteristic of a CRDI diesel engine. *Renewable Energy, 145*, 925–936.

Santhosh, K., & Kumar, G. N. (2020b). Impact of 1-Hexanol/diesel blends on combustion, performance and emission characteristics of CRDI CI mini truck engine under the influence of EGR. *Energy Conversion and Management, 217*, 113003.

Santhosh, K., & Kumar, G. N. (2021). Effect of injection time on combustion, performance and emission characteristics of direct injection CI engine fuelled with equi-volume of 1-hexanol/diesel blends. *Energy, 214*, 118984.

Saravanan, S. (2015). Effect of exhaust gas recirculation (EGR) on performance and emissions of a constant speed DI diesel engine fueled with pentanol/diesel blends. *Fuel, 160*, 217–226.

Sayed, E. T., Wilberforce, T., Elsaid, K., Rabaia, M. K. H., Abdelkareem, M. A., Chae, K. J., & Olabi, A. G. (2021). A critical review on environmental impacts of renewable energy systems and mitigation strategies: Wind, hydro, biomass and geothermal. *Science of the Total Environment, 766*, 144505.

Şen, M., Emiroğlu, A. O., & Keskin, A. (2019). Impact of pentanol addition and injection timing on the characteristics of a single-cylinder diesel engine. *Energy & Fuels, 33*(9), 9224–9231.

Shadidi, B., Alizade, H. H. A., & Najafi, G. (2021). The influence of diesel–ethanol fuel blends on performance parameters and exhaust emissions: Experimental investigation and multi-objective optimization of a diesel engine. *Sustainability, 13*(10), 5379.

Sharma, V., & Duraisamy, G. (2019). Production and characterization of bio-mix fuel produced from a ternary and quaternary mixture of raw oil feedstock. *Journal of Cleaner Production, 221*, 271–285.

Shi, W., Li, Z., Zhao, Z., Yu, X., Sun, P., Sang, T., ... & Li, M. (2024). A renewable clean energy application: Oxyhydrogen negative pressure inhalation for enhancing the combustion and emission characteristics of isopropanol/gasoline dual-fuel combined injection engine. *Energy, 290*, 130214.

Singh, D., Sharma, D., Soni, S. L., Sharma, S., Sharma, P. K., & Jhalani, A. (2020). A review on feedstocks, production processes, and yield for different generations of biodiesel. *Fuel, 262*, 116553.

Viswanathan, V. K., & Thomai, P. (2021). Performance and emission characteristics analysis of Elaeocarpus Ganitrus biodiesel blend using CI engine. *Fuel, 288*, 119611.

Wang, X., Wang, Y., & Bai, Y. (2021). Oxidation behaviors and nanostructure of particulate matter produced from a diesel engine fueled with n-pentanol and 2-ethylhexyl nitrate additives. *Fuel, 288*, 119844.

Wei, L., Cheung, C. S., & Huang, Z. (2014). Effect of n-pentanol addition on the combustion, performance and emission characteristics of a direct-injection diesel engine. *Energy, 70*, 172–180.

Wei, L., Yao, C., Wang, Q., Pan, W., & Han, G. (2015). Combustion and emission characteristics of a turbocharged diesel engine using high premixed ratio of methanol and diesel fuel. *Fuel, 140*, 156–163.

Wei, Z., Zhang, Y., Xia, Q., Liu, Y., & Xu, Y. (2022). A simulation of ethanol substitution rate and EGR effect on combustion and emissions from a high-loaded diesel/ethanol dual-fuel engine. *Fuel, 310*, 122310.

Xia, Q., Xia, Y., Xi, J., Liu, X., & Wang, Y. (2015). Energy-efficient production of 1-octanol from biomass-derived furfural-acetone in water. *Green Chemistry, 17*(8), 4411–4417.

Xu, C., Zhuang, Y., Qian, Y., & Cho, H. (2022). Effect on the performance and emissions of methanol/diesel dual-fuel engine with different methanol injection positions. *Fuel, 307*, 121868.

Yaman, H., Doğan, B., Yeşilyurt, M. K., & Erol, D. (2021). Application of higher-order alcohols (1-hexanol-C6 and 1-heptanol-C7) in a spark-ignition engine: Analysis and assessment. *Arabian Journal for Science and Engineering, 46*(12), 11937–11961.

Yaman, H., & Yeşilyurt, M. K. (2021). The influence of n-pentanol blending with gasoline on performance, combustion, and emission behaviors of an SI engine. *Engineering Science and Technology, an International Journal, 24*(6), 1329–1346.

Yang, Y., Zheng, X., & Sun, Z. (2020). Coal resource security assessment in China: A study using entropy-weight-based TOPSIS and BP neural network. *Sustainability, 12*(6), 2294.

Ye, J., Jiaqiang, E., & Peng, Q. (2023). Effects of porosity setting and multilayers of diesel particulate filter on the improvement of regeneration performance. *Energy, 263*, 126063.

Yilmaz, N., & Atmanli, A. (2017). Experimental evaluation of a diesel engine running on the blends of diesel and pentanol as a next generation higher alcohol. *Fuel, 210*, 75–82.

Yilmaz, N., Vigil, F. M., Atmanli, A., & Donaldson, B. (2022). Detailed analysis of PAH formation, toxicity and regulated pollutants in a diesel engine running on diesel blends with n-propanol, n-butanol and n-pentanol. *Energies, 15*(17), 6487.

Zameer, H., Wang, Y., Vasbieva, D. G., & Abbas, Q. (2021). Exploring a pathway to carbon neutrality via reinforcing environmental performance through green process innovation, environmental orientation and green competitive advantage. *Journal of Environmental Management, 296*, 113383.

Zhang, L., Li, H., Lee, W. J., & Liao, H. (2021). COVID-19 and energy: Influence mechanisms and research methodologies. *Sustainable Production and Consumption, 27*, 2134–2152.

Zhang, R., He, Y., Luo, Y., Lou, D., Zhu, R., Zhu, C., & Li, Q. (2023). Selective preparation of jet fuels from low carbon alcohols and ABE at atmospheric pressure. *Energy, 281*, 128246.

Zhang, Y., Gao, S., Zhang, Z., Li, W., Yuan, T., Tan, D., ... & Yang, G. (2023). A comprehensive review on combustion, performance and emission aspects of higher alcohols and its additive effect on the diesel engine. *Fuel, 335*, 127011.

Zhang, Z., Li, J., Tian, J., Zhong, Y., Zou, Z., Dong, R., ... & Tan, D. (2022). The effects of Mn-based catalysts on the selective catalytic reduction of NOx with NH_3 at low temperature: A review. *Fuel Processing Technology, 230*, 107213.

Zhang, Z., Tian, J., Li, J., Lv, J., Wang, S., Zhong, Y., ... & Tan, D. (2022). Investigation on combustion, performance and emission characteristics of a diesel engine fueled with diesel/alcohol/n-butanol blended fuels. *Fuel, 320*, 123975.

Zhang, Z., Ye, J., Tan, D., Feng, Z., Luo, J., Tan, Y., & Huang, Y. (2021). The effects of Fe_2O_3 based DOC and SCR catalyst on the combustion and emission characteristics of a diesel engine fueled with biodiesel. *Fuel, 290*, 120039.

Zhao, W., Yan, J., Gao, S., Lee, T. H., & Li, X. (2022a). The combustion and emission characteristics of a common-rail diesel engine fueled with diesel and higher alcohols blends with a high blend ratio. *Energy, 261*, 124972.

Zhao, W., Yan, J., Gao, S., Lee, T. H., & Li, X. (2022b). The combustion and emission characteristics of a common-rail diesel engine fueled with diesel, propanol, and pentanol blends under low intake pressures. *Fuel, 307*, 121692.

Zhou, L., Boot, M. D., & De Goey, L. P. H. (2012). The effect of the position of oxygen group to the aromatic ring to emission performance in a heavy-duty diesel engine. *SAE International Journal of Fuels and Lubricants, 5*(3), 1216–1239.

Zhou, L., Boot, M. D., Johansson, B. H., & Reijnders, J. J. E. (2014). Performance of lignin derived aromatic oxygenates in a heavy-duty diesel engine. *Fuel, 115*, 469–478.

Zhu, L., Xiao, Y., Cheung, C. S., Guan, C., & Huang, Z. (2016). Combustion, gaseous and particulate emission of a diesel engine fueled with n-pentanol (C5 alcohol) blended with waste cooking oil biodiesel. *Applied Thermal Engineering, 102*, 73–79.

3 Improving Solubility and Stability of Alcohol Fuels with Surfactants

E. A. Lohith, K. Keerthi, Sada Venkateswarlu, K. Siva Kumar, and N. V. V. Jyothi

3.1 INTRODUCTION

Interest in alcohol as a potential alternative to traditional fossil fuels for internal combustion engines (ICEs) has grown due to the global movement toward sustainable energy solutions. Alcohols such as methanol and ethanol offer numerous benefits, including higher octane ratings, cleaner combustion, and the potential for renewable production. However, their limited solubility in diesel fuel and issues with phase separation, particularly at low temperatures, present considerable obstacles to their use in diesel engines (Yudharaputra and Adisasmito 2023). These challenges make alcohol-diesel mixtures less stable, reducing their consistency and effectiveness as alternative fuels. Nicolaus Otto first used ethyl alcohol as fuel for the ICE in 1860. Since then, ethanol has become an important fuel for ICEs in cars, trucks, and other machinery. Methanol, ethanol, propanol, and butanol are used as fuels in ICEs; they can be synthesized chemically and biologically. Most methanol is derived from natural gas (Usman et al. 2023). While the ethanol is often produced through fermentation, a biological process. Alcohols produced biologically are known as bioalcohols. The combustion of alcohol in ICEs increases the power and thermal efficiency compared to gasoline while reducing greenhouse gas emissions (Salim and Saed 2023). Ethanol is a suitable fuel for gasoline engines due to its high-octane number (Amaral et al. 2021; Kurji et al. 2021; Taco-Vasquez et al. 2022). Fuel characteristics have a significant impact on the combustion process, which helps reduce engine noise. Studies have shown that the use of ring-shaped porous inert media (PIM) in a combustor can effectively reduce combustion noise (Krishnamoorthi et al. 2019). Noise levels can also be influenced by the design of fuel system components such as fuel rails and nozzles. For instance, modifying the fuel system's acoustic impedance by adjusting the size and placement of orifices (Meadows and Agrawal 2012). Engine noise can be affected by fuel attributes including fuel injection system design, lubrication, combustion characteristics, cleanliness, and octane rating. Cleaner, higher-quality fuels with regulated combustion processes and lubricant additives can result in quieter engine operation by reducing knocking, promoting complete combustion, decreasing friction, and improving fuel distribution.

DOI: 10.1201/9781003482277-3

Diesel engine, a type of ICE, ignites fuel through the elevated temperature of air in the cylinder caused by mechanical compression (Alaa and Mohamed 2014). Diesel engine development began in the 18th century, with early models being large and slow. After World War II, diesel-powered transport vehicles became popular in Europe (Weng and Ren 2018). Surfactants are known for their ability to lower surface tension and stabilize mixtures. As emulsifiers, surfactants create more stable alcohol-diesel emulsions, preventing phase separation and ensuring a homogenous blend suitable for diesel engines (Nguyen et al. 2023). This chapter explores the principles behind using surfactants in fuel blending, examines various types of surfactants, and evaluates their effectiveness in enhancing the solubility and stability of alcohols in diesel fuels.

The performance of diesel engines varies depending on speed and size. Diesel engines are categorized by power output into small, medium, and large sizes, with their success determined by design, reliability, efficiency, adaptability, and cost-effectiveness. This chapter also discusses current experimental research and practical applications, highlighting the advantages and potential drawbacks of using alcohol fuels stabilized with surfactants in diesel engines. It aims to support ongoing efforts to improve the efficiency and environmental impact of diesel engines through the use of alternative fuels by addressing these critical issues.

3.2 DIESEL ENGINES

A diesel engine refers to any ICE that uses compressed air heated to a high enough temperature to ignite diesel fuel in the cylinder, where combustion and expansion move a piston. This process converts the fuel's chemical energy into mechanical energy, which can be used to power trains, freight vehicles, large tractors, and maritime vessels. Some electric generator sets and a small number of cars are also diesel-powered.

The design of a diesel engine plays a crucial role in influencing its key characteristics. Diesel engines are equipped with start-assist systems that enhance ignition power, reducing starting noise, white smoke, and cranking time. The ignition quality of a fuel depends on factors such as misfire rate, smoothness of operation, smoke emissions, ease of starting, and noise levels. The noise produced by a diesel engine is a combination of combustion and mechanical noise, which is directly affected by the fuel's properties.

Starting a cold diesel engine is generally more laborious compared to starting a warm one. Typically, the average airborne noise level of a moderate-rated diesel engine is around 105 dB. Diesel and gasoline engines are similar in that they are both ICEs designed to convert chemical energy into mechanical energy through a series of combustions.

The addition of certain additives to diesel fuel can reduce friction and noise. Two such additives that work well in diesel engines are STP (Scientifically Treated Petroleum) oil additive and nylon. Fuel oil additives are chemical compounds that improve lubrication, alter viscosity, prevent sludge formation, and protect against corrosion.

3.2.1 Biofuels in Diesel Engines

Another oxygenated biofuel is alcohol, which is classified into higher and lower alcohols based on the type of carbon-to-carbon bond. The chemical and physical properties of pure diesel fuel and various alcohols are compared in Table 3.1. Blending oxygenated fuels, such as alcohols, with standard fuels like diesel and biodiesel is being explored worldwide as a way to reduce harmful emissions and meet growing energy demands. Methanol and ethanol are lower alcohols with one and two carbon atoms, respectively, while alcohols with four or more carbon atoms, such as butanol, pentanol, hexanol, octanol, decanol, and phytol, are considered higher alcohols.

However, due to their higher auto-ignition temperature, lower cetane number, and higher calorific value, the use of alcohols in compression ignition (CI) engines is limited. To address this, alcohols are utilized in CI engines through various methods, such as emulsification, direct injection, and blending. Numerous studies have explored these techniques for combining alcohol with diesel fuel. Enhancing the oxygen content of fuel depends largely on the oxygen concentration of the alcohol additives. Alcohol additives increase the atomic oxygen content, improving combustion quality in the chamber. It is believed that the oxygenated derivatives in the fuel raise the ignition temperature. One of the most pressing global issues today is the future of the environment. Young researchers have recently raised awareness about the need to protect the environment and the use of alcohol fuels or non-fossil fuels for engine combustion. Higher-octane alcohols can produce more power and fuel efficiency than gasoline, with reduced evaporative emissions. Alcohol fuels also contain less carbon than traditional fuels, resulting in lower ash emissions into the environment after combustion.

Countries that still depend on fossil fuels must put up enormous effort to achieve the aim of decarbonization. The yearly availability of biomass in India is close to 500 million tons, with a surplus of 120–150 million tons (Dey et al. 2022; Narwane et al. 2021). Timelines and various applications of biofuels are shown in Figures 3.1 and 3.2, respectively.

TABLE 3.1
Chemical and Physical Properties of Pure Diesel Fuel and Alcohols

Properties	Pure Diesel	Ethanol	n-Butanol	Methanol
Cinematic viscosity at 40°C ($cm^2/s.10^{-6}$)	3.35	1.2	3	0.75
Density at 20°C (kg/m^3)	837	788	810	790
Cetane number	50	5–8	25	3–5
Oxygen molecules	0	34.8	21.6	50
C/H	0.45	0.33	0.4	0.25
Flashpoint (°C)	45	13–14	35–37	11
Boiling point (°C)	180–360	78	118	64.7
Latent energy (kJ/kg)	250	840	585	1100

Source: Adapted with Permission from Fayyazbakhsh and Pirouzfar (2017). Copyright© 2017, Elsevier.

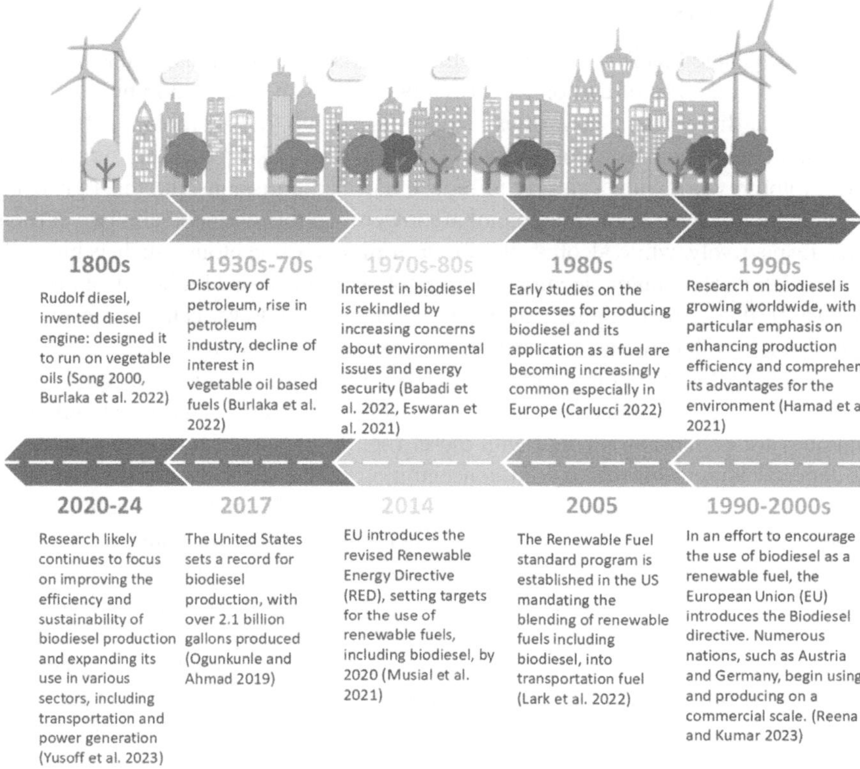

FIGURE 3.1 Timeline of significant events and research milestones in the development and use of biodiesel as fuel.

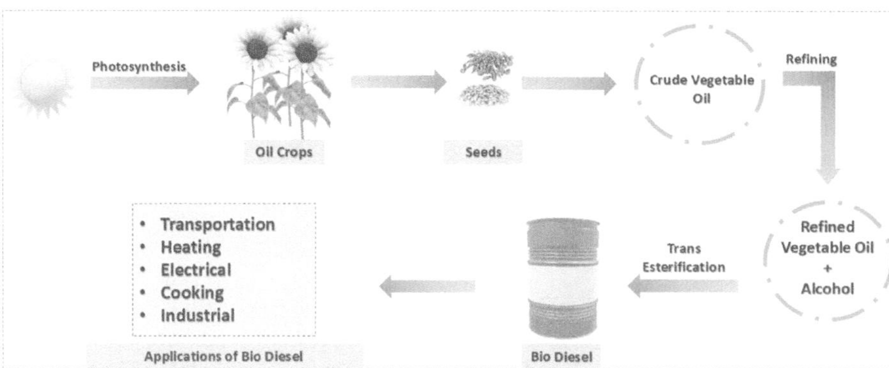

FIGURE 3.2 Various applications of biofuels in daily life. (Adapted with permission from Ogunkunle and Ahmed (2019). Copyright© 2019, Elsevier.)

According to recent research studies, biodiesel can improve a diesel engine's performance and decrease the amount of harmful exhaust (Goga et al. 2019). Compared to regular diesel, biodiesel significantly reduces carbon monoxide (CO) and unburned hydrocarbons (HC). Various researchers have focused on using biofuel as an alternate fuel in diesel engines. Using leftover vegetable oil (cocking oil) from the Tafila Technical University Restaurant, W.M. Adaileh et al. created a biofuel (Adaileh and AlQdah 2011). Testing results showed that the dynamic performance of the diesel engine remained normal under all conditions without any modifications. The B20 and B5 blend fuels, which contain 20% and 5% biodiesel, respectively, led to acceptable emissions at variable loads. When compared to regular diesel, there is an increase in NO_x emissions (Subramani et al. 2020). Due to its lower heating value, biodiesel results in a 5.95% increase in fuel consumption specifically for brake-specific fuel usage. However, for NO_x and brake-specific fuel usage, utilizing B20 and B5 diesel fuel produced better emission outcomes.

Thus, there has been significant interest in using vegetable oils, such as jatropha, palm, algae, and leftover cooking oil, as alternative fuels in diesel engines. K.A. Abed et al. utilized the trans-esterification process to create biodiesel from these sources. Diesel oil and biodiesel from various feedstocks are combined in different ratios, such as B10 and B20. The physical and chemical characteristics of biodiesel are measured in compliance with ASTM guidelines. The test engine used in this work is a single-cylinder diesel engine. Diesel oil and exhaust emissions, including CO, CO_2, NO_x, HC, and smoke, are measured and compared. The biodiesel blends emit less CO, HC, CO_2, and smoke when compared to diesel fuel; however, CO_2 emissions from the B10 and B20 blends made from leftover cooking oil are higher. For any biodiesel blend, including B10 and B20, NO_x emissions are greater than those of diesel fuel (Nunes and Teixeira 2022). The percentage of emission reduction from biofuels is shown in Figure 3.3.

3.2.2 Alcohols as Fuel in Diesel Engines

Another oxygenated biofuel is alcohol, which is divided into higher and lower alcohol fractions according to the type of carbon-to-carbon bond (Loyte et al. 2023). The chemical and physical properties of pure diesel fuel and various alcohols are comparatively shown in Table 3.1. Blending oxygenated fuels, such as alcohols, with standard fuels like diesel and biodiesel, is being explored globally as a possible way to reduce harmful emissions and meet the world's growing energy needs. Methanol and ethanol are primary/lower alcohols with one and two carbon atoms, whereas alcohols with four or more carbon atoms can be considered higher alcohols, such as butanol, pentanol, hexanol, octanol, decanol, and phytol (Nanthagopal et al. 2020). However, due to its higher auto-ignition temperature, lower cetane number, and higher calorific value, alcohol application in CI engines is limited. Because of this, alcohol is used in CI engines in a variety of ways, such as emulsification, direct injection, and blending. Numerous studies have examined these techniques for combining alcohol with diesel fuel (Nour et al. 2021). Enhancing the oxygen content of the fuel is mostly dependent on the oxygen concentration of alcohol additions. An alcohol additive fuel increases

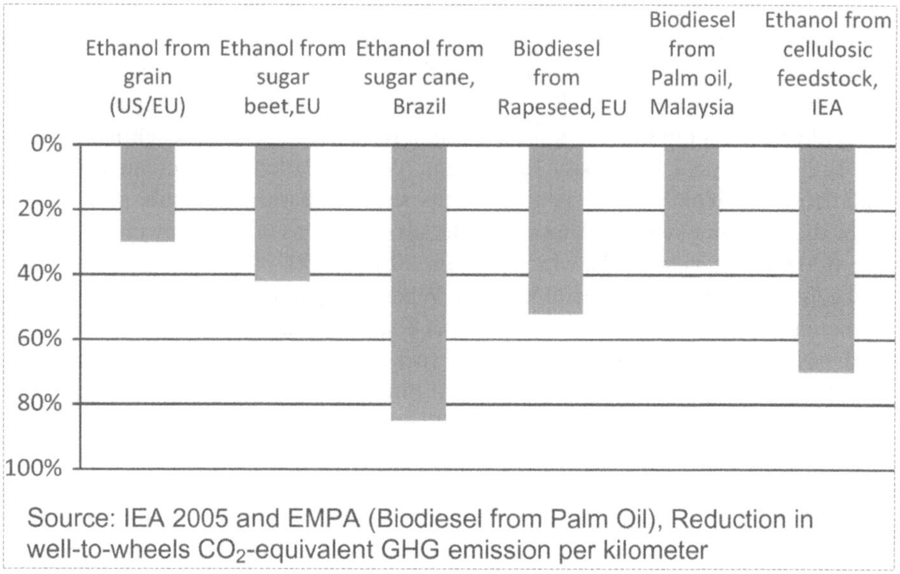

FIGURE 3.3 Percentage of emission reduction from biofuels. (Adapted with permission from Mofijur et al. (2015). Copyright© 2015, Elsevier.)

atomic oxygen content that improves a combustion chamber's quality of combustion. It is thought that the fuel's oxygenated derivatives cause an increase in ignition temperature (Fayyazbakhsh and Pirouzfar 2017). The most pressing issue at present is undoubtedly how the global ecology will develop in the future. Our young researchers have recently raised awareness of the need to protect the environment and use alcohol fuels or non-fossil fuels for engine combustion. Higher average octane alcohols can produce more power and fuel economy than gasoline. They also have reduced evaporative emissions. Alcohol fuel has less carbon compared to normal fuel, so less amount of ash is emitted into the environment after combustion.

3.3 SURFACTANTS

Surface-active agents, commonly known as surfactants, are substances that reduce the surface tension between two phases, most frequently between two liquids or between a liquid and a solid. They interact with both polar and non-polar substances because they have a lipophilic (oil-attracting) tail and a hydrophilic (water-attracting) head. Surfactants are essential for a wide range of applications across different sectors. By lowering the interfacial tension between immiscible liquids, such as oil and water, surfactants can create stable emulsions (Griffin 1954), which are extensively used in the culinary, cosmetics, and pharmaceutical. Surfactants are necessary ingredients in cleaning supplies and detergents, facilitating the dissolution and breakdown of oils and greases so that water can wash them away. They also decrease the contact angle between a liquid and a solid surface, improving surface wetting, which is crucial for applications in printing, painting, and agriculture (pesticide spreading). Surfactants can stimulate or prevent foam production, and they are selected for industrial operations, cleaning agents, and shampoos based on their ability to foam or defoam. Various applications of surfactants have been shown in Figure 3.4. To prevent particles in liquids from clumping together, surfactants are employed as dispersion agents, which is useful in the paint, ink, and pharmaceutical industries (Ramprasad and Philip 2016). Poorly soluble compounds can become more soluble in a given solvent when surfactants are added, a property that is exploited in some chemical processes and drug formulations. They stabilize colloidal systems by preventing particles from settling or agglomerating, which is crucial in the production of paints, inks, and other industrial products. Pharmaceutical companies use surfactants for drug delivery, solubilization, and as constituents of lotions and ointments. In the oil and gas sector, surfactants modify the characteristics of crude oil, enhancing its flow and enabling improved oil recovery. In agricultural formulations, surfactants are added as adjuvants to enhance the efficacy and dispersal of insecticides and herbicides. Due to their ability to cleanse and

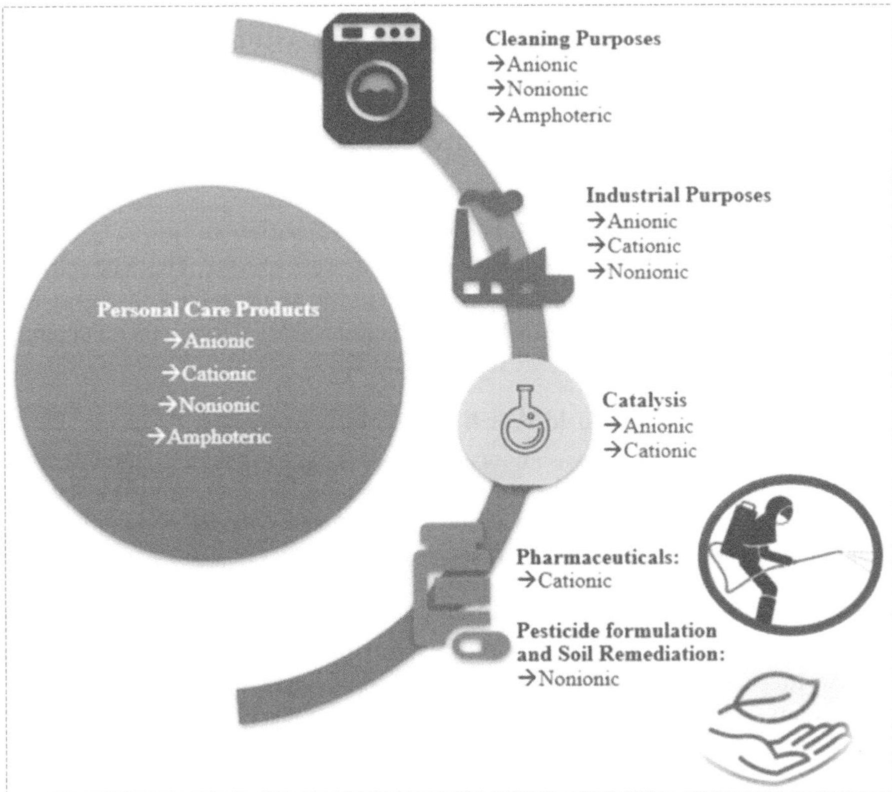

FIGURE 3.4 Applications of various classes of surfactants. (Adapted with permissions from Nunes and Teixeira (2022). Copyright© 2022, Elsevier.)

emulsify, surfactants are frequently found in shampoos, body washes, and lotions in cosmetics and personal care products. Their versatility makes surfactants essential to many applications, contributing to the performance and usefulness of everyday items as well as industrial operations.

3.3.1 History of Surfactants

3.3.1.1 Pre-20th Century (Early Discoveries)

For ages, people have been using natural surfactants without realizing it. Materials such as minerals from clay and plants called soapwort were employed as cleaning agents. However, nothing was known about the science underlying these mechanisms (Kim and Park 2022).

3.3.1.2 Ancient Civilizations (Soap Production)

From ancient civilizations including the Babylonians, Egyptians, and Romans, soaps are among the first examples of surfactants. They combined animal or vegetable fats with alkaline materials like ash to make soap (Marquez et al., 2024).

3.3.1.3 18th–19th Century (Industrial Revolution)

Chemistry has advanced significantly since the Industrial Revolution. The French chemist Michel Eugene Chevreul discovered the structure of fats and proved that fatty acids and glycerol were related early in the 19th century. His research established the groundwork for our current understanding of surfactant structure (McNamara et al. 2006).

3.3.1.4 Early to Mid-20th Century (Synthetic Surfactants)

In the early 1900s, the first synthetic surfactants were created. German scientists created "syndet", the first synthetic detergent, in 1913 in response to the scarcity of fats and oils during World War I (Ciccone et al. 2022). The synthetic detergent industry got its start with this.

3.3.1.5 1930s–1950s (World War II and Detergent Boom)

During World War II, there was an increase in demand for cleaning supplies, which prompted more advancements in synthetic detergent technology. Alkylbenzene sulfonates were introduced during this period and went on to become an essential ingredient in many household detergents (Sharipovich 2021).

3.3.1.6 1950s–1970s (Post-War Innovations and Specialized Surfactants)

Surfactant chemistry continued to progress during the post-war era. For particular uses, researchers created a variety of specialty surfactants, such as nonionic surfactants, which are frequently found in contemporary formulations (Blunk et al. 2007).

3.3.1.7 1970s–Present (Environmental Concerns and Regulations)

Concerns concerning the effects of some surfactants on the environment arose in the late 20th century as environmental consciousness grew. For instance, phosphate-based surfactants were linked to water pollution, which prompted laws and the creation of more ecologically friendly substitutes (Vasylenko et al. 2022).

3.3.1.8 Late 20th Century–Present (Modern Applications and Research)

Even now, surfactant technology is still advancing. Researchers concentrate on creating surfactants with certain qualities for use in various sectors, including nanotechnology, agriculture, and pharmaceuticals. The evolution of surfactants has always been intimately related to scientific knowledge, industrial advancement, and societal demands. Surfactants are now essential to many goods and procedures, benefiting everything from sophisticated industrial uses to home cleaning. It has been shown that adding surfactants to engine fuel improves both environmental and performance aspects.

3.3.2 Types of Surfactants

Based on their chemical structure, charge, and function, surfactants can be divided into multiple categories (Nakama 2016). Examples of various surfactants are shown in Figure 3.5. The major categories include the following.

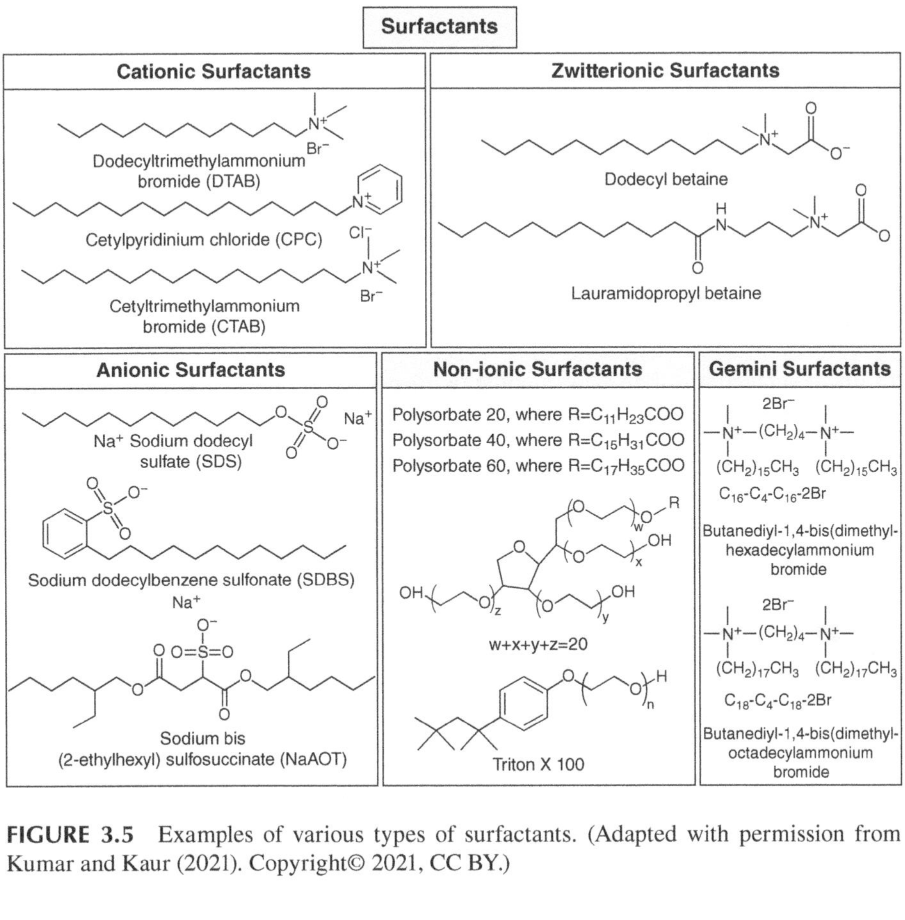

FIGURE 3.5 Examples of various types of surfactants. (Adapted with permission from Kumar and Kaur (2021). Copyright© 2021, CC BY.)

3.3.2.1 Based on Charge

3.3.2.1.1 Anionic Surfactants

Their hydrophilic head is negatively charged. Examples include sulfonates (alkylbenzene sulfonates in detergents) and sulfates (like sodium lauryl sulfate in shampoos).

3.3.2.1.2 Cationic Surfactants

Their hydrophilic head is positively charged. Common in disinfectants (such as cetyltrimethylammonium bromide) and fabric softeners.

3.3.2.1.3 Nonionic Surfactants

These have no net charge. Common in products for sensitive skin, such as baby shampoos, e.g., alcohol ethoxylates and alkylphenol ethoxylates.

3.3.2.2 Based on Hydrophilic Head: Amphiphilic Surfactants

These are composed of both hydrophilic and hydrophobic parts. The majority of surfactants are included in this group.

3.3.2.3 Based on Hydrophobic Tail

3.3.2.3.1 Alkyl Surfactants

The hydrophobic tail is typically a long alkyl chain. e.g., alkyl sulfates, alkylbenzene sulfonates, and alkylphenol ethoxylates.

3.3.2.3.2 Fluorinated Surfactants

Atoms of fluorine can be found in the hydrophobic tail. Because of their specific qualities, they are employed in niche applications.

3.3.2.4 Based on Molecular Structure

3.3.2.4.1 Linear Surfactants

In linear surfactants, their hydrophobic tail is a straight chain. A commonly known example is linear alkylbenzene sulfonates (LAS).

3.3.2.4.2 Branched Surfactants

There are branches on the hydrophobic tail. Compared to linear surfactants, branched surfactants could have different characteristics.

3.3.2.5 Based on Function

Surfactants based on function are shown in Table 3.2.

3.3.2.6 Specialty Surfactants

3.3.2.6.1 Gemini Surfactants

Gemini surfactants consist of two hydrophilic heads and two hydrophobic chains connected by a spacer at or close to the head groups (Guerrero-Hernandez et al. 2022).

TABLE 3.2
Surfactants Based on Function

Types	Function
Emulsifying surfactants	They assist mixes of immiscible liquids to stay mixed by stabilizing emulsions. Prevalent in cosmetic and culinary items.
Wetting surfactants	To improve wetting, decrease the liquid's contact angle with the solid surface. Utilized in printing, coatings, and agriculture.
Foaming and defoaming surfactants	Manage the production of foam or disintegrate it. Discovered in industrial operations, shampoos, and detergents.
Solubilizing surfactants	Boost a substance's solubility in a specific solvent. Utilized in chemical and medicinal operations.
Corrosion inhibiting surfactants	Reduce corrosion on metal surfaces by forming protective layers. Used in some industrial processes and metalworking fluids.

3.3.2.6.2 Polymeric Surfactants

Macromolecules that contain both hydrophilic and hydrophobic segments, as well as polymer chains, are called polymeric surfactants (Raffa et al. 2015), providing unique properties used in specific industrial applications.

3.4 ROLE OF SURFACTANTS IN DIESEL ENGINES

Surfactants are essential for lowering cold-start problems in diesel engines by enhancing the functionality of a mixture of biodiesel. Research indicates that the incorporation of surfactants such as octanol and sorbitan monooleate (SMO) can effectively reduce the cloud point (CP) and cold filter plugging point (CFPP) of biodiesel, hence impeding the agglomeration of saturated monoglycerides (SMG) during low-temperature conditions. Moreover, diesel and water emulsions made with a combination of surfactants have demonstrated potential in lowering NO_x, SO_x, CO, and particulate matter emissions without reducing engine efficiency. Because they help disperse and stabilize water droplets in the fuel, surfactants are essential to diesel engines. Diesel fuel is prone to water contamination from condensation and other sources, especially while it's being transported or stored. Diesel fuel containing water can cause several problems, including microbiological development, corrosion, and reduced combustion efficiency. Lowering the surface tension separates water from diesel fuel, enabling the former to spread and condense into smaller droplets (Vellaiyan and Amirthagadeswaran 2016). Because of this dispersion, water droplets are kept from coalescing and stay finely disseminated inside the fuel. Surfactants lessen the negative effects of water pollution in diesel engines by preserving water in this state (Morozumi and Saito 2010). Furthermore, surfactants can stabilize fuel-water emulsions. Diesel fuels that have been emulsified have been investigated because they may offer advantages like better combustion and lower emissions. By stopping the separation of the diesel and water phases, surfactants aid in the stabilization of these emulsions. Emulsified diesel fuel is a combination of diesel fuel and water that is often stabilized with an emulsifying agent. This fuel blend has various advantages over normal diesel fuel, most notably improved combustion efficiency and lower pollutants. This is a thorough description of how using emulsified diesel fuel results in combustion and less emissions. When using alcoholic blends, surfactants are essential for strengthening the cold-start performance of CI engines. This is because they improve the solubility and stability of ethanol in diesel fuel. By enhancing fuel properties, surfactants make it easier to employ alcoholic mixes in CI engines and successfully solve the cold-start issue.

3.4.1 Enhanced Efficiency of Combustion by Surfactants

The high temperature of the emulsified diesel fuel causes the water droplets to evaporate and expand quickly when it is introduced into the combustion chamber. Micro-emulsions are a phenomenon that improves the fuel-air mixing process by breaking the fuel into smaller droplets. Completer and more effective combustion results from improved fuel atomization. When fuel vaporizes, the fuel's total surface area increases due to the water content. As a result, the diesel burns more thoroughly, and the fuel is more evenly distributed throughout the combustion chamber.

3.4.2 Lowering of Emissions by Surfactants

Emulsified fuel reduces NO_x emissions by lowering peak combustion temperature due to water's high latent heat of vaporization. It also reduces particulate matter (PM) and carbon monoxide (CO) emissions due to improved atomization and mixing from the micro-explosion effect, resulting in lower concentrations in the exhaust. Emulsified diesel offers reduced smoke and potential fuel savings due to improved combustion efficiency. Emulsifying agents maintain a stable mixture of diesel and water, preventing separation. Modern diesel engines can run efficiently on emulsified diesel fuel, but their use may be limited by engine design and storage practices. An effective way to raise combustion efficiency and lower hazardous emissions from diesel engines is to use emulsified diesel fuel. This fuel blend can assist in complying with rigorous environmental requirements and lessen the overall environmental effect of diesel-powered cars and equipment by utilizing the qualities of water to improve the combustion process. It is crucial to remember that choosing the right surfactants for diesel engines is crucial since using the wrong ones can have unforeseen ramifications that could negatively impact fuel stability and engine performance.

3.4.3 Effect of Surfactant on the Environment

Surfactants are widely used, and as a result, a significant amount of their concentration from largely urban or industrial and domestic wastewater might wind up in municipal wastewater either through direct environmental discharge from treatment plants (MWTPs) (Bautista-Toledo et al. 2014). Various environmental parameters along with positive and negative impacts are given in Table 3.3.

TABLE 3.3
Environmental Impact of Surfactants

Environmental Impact of Surfactants	Positive Impact	Negative Impact
Biodegradability	Biodegradable surfactants decompose into elements that are not harmful	Long-term ecological hazards might arise from the accumulation of non-biodegradable surfactants
Aquatic toxicity	Surfactants that are good for the environment are not harmful to aquatic life	Certain surfactants can be hazardous to aquatic life, which could be detrimental to ecosystems
Foam formation	Controlled foam in specific applications (e.g., wastewater treatment)	An overabundance of foam can damage aquatic life and disturb aquatic ecosystems
Groundwater contamination	Surfactants that are less mobile and biodegradable have a lower potential to contaminate groundwater	Some surfactants may leach into groundwater posing contamination risks
Bioaccumulation	Biodegradable surfactants are less likely to bioaccumulate	Ecosystems may be at risk due to the accumulation of persistent surfactants in organisms.
Impact on soil	Certain surfactants enhance the qualities of soil and boost agricultural productivity	High amounts or improper application could hurt soil health
Air quality	By increasing combustion efficiency, surfactants can help lower air pollution	Certain surfactants might be involved in air pollution and VOC (volatile organic compounds) emissions
Eutrophication	–	Surfactants in wastewater can contribute to nutrient loading and eutrophication

3.5 ENHANCING THE ROLE OF SURFACTANTS IN ALCOHOL-BASED FUELS

Fossil fuel burning is the most well-known source of environmental pollution, and it is thought to be the main energy source in use today. Finding fossil fuel substitutes and replacing them is therefore of interest to maintain sustainable economic growth and adhere to stringent environmental emission requirements (ENTERING A, GAS AG, 2011). The blended use of ethanol and diesel fuel is an excellent strategy to address these problems. Many studies blending ethanol with diesel fuel found that the greatest results come from blending only a small amount of ethanol (less than 5%) since ethanol and diesel fuel are not miscible (Mofijur et al. 2015). Very recently M. M. El-Sheekh et al. performed an extensive experiment to ascertain the effects of altering the diesel and ethanol fuel blends (10%, 20%, 30%, and 40% by weight). For the first time, it was found that adding surfactant (tri-n-butyl phosphate) enhanced

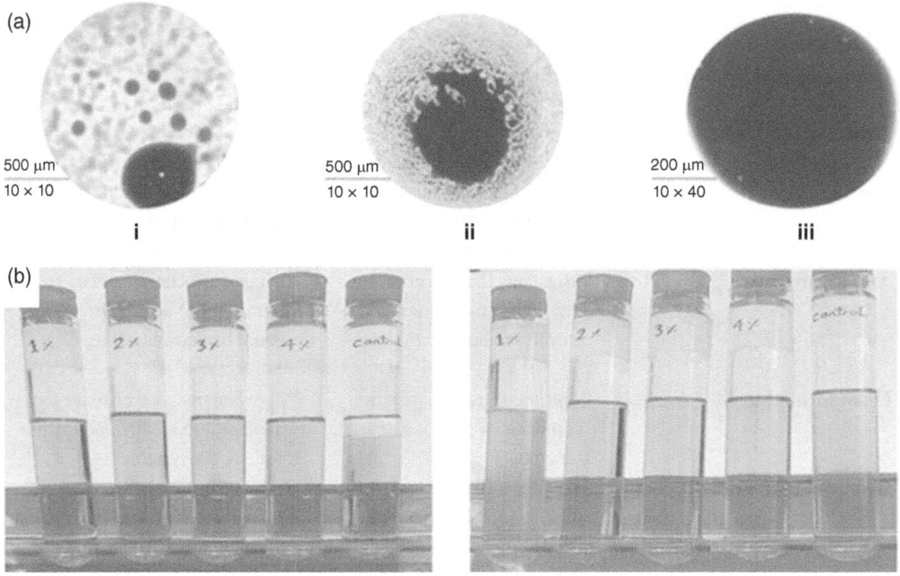

FIGURE 3.6 (a) Miscibility of fuel systems observing using a microscope: (i) heavy fuel oil-methanol: (ii) heavy fuel oil-ethanol: (iii) heavy fuel oil-ethanol-surfactant. (b) Stability of ethanol inside diesel with 20%–30% of ethanol and 80%–70% diesel. ((a) Adapted with permission from Jin et al. (2022). Copyright© 2022, Elsevier and (b) adapted with permission from El-Sheekh et al. (2023). Copyright© 2023, CC BY 4.0.)

the stability of high-ethanol diesel fuel. Throughout the series of tests, every biofuel blend is used in varied load settings with a diesel engine running at 1500 rpm. In every test, emissions were gathered and assessed using an exhausting gas analyzer equipment. The stability of ethanol inside diesel is shown in Figure 3.6b. At varying engine load circumstances, the exhaust gas temperature, brake-specific fuel consumption (BSFC), and brake thermal efficiency of the engine are all measured. They concluded that a 40% increase in the base volume of bioethanol caused an increase in the combustion of unburned HC and the emission of CO, while O_2 emission increased with a gradual decrease in CO_2 emission. The effect of ethanol+surfactant on CO, HC, NO_x, and O_2 is shown in Figure 3.7 (El-Sheekh et al. 2023). Similarly, S. Nanadakumar et al. investigated the properties of diesel-ethanol blended fuels with surfactant. Based on the "Hydrophilic-Lipophilic Balance" and the "Theory of similarity", a surfactant solution is chosen. Similar outcomes are seen when cetyl trimethyl ammonium bromide, a surfactant is introduced. The concentration of carbon and hydrogen dropped as the amount of ethanol added increased. This will result in a decrease in the specific energy content of the various fuel blends. When oxygenated fuels like ethanol are introduced, the emission rates go down (Nanadakumar and Malayalamurthy 2015).

To reduce the use of fossil fuels and engine emissions, respectively, A. Molea et. al. emphasized the potential use of diesel-ethanol blends as an alternative fuel in ICEs. Ethanol was used in the blends at concentrations of up to 15% v/v. Additionally,

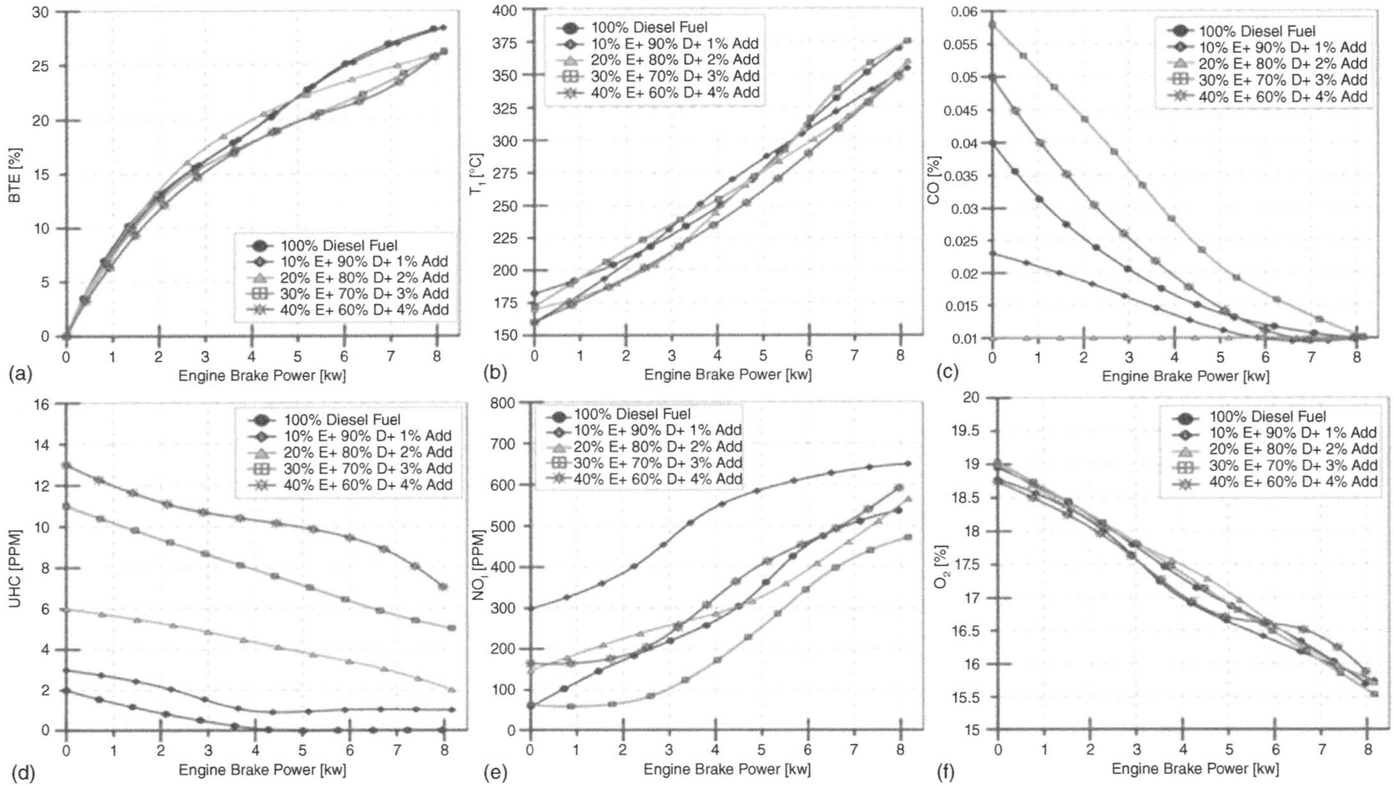

FIGURE 3.7 (a) Effect of ethanol+surfactant on engine thermal efficiency. (b) Effect of ethanol+surfactant on exhaust gas temp. (c) Effect of ethanol+surfactant on CO emissions. (d) Effect of ethanol+surfactant on HC emissions. (e) Effect of ethanol+surfactant on NO_x emissions. (f) Effect of ethanol+surfactant on O_2 emissions. (Adapted with permission from El-Sheekh et al. (2023). Copyright© 2023, CC BY 4.0.)

stability studies revealed that at low ethanol concentrations, the blends could be stabilized at 22°C in two weeks with 2.4% v/v THF (tetrahydrofuran); however, at high ethanol concentrations, diesel and ethanol were miscible and could only be stabilized with the addition of up to 7% v/v THF. The density values of the diesel-ethanol-THF blends were found to be within EN 590 limits based on their physicochemical analysis. The kinematic viscosity values of the blends containing up to 5% v/v ethanol were found to be close to the standard inferior limit, while the blends containing more than 5% v/v ethanol had kinematic viscosity values below the inferior limit. The surface tension values exhibit a modest drop as the proportion of ethanol and THF increases. However, the blend's three properties remain within acceptable limits when used as fuels in diesel engines. However, the mixtures show a flash point lower than 25°C, necessitating additional safety precautions due to the low flash points of both THF and ethanol. Additionally, it was highlighted using UV-Vis spectroscopy that tetrahydrofuran compounds function as a surface-active agent, establishing a bridge between diesel and ethanol that facilitates the creation of a miscible phase. There are no chemical interactions between diesel and ethanol or THF compounds. The engine simulation demonstrates that, as compared to diesel operation, the engine running with the blends produced slightly lower cylinder pressures and temperatures but superior emission parameters (Molea et al. 2017).

The potential to create molecules with different architectures and customized emulsification properties is provided by the comparatively new materials known as ionic liquid Gemini surfactants (ILGS) molecules, which have two head groups joined together on each molecule. Through titration using 1-butanol as a co-surfactant, the efficacy of Tween 80, Span 80, and ILGS in the formation of single-phase fuel micro-emulsions was evaluated as shown in Figure 3.8. [C_{12}-5-im] [2Ac] ILGS with a longer spacer length has performed better than [C_{12}-5-im] [2Ac] with a shorter space length, leading to reduced 1-butanol consumption, explained by the creation of more stable micelles. The ideal Tween80:Span80:ILGS ratio is 70:27:3, producing samples that have stayed stable as a single phase for as long as six months (Ghazal and Najjar 2024).

On the one hand, research is concentrated on alternative fuels; on the other hand, fuel is altered by additives. The addition of metal oxide nanoparticles remains a viable approach when emissions decrease, as size confinement enhances catalytic surface area. Cerium oxide is prepared via the precipitation process and chosen as a catalytic nanoparticle. Oleic acid surfactant is used to modify nanoparticles so that their dispersion stabilities can be assessed in comparison to their unmodified states. It is cost-effective and exhibits excellent stability and morphological qualities. Different quantities of cerium oxide (CeO_2), nanoparticles are used in a DI-CI engine to test engine performance and emission characteristics. By adding nanoparticles, the emissions are reduced; the ideal dose of nanoparticles for engine performance and emission patterns is 35 ppm. The best mixture for creating nano-fuel is found to be 20% diesel and biodiesel. In the load test, efficiency rose to 5%. At greater loads, emissions of HC dropped by 45% and emissions of NO_x by 30%, which is shown in Figures 3.9 and 3.10 (Roy et al. 2024).

To improve the stability and solubility of alcohol-diesel mixtures, surfactants are essential. Studies have demonstrated that adding surfactants to ethanol-diesel

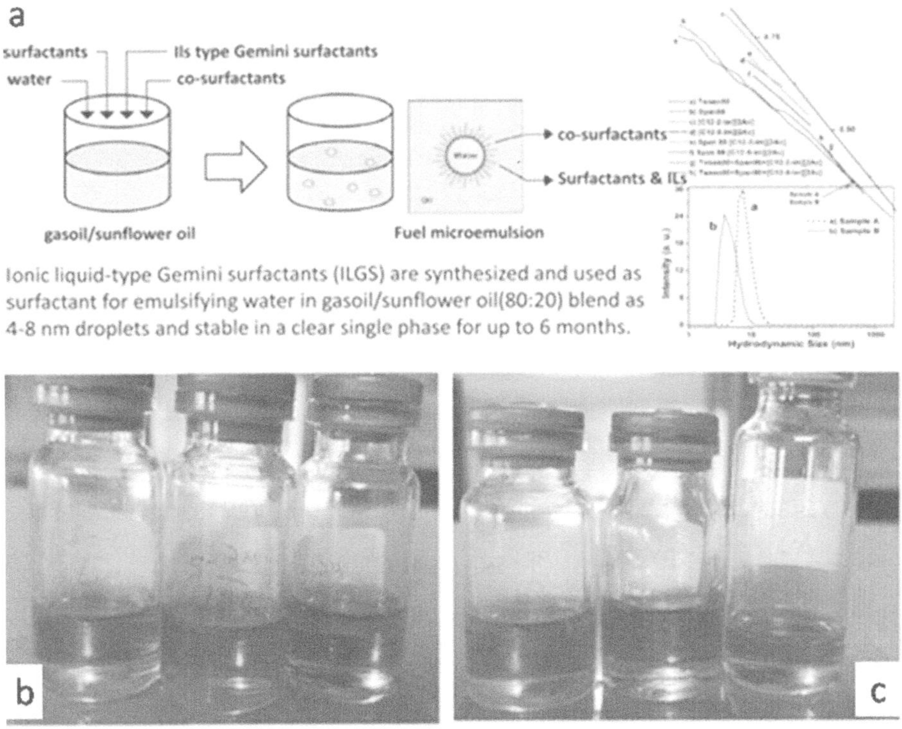

FIGURE 3.8 Schematic illustration of ILGS surfactants. (a) Picture of gas oil/sunflower oil (80:20) blend, prepared by using [C_{12}-2-im] [2Ac] (b, c) prepared by using [C_{12}-2-im] [2Ac] and various surfactants with $\gamma = 0.05$, including (Tween 80 and Span 80) and 1-butanol as co-surfactant at room temperature. (Adapted with permission from Ghazal and Najjar (2024). Copyright© 2024, Elsevier.)

blends, such as 1-octanal, greatly increases stability and prevents phase separation across a broad temperature range (Shaikh and Patel 2022). Similar research has shown that in the absence of surfactants, variables such as ultrasonication, stirring speed, and time affect how soluble ethanol is in diesel fuel, under ideal circumstances, ethanol is more soluble in diesel fuel without the need for surfactants (Parray and Bhattacharya 2022). Furthermore, it has been discovered that adding surfactants to biodiesel-diesel blends, such as Span 80 and Tween 80, improves stability over time by reducing phase separation through the utilization of larger droplet sizes and higher viscosities (Maji et al. 2022). In general, surfactants are necessary to ensure homogenous mixes and extended storage stability by increasing the stability and solubility of alcohol-diesel blends.

Surfactants improve fuel atomization, mixing, and combustion efficiency, which is essential for creating different types of low-temperature combustion (LTC). The ability of surfactants to enhance fuel evaporation and mixing properties is advantageous for several forms of LTC, including homogeneous charge CI (HCCI), partially premixed CI (PPCI), and reactivity-controlled CI (RCCI) (Gwalwanshi et al. 2023;

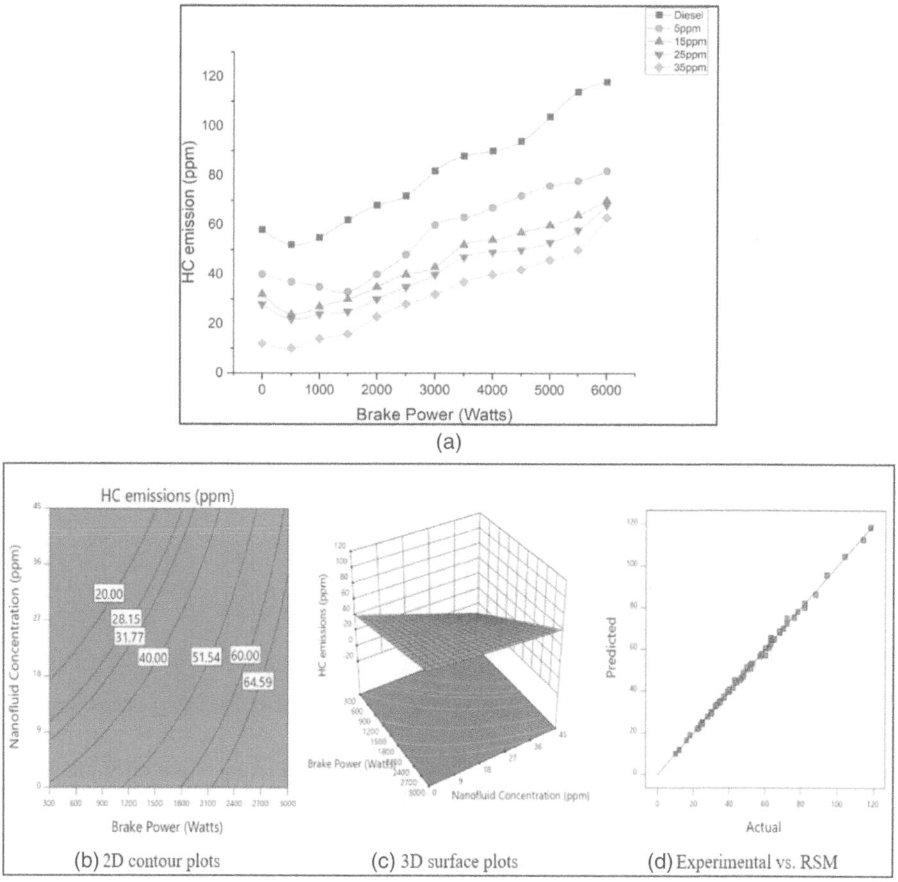

FIGURE 3.9 (a) Variation in HC productions for both regular and modified diesel under load. (b) 2D contour plots for Variation in HC productions for both regular and modified diesel under load. (c) 3D surface plots for Variation in HC productions for both regular and modified diesel under load. (d) Experimental vs. RSM for Variation in HC productions for both regular and modified diesel under load. HC emissions (ppm) = 112.185 + 0.199088. Brake Power (Watts)+ −4.41278. CR+ −14.8935. Nanofluid Concentration (ppm)+ −0.010959. Brake Power (Watts). CR + 0.000333353. Brake Power (Watts). Nanofluid Concentration (ppm) + 0.798619. CR. Nanofluid Concentration (ppm). (Adapted with permission from Roy et al. (2024). Copyright© 2024, Elsevier.)

Singh and Agarwal, 2017). By encouraging the auto-ignition of strongly premixed charges, surfactants can help achieve LTC conditions. This is a fundamental idea in LTC engines such as dual-fuel-based RCCI, premixed charge CI (PCCI), and gasoline-controlled auto-ignition (GCAI) (Biswas and Gururajan 2021). Moreover, surfactants can improve thermal efficiency and help reduce emissions such as CO and oxides of nitrogen (NO_x), which is in line with the LTC engine's objectives of achieving ultra-low emissions and high-efficiency levels (Maurya 2018).

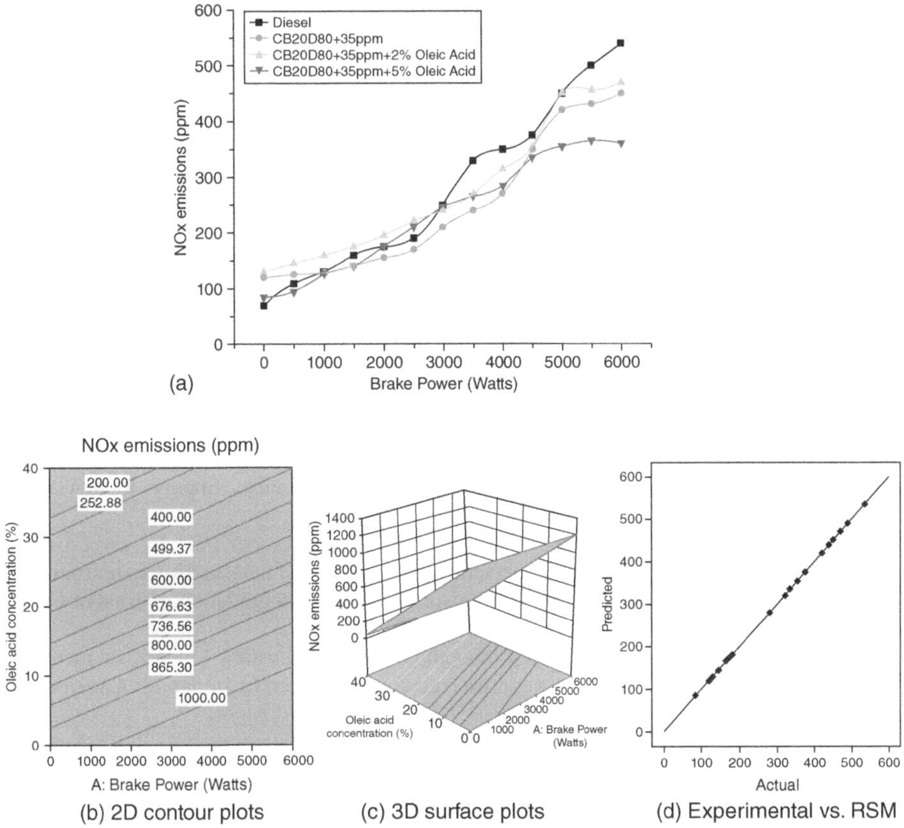

FIGURE 3.10 (a) BTE variation with load for modified diesel and diesel that has different surfactant volume percentages (oleic acid). (b) 2D contour plots for BTE variation with load for modified diesel and diesel that has different surfactant volume percentages (oleic acid). (c) 3D surface plots for BTE variation with load for modified diesel and diesel that has different surfactant volume percentages (oleic acid). (d) Experimental vs. RSM for BTE variation with load for modified diesel and diesel that has different surfactant volume percentages (oleic acid). BTE (%) = 112.185 + 0.199088. Brake Power (Watts) + 94.5825. CR + 37.4846. Oleic acid Concentration (%) + 0.00193014. Brake Power (Watts). CR+ −1.52239E-05. Brake Power (Watts). Oleic acid Concentration (ppm)+ −2.35064. CR. Oleic acid Concentration (%). (Adapted with permission from Roy et al. (2024). Copyright© 2024, Elsevier.)

3.6 CONCLUSION

In conclusion, surfactants help to stabilize fuel-water emulsions and reduce the risk of water contamination in diesel engines by spreading out water droplets in the fuel. Their application is a component of initiatives to guarantee the dependable and effective operation of diesel engines. One interesting approach to improving the fuel stability of alcohol in diesel engines is the use of surfactants. Surfactants can improve the overall stability of alcohol-diesel blends by modifying the interfacial characteristics between alcohol and diesel fuel, hence mitigating difficulties like phase separation and sedimentation. This stability improvement is a workable way to increase the use of alcohol fuels in diesel engines because it not only encourages increased combustion efficiency but also lowers emissions and wear on the engine. Furthermore, a growing number of surfactants are now accessible for a wider range of applications owing to continuous research and development.

REFERENCES

Adaileh, W. M., and AlQdah, K. S. 2011. Performance of diesel engine fuelled by a biodiesel extracted from a waste cooking oil. Energy Procedia. 18: 1317–1334.

Alaa, E.-D., and Mohamed, M. 2014. Conventional and insulated pulse diesel engines performance and combustion. Middle-East Journal of Scientific Research. 22(8): 1127–1131.

Amaral, L. V., Santos, N. D. S. A., Roso, V. R., De Cassia De Oliveira Sebastiao, R., and Pujatti, F. J. P. 2021. Effects of gasoline composition on engine performance, exhaust gases and operational costs. Renewable and Sustainable Energy Reviews. 135: 110196.

Bautista-Toledo, M. I., Rivera-Utrilla, J., Mendez-Díaz, J. D., Sanchez-Polo, M., and Carrasco-Marín, F. 2014. Removal of the surfactant sodium dodecyl benzenesulfonate from water by processes based on adsorption/bioadsorption and biodegradation. Journal of Colloid and Interface Science. 418: 113–119.

Biswas, S., and Gururajan, V. 2021. Low-temperature combustion kinetics of methanol-blended gasoline and methanol synthesized dimethyl ether. In *Methanol: A Sustainable Transport Fuel for SI Engines* (pp. 193–219). Springer Singapore.

Blunk, D., Bierganns, P., Bongartz, N., Tessendorf, R., and Stubenrauch, C. 2007. New speciality surfactants with natural structural motifs. ChemInform. 38(14).

Ciccone, G., Skopalik, S., Smart, C., Gezgin, S., Ridland, D., Paul, M. C., Noriega Escobar, D. P., and Tassieri, M. 2022. A rheological characterization of synthetic detergent formulations. Physics of Fluids. 34(9).

Dey, S., Sreenivasulu, A., Veerendra, G. T. N., Rao, K. V., and Babu, P. S. S. A. 2022. Renewable energy present status and future potentials in India: An overview. Innovation and Green Development. 1(1): 100006.

El-Sheekh, M. M., El-Nagar, A. A., ElKelawy, M., and Bastawissi, H. A. 2023. Solubility and stability enhancement of ethanol in diesel fuel by using tri-n-butyl phosphate as a new surfactant for CI engine. Scientific Reports. 13(1).

ENTERING A, GAS AG 2011. World Energy Outlook. International energy Agency: Paris, France.

Fayyazbakhsh, A., and Pirouzfar, V. 2017. Comprehensive overview on diesel additives to reduce emissions, enhance fuel properties and improve engine performance. Renewable and Sustainable Energy Reviews. 74: 891–901.

Ghazal, Y., and Najjar, R. 2024. Gasoil/sunflower oil blend/water fuel microemulsions prepared with ionic liquids-type Gemini surfactants. Alexandria Engineering Journal. 90: 161–169.

Goga, G., Chauhan, B. S., Mahla, S. K., and Cho, H. M. 2019. Performance and emission characteristics of diesel engine fueled with rice bran biodiesel and n-butanol. Energy Reports. 5: 78–83.

Griffin, W. C. 1954. Calculation of HLB values of non-ionic surfactants. Journal of the Society of Cosmetic Chemists. 5(4): 249–256.

Guerrero-Hernandez, L., Melendez-Ortiz, H. I., Cortez-Mazatan, G. Y., Vaillant-Sánchez, S., and Peralta-Rodriguez, R. D. 2022. Gemini and bicephalous surfactants: A review on their synthesis, micelle formation, and uses. International Journal of Molecular Sciences. 23(3): 1798.

Gwalwanshi, M., Chauhan, M. K., and Kumar, R. 2023. A review of low temperature combustion techniques and the evolution of combustion strategies. Emerging Trends in Mechanical and Industrial Engineering: Select Proceedings of ICETMIE 2022., pp. 75–83.

Jin, C., Sun, T., Ampah, J. D., Liu, X., Geng, Z., Afrane, S., Yusuf, A. A., and Liu, H. 2022. Comparitive study on synthetic and biological surfactants role in phase behaviour and fuel properties of Marine heavy fuel oil-low carbon alcohol blends under different temperatures. Renewable Energy. 195: 841–852.

Kim, G. S., and Park, C. H. 2022. Research trends of natural and natural-derived surfactant in raw materials. Journal of the Korean Society of Cosmetology. 28(5): 1124–1134.

Krishnamoorthi, M., Malayalamurthi, R., He, Z., and Kandasamy, S. 2019. A review on low temperature combustion engines: Performance, combustion and emission characteristics. Renewable and Sustainable Energy Reviews. 116: 109404.

Kumar, H., and Kaur, G. 2021. Scrutinizing self-assembly, surface activity and aggregation behavior of mixtures of imidazolium based ionic liquids and surfactants: A comprehensive review. Frontiers in Chemistry. 9: 667941.

Kurji, H. J., Imran, M. S., and Bded, A. S. 2021. The impact of using pure ethanol additives on gasoline fuel with respect to SI engine emissions. IOP Conference Series Materials Science and Engineering. 1067(1): 012090.

Loyte, A., Suryawanshi, J., Bhiogade, G., Devarajan, Y., Thandavamoorthy, R., and Mishra, R. 2023. Influence of injection strategies on ignition patterns of RCCI combustion engine fuelled with hydrogen enriched natural gas. Environmental Research. 234: 116537.

Maji, N. C., Rastogi, P., Krishnasamy, A., Aidhen, I. S., Kaisare, N. S., and Basavaraj, M. G. 2022. Storage and temperature stability of emulsified biodiesel-diesel blends. ACS Omega. 7(49): 44762–44771.

Marquez R, Ontiveros JF, Barrios N, Tolosa L, Palazzo G, Nardello-Rataj V, and Salager JL. 2024. Advantages and limitations of different methods to determine the optimum formulation in surfactant–oil–water systems: A review. Journal of Surfactants and Detergents. 27(1):5–36.

Maurya, R. K. 2018. *Characteristics and Control of Low Temperature Combustion Engines. Mechanical Engineering Series.* Springer

McNamara, J. R., Warnick, G. R., and Cooper, G. R. 2006. A brief history of lipid and lipoprotein measurements and their contribution to clinical chemistry. Clinica Chimica Acta. 369(2): 158–167.

Meadows, J., and Agrawal, A. 2012. Passive mitigation of combustion noise in liquid fuel combustion using porous inert media. In 50th AIAA Aerospace Sciences Meeting Including the New Horizons Forum and Aerospace Exposition, p. 206.

Mofijur, M., Rasul, M., Hyde, J., and Bhuyia, M. 2015. Role of biofuels on IC engines emission reduction. Energy Procedia. 75: 886–892.

Mofijur, M., Rasul, M., Hyde, J., Azad, A., Mamat, R., and Bhuiya, M. 2015. Role of biofuel and their binary (diesel-biodiesel) and ternary (ethanol-biodiesel-diesel) blends on internal combustion engines emission reduction. Renewable and Sustainable Energy Reviews. 53: 265–278.

Molea, A., Visuian, P., Barabás, I., Suciu, R. C., and Burnete, N. V. 2017. Key fuel properties and engine performances of diesel-ethanol blends, using tetrahydrofuran as surfactant additive. IOP Conference Series Materials Science and Engineering. 252: 012077.

Morozumi, Y., and Saito, Y. 2010. Effect of physical properties on microexplosion occurrence in water-in-oil emulsion droplets. Energy & Fuels. 24(3): 1854–1859.

Nakama, Y. 2016. Surfactants. Cosmetic Science and Technology. 231–244.

Nanadakumar, S., and Malayalamurthy, R. 2015. Properties of diesel – ethanol blended fuels with surfactant. International Journal of Engineering Research and Technology. V4(11).

Nanthagopal, K., Kishna, R. S., Atabani, A. E., Al-Muhtaseb, A. H., Kumar, G., and Ashok, B. 2020. A compressive review on the effects of alcohols and nanoparticles as an oxygenated enhancer in compression ignition engine. Energy Conversion and Management. 203: 112244.

Narwane, V. S., Yadav, V. S., Raut, R. D., Narkhede, B. E., and Gardas, B. B. 2021. Sustainable development challenges of the biofuel industry in India based on integrated MCDM approach. Renewable Energy. 164: 298–309.

Nguyen, V. G., Pham, M. T., Le, N. V. L., Le, H. C., Truong, T. H., and Cao, D. N. 2023. A comprehensive review on the use of biodiesel for diesel engines. IJRED (International Journal of Renewable Energy Development. 12(4): 720–740.

Nour, M., Sun, Z., El-Seesy, A. I., and Li, X. 2021. Experimental evaluation of the performance and emissions of a direct-injection compression-ignition engine fueled with n-hexanol-diesel blends. Fuel. 302: 121144.

Nunes, R. F., and Teixeira, A. C. S. C. 2022. An overview on surfactants as pollutants of concern: Occurrence, impacts and persulfate-based remediation technologies. Chemosphere. 300: 134507.

Ogunkunle, O., and Ahmed, N. A. 2019. A review of global current scenario of biodiesel adoption and combustion in vehicular diesel engines. Energy Reports. 5: 1560–1579.

Parray, R. A., and Bhattacharya, T. K. 2022. Effectiveness of ethyl acetate, 1-octanol, and soy biodiesel in stabilizing ethanol-diesel fuel blends and performance of compression ignition engine on stabilized fuel blends. ACS Omega. 7(2): 1648–1657.

Raffa, P., Wever, D. a. Z., Picchioni, F., and Broekhuis, A. A. 2015. Polymeric surfactants: Synthesis, properties, and links to applications. Chemical Reviews. 115(16): 8504–8563.

Singh, A. P. and Agarwal A.K. 2017. Low-temperature combustion engine control. Advances in internal combustion engine research, pp. 9–41.

Ramprasad, C., and Philip, L. 2016. Surfactants and personal care products removal in pilot scale horizontal and vertical flow constructed wetlands while treating grey water. Journal of Chemical Engineering. 284: 458–468.

Roy, R. G., Ağbulut, M., Koshy, C. P., Alex, Y., Sailesh, K., Khan, S. A., Jilte, R., Linul, E., and Asif, M. 2024. Impact of synthesizing surfactant-modified catalytic ceria nanoparticles on the performance and environmental behaviours of coconut oil/diesel-fueled CI engine: An optimization attempt. Energy. 295: 130825.

Salim, M. A., and Saed, U. A. 2023. Production of high-octane number gasoline from Basra low octane-number gas condensate and ethanol over modified Zn/ZSM5 zeolite catalyst. Bulletin of Chemical Reaction Engineering & Catalysis. 18(1): 59–70.

Shaikh, M. A., and Patel, V. R. 2022. Experimental studies on ethanol solubility and nanoparticle (NP) stability in diesel fuel. Chemical Engineering Research and Design. 188: 105–129.

Sharipovich, A. M. 2021. Application of Benzol Sulfonates in the Preparation of Detergents. Zenodo (CERN European Organization for Nuclear Research).

Subramani, S., Govindasamy, R., and Rao, G. L. N. 2020. Predictive correlations for NOx and smoke emission of DI CI engine fuelled with diesel-biodiesel-higher alcohol blends-response surface methodology approach. Fuel. 269: 117304.

Taco-Vasquez, S., Salinas, M., Murillo, H., and Oñate, W. 2022. Ethanol to high-octane hydrocarbons using HZSM-5 as a catalyst. IOP Conference Series Earth and Environmental Science. 1094(1): 012005.

Usman, M., Malik, M. a. I., Chaudhary, T. N., Riaz, F., Raza, S., Abubakar, M., Malik, F. A., Ahmad, H. M., Fouad, Y., Abbas, M. M., and Kalam, M. A. 2023. Comparative assessment of ethanol and methanol–ethanol blends with gasoline in SI engine for sustainable development. Sustainability. 15(9): 7601.

Vasylenko, L., Bereznytska, Y., Kravchenko, A., Shevchenko, O., and Tsema, T. 2022. Pollution of surface waters with phosphates and heavy metals. Problems of Water Supply Sewerage and Hydraulic. 38: 4–17.

Vellaiyan, S., and Amirthagadeswaran, K. 2016. The role of water-in-diesel emulsion and its additives on diesel engine performance and emission levels: A retrospective review. Alexandria Engineering Journal. 55(3): 2463–2472.

Weng, L., and Ren, J. 2018. Optimization of diesel engine based on digital design technology and finite element simulation analysis. Proceedings of the 2018 8th International Conference on Mechatronics, Computer and Education Informationization (MCEI 2018). 274–277.

Yudharaputra, D. L., and Adisasmito, S. 2023. Determination of the optimum operating conditions for integrated methanol and ethanol plant from natural gas reactors. International Journal of Biological, Physical and Chemical Studies. 5(1): 09–17.

4 Optimization of Methanol/WCO Biodiesel/Diesel Ternary Blend in CI Engine

Ananda Murugan M. and Nataraj G.

4.1 INTRODUCTION

Energy is a fundamental requirement for global progress. Researchers are increasingly concerned about the rapid depletion of fossil fuels and the ever-increasing need for energy supplies, prompting them to seek new energy sources that can quickly replace them (Verma et al., 2021). About 3 kg of greenhouse gases are discharged into the atmosphere when 1 l of diesel fuel is burnt (Ağbulut et al., 2020). The energy security and environmental pollution caused by engines that run on conventional fuels have prompted a lot of experimental studies into finding renewable, sustainable, and alternative fuels (Mohammed et al., 2023). Because of dwindling oil supplies and rising climate concerns, there has been a dramatic increase in research efforts to discover alternative fuels for internal combustion engines in the last several years (Adhikesavan et al., 2022). Biodiesel is a kind of biofuel that shows promise as a tool to improve engine performance while decreasing harmful emissions (Tuan Hoang et al., 2021). For compression ignition engines, biodiesel is a viable alternative to diesel fuel that is both environmentally beneficial and functionally equivalent.

The lesser soot formation in biodiesel than in petro-diesel results in fewer engine deposits and carbon deposits. The methyl esters produce less soot formation than the ethyl esters due to their longer hydrocarbon chains and lower aromatic compounds than the neat diesel. As the double bond in the structure increases, soot emissions also increase (Feng et al., 2012). NO_x emissions are one of the harmful emissions from CI engines that affect humans as well as the environment. It is mainly the byproduct of combustion, with higher brake thermal efficiency (BTE) and fewer CO emissions. Most of the NO is converted into NO_2 when it reacts with the atmospheric ozone. At the same time, NO_2 is further decomposed to NO due to the heat transfer from the sun in the atmosphere. The equations for the NO_x decomposition are as follows (Lozhkina & Lozhkin, 2016):

$$NO + O_3 \rightleftharpoons NO_2 + O_2$$

$$NO_2 + h\nu \rightleftharpoons NO + O_2$$

DOI: 10.1201/9781003482277-4

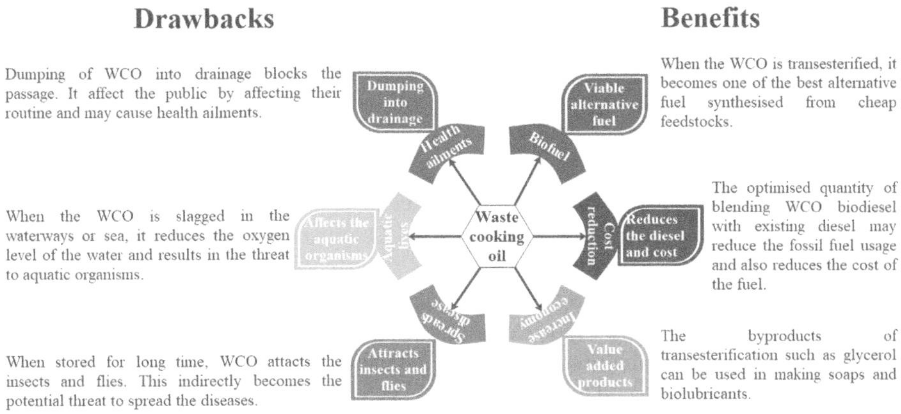

FIGURE 4.1 Drawbacks and the potential benefits of waste cooking oil.

Edible vegetable oil is used for frying food items and undergoes several physiochemical changes. When the food is fried in the oil, the water content of the food mixes with the oil, increasing the viscosity and acidity level. This happens due to the hydrolysis process occurring during high-temperature cooking, which results in excess free fatty acids (FFAs) (Zhang, 2003). Also, there will be excess food sludge suspended in the oil. So, it is mandatory to do a pretreatment process to purify the waste cooking oil (WCO). There are different methods of pretreatment, such as vacuum filtration, film vacuum evaporation, injection of steam, and sedimentation methods (Kulkarni & Dalai, 2006). Also, the water content can be removed by heating oil to 100°C. Some of the drawbacks and potential benefits of waste cooking to the environment is illustrated in Figure 4.1.

After the pretreatment process, the oil can be converted into fuel by either the transesterification process, micro-emulsification, pyrolysis, or direct blending process (E et al., 2017). Among all, transesterification is a popular and affordable method. Since the FFA in the WCO is higher, there is a two-step transesterification process. The first process is called esterification, where the FFA is reduced to less than 1%. The next step that follows is called transesterification. There are some drawbacks involved with the two-step transesterification process. The acid catalyst content will be higher in the fatty acid methyl ester, which may harden the engine components when this fuel is used. To overcome this issue, a base catalyst can be used instead of an acid catalyst. The direct blending of methanol with conventional fuel has more advantages in terms of reducing exhaust emissions. But the blending of methanol fuel with conventional diesel causes an immediate phase separation due to its immiscibility. The immediate phase separation can be avoided by adding emulsifiers or cosolvents to improve stability. Also, high-carbon alcohols, such as n-butanol, n-decanol, and n-octanol, can be blended with diesel and methanol to prevent phase separation. A higher cetane number and a lower latent heat of vaporization characterize the high-carbon alcohols in comparison to methanol (El-Seesy et al., 2021).

Lei Zhu et al. (2010) conducted an emission study using WCO biodiesel blended with methanol and ethanol in binary blended form. The study showcased that increasing the ethanol and methanol ratio up to 15% in the blend increased the HC and CO emissions, whereas the ratio up to 5% reduced the HC and CO emissions. When the ratio of biodiesel in the blend is lower than that of neat diesel and replaced by alcohol, CO emissions tend to increase. At lower engine loads, the CO emissions are higher, whereas at higher loads, CO emissions tend to reduce. In another study, the author compared the B20 blend and 5% methanol-added B20 blend with neat diesel and concluded that B20 and B20M5 produced 18.3% and 13.4% lower CO emissions at lower loads, 19.8% and 17.2% lower CO emissions at midloads, and 18.9% and 11.9% lower CO emissions at high loads than the neat diesel (Mat Yasin et al., 2014). Methanol has the greatest ability to reduce NO_x and PM emissions because of its higher latent heat of vaporization and excess oxygen content. However, when blended with diesel, there are several disadvantages, such as phase separation, a lower cetane number, and reduced lubricity. To overcome these disadvantages, several studies focused on adding the cetane enhancers to improve the blends' properties. In a study, Li Ruina et al. (Li et al., 2014) conducted experiments with 2-ethylhexyl nitrate, 2-methoxyethyl ether, and cyclohexyl nitrate as the cetane enhancers in methanol-biodiesel blends. The results revealed that the addition of cetane enhancers shortened the ignition duration by 1–3 deg. crank angle and retarded the combustion duration by 1–6 deg. crank angle. In terms of emissions, the cetane enhancers in the blend reduced NO_x and smoke emissions, whereas HC and CO emissions increased.

In a study (Qi et al., 2011), the authors conducted the experiment by taking two ternary fuel blends, i.e., biodiesel/diesel/diethyl ether and biodiesel/diesel/ethanol blends. The results proved that the brake-specific fuel consumption (BSFC) of the binary blend, i.e., the B30 blend, is higher than the two ternary blends. This is due to the volatility properties of the ethanol and diethyl ether that aid in the mixture of the fuel with the compressed air inside the cylinder. This increased the combustion efficiency. The CO emissions of the ternary blends are lower than those of the B30 blend, whereas the HC emissions are higher. The NO_x emissions of the blend with diethyl ether are similar to those of B30, whereas the blend with ethanol produced higher NO_x. Finally, the author concluded that oxygenated additives such as ethanol and diethyl ether are the most promising solution to form ternary blends, which result in balancing results in performance, emissions, and combustion without modifying the engine. In another study (Kumar et al., 2022), the authors conducted experiments by formulating ternary fuel blends with palm biodiesel and diesel using higher alcohols such as 1-decanol and n-hexanol. The addition of 20% and 30% higher alcohol in the blend produced excellent BSFC and BTE results compared to diesel. It is said that BSFC was reduced by 2.2%–10.87% by using ternary fuel blends. The BTE of the ternary fuel blends is 31.68% and 33.35% at higher loading conditions. The compiled information in the literature states that low-carbon alcohols such as ethanol and methanol are colourless and miscible with organic solvents and water. They are very hygroscopic in nature. When low-carbon alcohols are blended with fossil diesel, the blends become unstable. Hence, the further addition of biodiesel to that blend; the stratum of the blend is avoided.

From an extensive literature survey, there are many research works focused on ternary fuel blends with methyl ester, diesel, and high carbon alcohols (Coughlin & Hoxie, 2017; Gülüm, 2023; Krishnamoorthy et al., 2018; Swarna et al., 2022). But the utilization of WCO biodiesel, diesel, and low-carbon alcohol as ternary blends is found to be lacking. Since methanol has a low carbon content and is cost-effective, it can be used to formulate ternary fuel blends. But the problem here is the blend stability. Though the WCO and methanol are miscible, the addition of diesel to the blends causes stability issues, but not immediately. So, in order to avoid blend instability, polyglycerol polyricinoleate (PGPR), a novel surfactant, is added to make the blend stable. The PGPR surfactant is added at 5 ml. To improve the stability of the methanol blends in the literature, the researchers used only the higher alcohols to stabilize the blend. Though the PGPR in this study did not aid in the combustion or emission aspects, it was very helpful in the stability of the blends. In this study, the authors used the PGPR emulsifier to stabilize the blend, which is a newer part of this study. Also, the application of the entropy weighted Technique for Order Preference by Similarity to Ideal Solution (TOPSIS), a multicriteria decision-making method, in this distinct blend selection is not much explored in the literature.

4.2 MATERIALS AND METHODS

4.2.1 Preparation of Biodiesel and Its Blends

In the current study, neat diesel procured from local fuel stations was used as the baseline fuel. The methanol fuel used in this study is supplied by Central Drug House Pvt. Ltd., Gujarat. The WCO is collected from local restaurants and snack shops and undergoes esterification as well as the transesterification process due to the presence of higher FFAs and excess acidic content. In the esterification process, the ratio of methanol to WCO is 6:1, with 1% sulphuric acid at 60°C. After the esterification process, the oil is water-washed multiple times to remove the alcohol stain. Then the oil is introduced into the transesterification process, where the ratio of oil to methanol is 1:3, and it reacts with 1% potassium hydroxide at 60°C for 1 hour. Following the transesterification process, the oil enters the separating funnel and remains undisturbed for some time. The biodiesel will come to the top layer, while the glycerol will settle at the bottom of the funnel. Again, the top layer of biodiesel is water-washed with warm water at 40°C to remove the excess catalyst, alcohol, and the remaining glycerol stains. The water washing process was repeated three times, and the finally obtained biodiesel was heated at 100°C for half an hour. After the water evaporates, the final WCO biodiesel is obtained. The flowchart of the biodiesel preparation process in this study is described in Figure 4.2.

The WCO is taken into the burette and placed near the magnetic stirrer, where the beaker filled with diesel is placed over it. For preparing the B30M20D50 blend, 30% by volume of WCO biodiesel is taken in the burette, and 50% by volume of diesel is placed in the beaker. When the magnetic stirrer is switched on, the WCO biodiesel is allowed to drop by drop into the beaker containing diesel. After the WCO biodiesel is fully poured into the diesel, the mixture is allowed to stir for about 30 minutes. After that, 20% by volume of methanol is added into the binary mixture of diesel and biodiesel using the burette. Now the ternary blend is allowed to stir for about 30 minutes. To improve the stability of the ternary blend, 5 ml of PGPR is added to the ternary blend and allowed to stir for 20 minutes to form

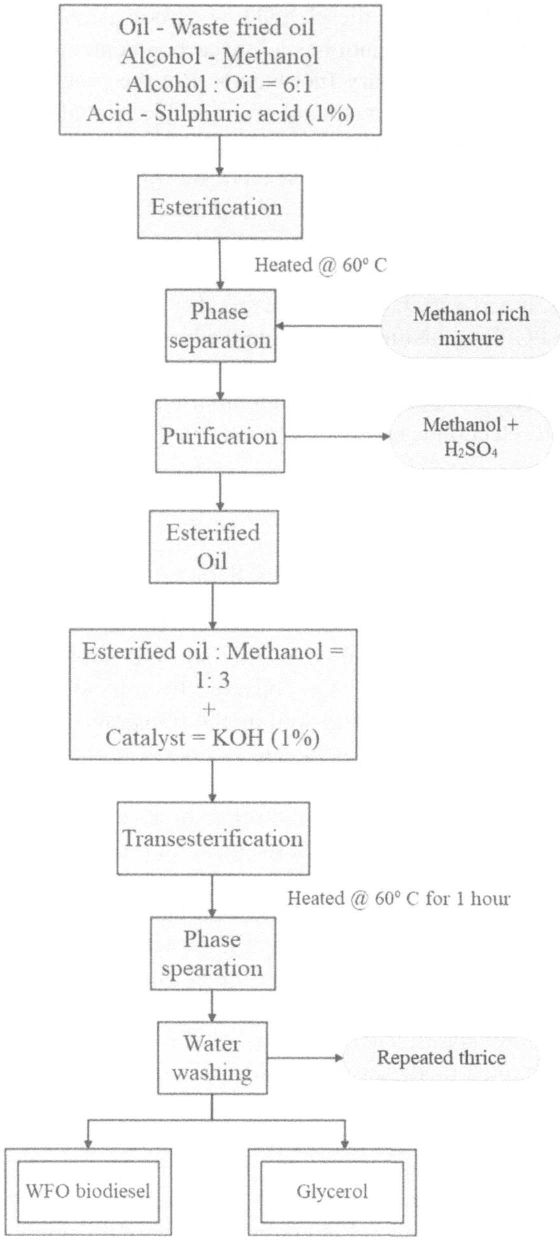

FIGURE 4.2 Flowchart of the WCO biodiesel conversion process.

a stable mixture. A similar procedure is repeated for preparing the B40M10D50 blend by varying the proportion of the fuels. The properties of the test fuels are given in Table 4.1. The fatty acid composition of the WCO biodiesel prepared was measured using gas chromatography and can be found in previous publications (Ananda Murugan & Nataraj, 2024).

TABLE 4.1
Properties of Test Fuels in the Study

Property	Apparatus Used	ASTM Standards	Diesel	Methanol	B30M20D50	B40M10D50
Lower heating value, MJ/kg	Bomb calorimeter	ASTM D240	42.5	20.1	37.105	39.04
Density, kg/m^3	Hydrometer	ASTM D1298	830	792	835.6	843.8
Kinematic viscosity, mm^2/s	Digital viscometer	ASTM D445	2.7	0.5	2.84	2.937
Flash point, °C	Pensky Martens Apparatus	ASTM D93	59	10	87	102
Fire point, °C	Pensky Martens Apparatus	ASTM D93	68	12	99	112

4.2.2 Experimental Setup

A stationary engine with one cylinder and variable speed with four strokes is used for the experiment in this study. The load is applied by coupling the engine's flywheel with the modern air-cooled eddy current dynamometer. A fully automatic optical liquid level sensor measures the fuel flow, while a differential pressure sensor with an inbuilt transmitter measures the air flow to the engine. Based on data obtained from the crankshaft position sensor, the combustion data is measured by the piezoelectric pressure sensor, which is positioned at the cylinder head. The engine conditions are varied using an open ECU from Bosch. Table 4.2 contains

TABLE 4.2
Specifications of Engine Test Rig

Engine Description	Specifications
Engine type	Single cylinder, water-cooled, four stroke, naturally aspirated engine
Compression ratio	17:1
Bore × stroke	87.5 mm × 110 mm
Length of connecting rod	235 mm
Rated power	3.7 kW
Fuel injection timing	23 deg. bTDC
Fuel injection pressure	250 bar
Intake valve timing	4.5 deg. before TDC (open) and 35.5 deg. after BDC (close)
Exhaust valve timing	35.5 deg. after BDC (open) and 4.5 deg. before TDC (close)
Loading device	Air-cooled eddy current dynamometer
No. of injector	1
Number of nozzles and orifice diameter	3 and 0.2 mm

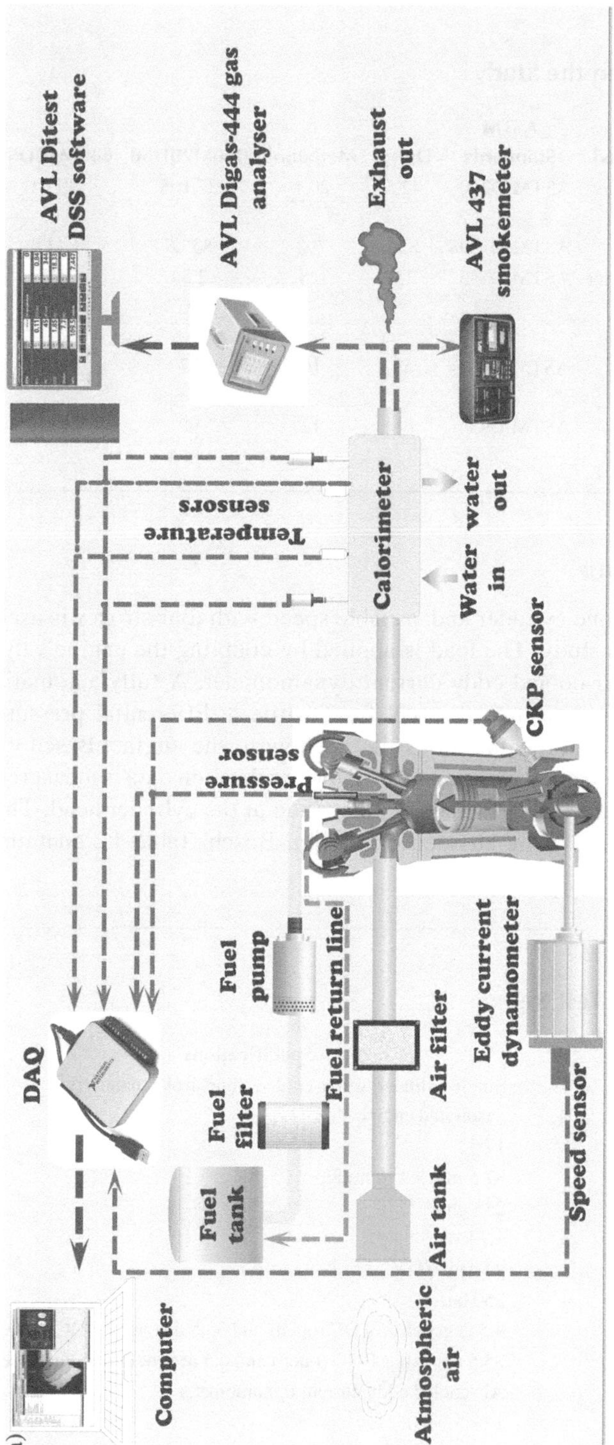

FIGURE 4.3 (a) Graphical layout of the experimental setup; (b) actual experimental setup. *(Continued)*

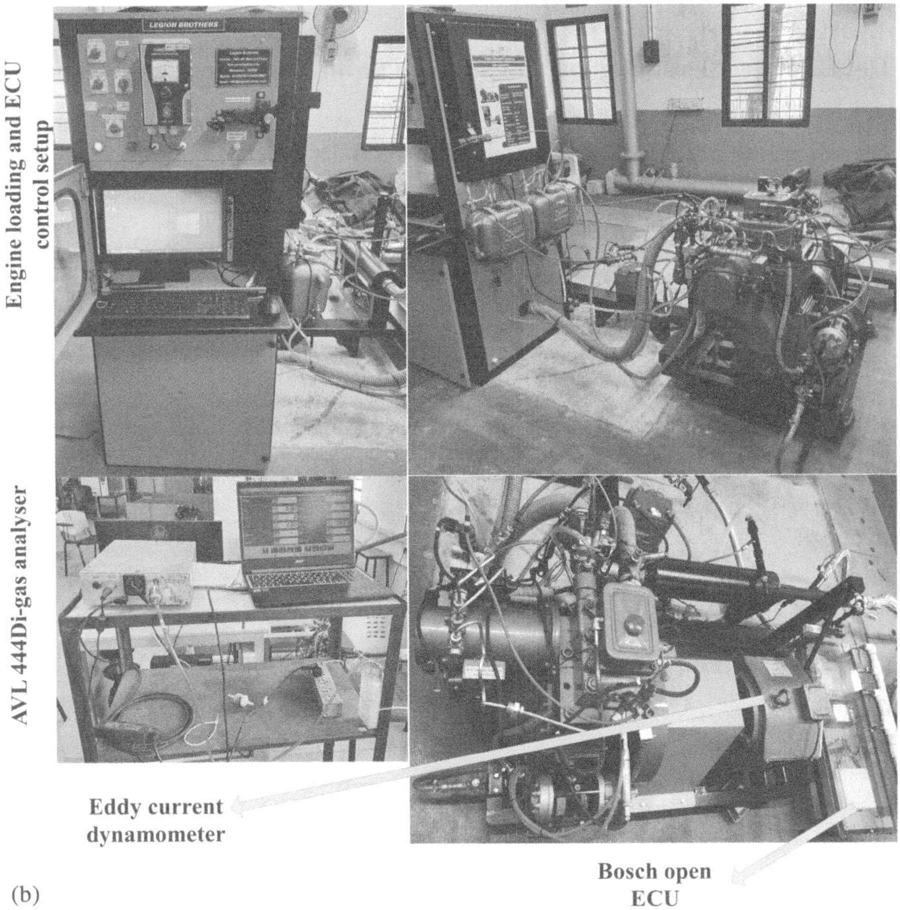

(b)

FIGURE 4.3 *(Continued)*

the test engine's complete specifications. By inserting the probe into the exhaust pipe, the AVL DiGas 444 gas analyser and the AVL 437c smoke meter measure the HC, CO, NO_x, and smoke emissions in the exhaust gas. At 1000 and 1500 rpm, the experiment is conducted under maximum loading conditions. Additionally, injection timings of 21, 23, and 25 deg. bTDC are varied throughout the experiment. In order to reduce the margin of error, the experiment is carried out three times for each fuel sample. Figure 4.3 (a & b) represents the graphical and actual representation of the experimental setup.

4.2.3 Uncertainty Analysis

While conducting experiments, the possibility of error occurrence may arise for different reasons. One of the major reasons for the error is the measuring instruments. Table 4.3 describes the range of the measuring instruments along with their accuracy

TABLE 4.3
Details of the Measuring Instruments along with Their Range and Accuracy

Equipment	Model	Measuring Range	Resolution
Emission analyser	AVL DiGas-444 gas analyser	CO – 0–10 (vol.%)	CO – ±0.01 vol.%
		NO_x – 0–5000 (ppm)	NO_x – ±5 ppm
		HC – 0–20000 (ppm)	HC – ±4.5 ppm
		O_2 – 0–22 (vol.%)	O_2 – ±0.01 vol.%
		Lambda – 0–9.99	Lambda – 0.001
Smoke meter	AVL 437C	0–10 FSN	±0.2 FSN

scale. The uncertainty analysis of the measured parameters is calculated using the following equation (Rajak et al., 2022):

$$E_t = \sqrt{\left(\frac{\partial R}{\partial x_1}E_1\right)+\left(\frac{\partial R}{\partial x_2}E_2\right)+.....+\left(\frac{\partial R}{\partial x_n}E_n\right)} \tag{4.1}$$

where E_t is the overall error in %; E_1, E_2.... E_n are the errors of measured parameters:

$$\begin{aligned}\text{Overall uncertainty} &= \sqrt{(\text{BTE})^2+(\text{SFC})^2+(\text{Speed})^2+(\text{NO}_x)^2+(\text{smoke})^2+(\text{CO}_2)^2}\\ &= \sqrt{(2.29)^2+(0.007)^2+(1.5)^2+(1.2)^2+(1.5)^2+(1.3)^2}\\ &= \pm 3.588\end{aligned} \tag{4.2}$$

4.3 RESULTS AND DISCUSSIONS

4.3.1 Brake Thermal Efficiency

The diesel blend produced a higher BTE than the ternary blends B30M20D50 and B40M10D50 due to its higher energy content. The ternary blends had higher density and viscosity and a lower calorific value compared to neat diesel. These properties must have affected the fuel spray and atomization processes inside the cylinder. This might directly adversely affect the evaporation, air, and fuel mixing inside the cylinder. The addition of methanol to fuel blends reduces the temperature of combustion due to its higher latent heat of vaporization (Cheng et al., 2008). Thus, the combustion efficiency is deteriorating in both ternary fuel blends. This might be another reason for the reduced BTE in the B30M20D50 and B40M10D50 blends. Figure 4.4 represents the BTE vs. injection timings at different speed ranges.

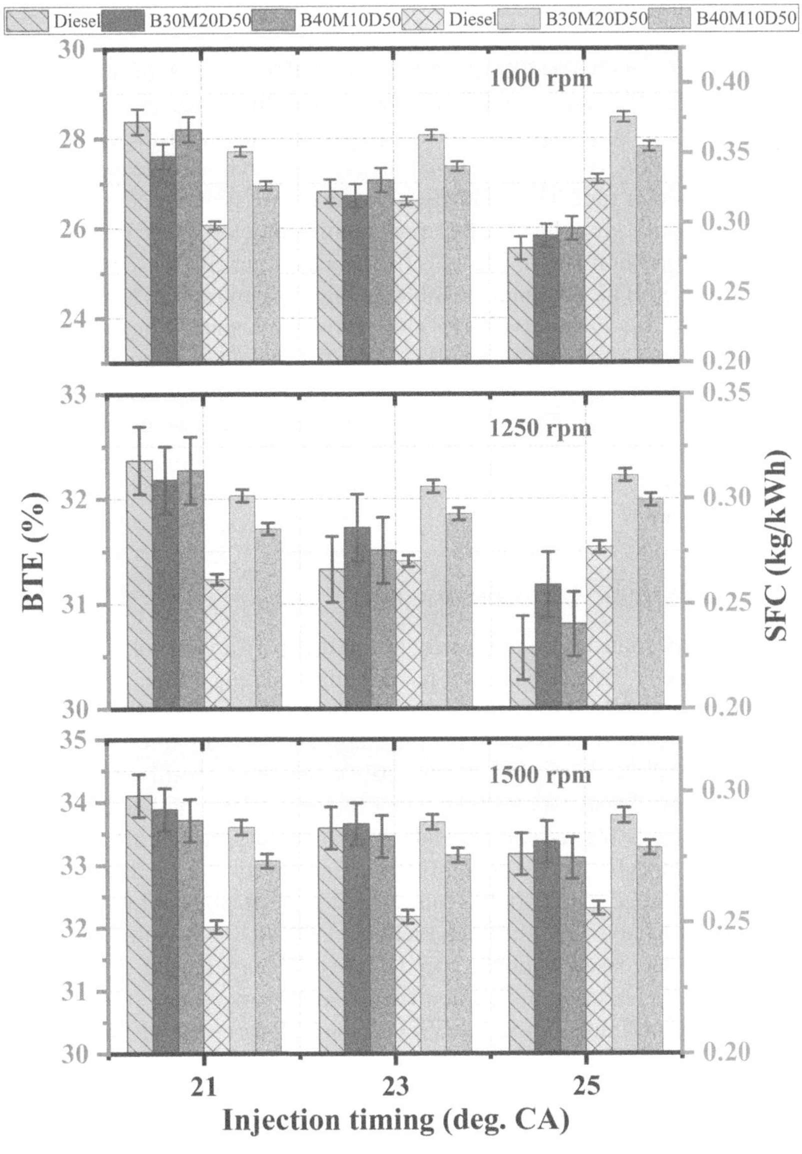

FIGURE 4.4 Variations of BTE and SFC with respect to injection timing at different speed ranges.

In the BTE, the injection time effect was crucial. In contrast to the default map condition of 23 deg. bTDC, the delayed injection timing of 21 deg. bTDC resulted in greater BTE, while the advanced injection timing of 25 deg. bTDC resulted in lower BTE. At 1000 rpm, the diesel, B30M20D50, and B40M10D50 blends of advanced injection timing produced 5.62%, 3.25%, and 4.10% higher BTE than the default map condition. When compared to the default map condition at 1250 rpm, the diesel,

B30M20D50, and B40M10D50 blends with advanced timing generated 3.26%, 1.42%, and 2.40% higher BTE, respectively. When compared to default map condition at 1500 rpm, the diesel, B30M20D50, and B40M10D50 blends with advanced timing injection produced BTE that was 1.52%, 0.68%, and 0.762% higher, respectively.

Compared to default map condition, the BTE was lower in the advanced injection timing condition. Before the piston hits TDC, the cylinder pressure rises due to the early injection of fuel. This indicates that the combustion process began prior to the expansion stroke, and thus the thermodynamic outcome of the pressure increase during the compression stroke is a negative work output. Increased fuel consumption and decreased BTE are made possible by this unfavourable effort. At 1000 rpm, the diesel, B30M20D50, and B40M10D50 blends of advanced injection timing produced 4.88%, 3.43%, and 4.09% less BTE than the default map condition. The diesel, B30M20D50, and B40M10D50 advanced injection timing blends reduced BTE by 2.44%, 1.74%, and 2.27% compared to the default map condition at 1250 rpm, respectively. The diesel, B30M20D50, and B40M10D50 advanced injection timing blends reduced BTE by 1.25%, 0.86%, and 1.03% compared to the default map condition at 1500 rpm, respectively.

4.3.2 Brake-Specific Fuel Consumption

The heating value, density, and viscosity of the fuel play a vital role in the fuel consumption parameter. Since the diesel has a higher heating value, the BSFC of the diesel was very much lower at all experimental conditions. The BSFC of biofuels is always higher than that of conventional diesel fuel due to its lower energy content (El-Seesy et al., 2021; Emiroğlu & Şen, 2018). Likewise, in this study, the BSFC of both the ternary blends B30M20D50 and B40M10D50 is higher than that of the neat diesel due to its higher density and higher viscosity, which affected the spray droplets inside the cylinder. By increasing the concentration of methanol, BSFC also increases (Hasan et al., 2021). In the study, the B30M20D50 blend with 20% methanol resulted in higher BSFC than the fuel blend B40M10D50 with 10% methanol. Also, the BSFC decreased when the speed of the engine increased. Figure 4.4 represents the BSFC vs. injection timing at different speed ranges.

Advancing the injection timing, the BSFC of the fuel increases, whereas by retarding the injection timing, the BSFC of the fuel decreases. From the discussion in the previous section, BTE decreased under advanced injection timing and increased under advanced injection timing. Since BSFC is inversely proportional to BTE, BSFC is opposite in the injection timing trend. By retarding the injection timing, the delay period increases along with the increase in brake mean effective pressure and thus reduces the BSFC. While advancing the injection timing, the delay period is shortened with reduced brake mean effective pressure, which thus increases the BSFC (Jindal, 2011).

When compared to default map condition, the diesel, B30M20D50, and B40M10D50 engines produced 1.53%, 0.68%, and 0.76% less BSFC at 1500 rpm, respectively, when the injection timing was retarded. Similarly, compared to default map condition, diesel, B30M20D50, and B40M10D50 produced 3.26%, 1.42%, and 2.40% fewer BSFC at 1250 rpm. Diesel, B30M20D50, and B40M10D50 all generated 5.62%, 3.25%, and 4.10% less BSFC at 1000 rpm, respectively, compared to the default map condition.

At 1500 rpm, diesel, B30M20D50, and B40M10D50 all produced higher BSFC at advanced injection timing compared to default map condition by 1.25%, 0.86%, and 1.03%, respectively. Diesel, B30M20D50, and B40M10D50 all obtained a 2.43%, 1.74%, and 2.26% increase in BSFC at 1250 rpm, respectively, compared to the default map condition. Compared with default map condition, diesel, B30M20D50, and B40M10D50 produced 4.88%, 3.43%, and 4.09% higher BSFC at 1000 rpm, respectively.

4.3.3 Oxides of Nitrogen

The NO_x emission increases mainly in biodiesel blends due to their higher oxygen content, longer residence duration inside the cylinder, and the resulting higher cylinder temperature. Compared to diesel fuel, NO_x emissions of both the ternary fuel blends increase with an increase in engine speed and also at all injection timings. At higher engine speeds, less time is provided for combustion, and their higher oxygen content contributes to the increased in-cylinder temperature. The addition of methanol to the ternary blends further increases the oxygen concentration and increases the temperature of the cylinder, thus contributing to higher NO_x emissions. Figure 4.5 represents the NO_x vs. injection timing at different speed ranges.

With respect to injection timing, the default map condition of 23 deg. bTDC produced medium NO_x emissions compared to advanced and retarded injection timing. Since more efficient combustion takes place at an advanced injection timing of 25 deg. bTDC, NO_x emissions are higher under these conditions. Thermal NO_x levels surged during the advanced injection phase because the combustion temperature was raised due to the reduced ignition delay period (Kumar & Dinesha, 2021). More heat is generated during combustion because the ternary blends have a higher oxygen concentration due to the higher methanol percentage. Compared to the B40M10D50 blend, the NO_x emissions from the B30M20D50 were higher due to the 20% methanol content. The NO_x output was 9.62% higher at 1500 rpm for diesel, 5.51% higher for B30M20D50, and 7.73% higher for B40M10D50 compared to the default map condition of 23 deg. bTDC. The NO_x output was 9.88% higher at 1250 rpm for diesel, 5.81% higher for B30M20D50, and 7.19% higher for B40M10D50 compared to the default map condition. Diesel, B30M20D50, and B40M10D50 emissions of NO_x were 6.26%, 1.60%, and 3.98% higher than default map condition, respectively, at 1000 rpm.

The reduction in NO_x emissions was mostly attributable to the shorter delay period and lower fuel injection prior to combustion, starting at a retarded injection timing of 21 deg. bTDC. As the afterburning stage burns the most, the pressure and temperature inside the cylinder are reduced during the delayed injection timing (Fayad, 2019). Therefore, NO_x levels are lower with delayed injection. Diesel, B30M20D50, and B40M10D50 all reduced NO_x emissions by 13.39%, 4.21%, and 10.96%, respectively, as compared to the default map condition at 1500 rpm. The NO_x emissions were 10.34%, 4.86%, and 7.78% lower at 1250 rpm for diesel, B30M20D50, and B40M10D50, respectively, compared to the default map condition. Diesel, B30M20D50, and B40M10D50 all reduced NO_x emissions by 9.38%, 3.64%, and 6.10%, respectively, as compared to the default map condition at 1000 rpm. In terms of the emission reduction aspect, the retarded injection timing is preferred for lesser NO_x emissions.

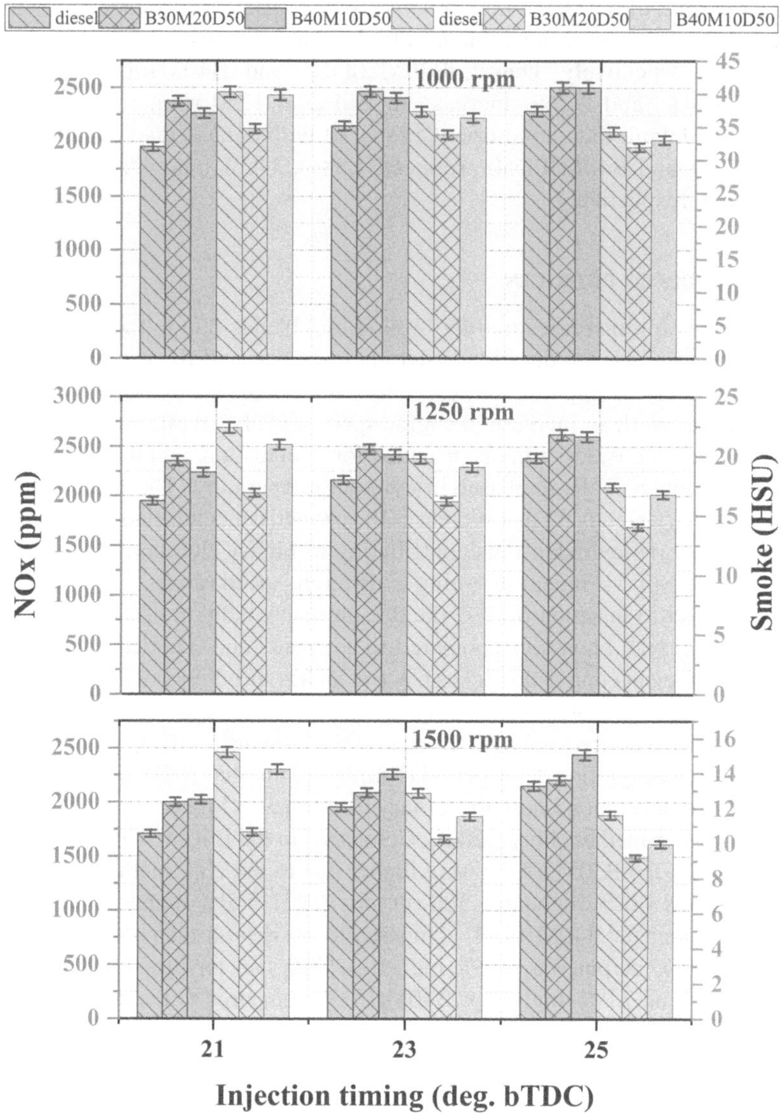

FIGURE 4.5 Variations of NO_x and smoke with respect to injection timing at different speed ranges.

4.3.4 Smoke

Incomplete combustion results in the generation of smoke. Soot particles are the main contents of the particulate matter, which is responsible for the smoke emissions due to the deficiency of air. In the absence of oxygen, long-chain molecules undergo thermal cracking, resulting in soot (Canakci, 2007). Compared to both the ternary fuel blends, diesel emitted more smoke emissions due to the insufficient oxygen to burn completely. The oxygen content in WCO biodiesel and methanol was higher in the ternary fuel blends, and thus the soot content after combustion is oxidized and releases very limited smoke (Wei et al., 2021). Especially, B30M20D50 released less smoke than the other two fuels due to the presence of a high concentration of methanol. Thus, the result justifies that the increasing concentration of methanol in the ternary fuel blends favours the reduction of smoke emissions. The lower carbon-to-hydrogen ratio of methanol inhibits smoke formation when its concentration is increased during the blending with diesel and WCO biodiesel. This happens by reducing the carbon chain (Ge et al., 2022) of the ternary fuel blends. Mixing oxygenated fuels like methanol and WCO biodiesel with non-oxygenated diesel lowers smoke emissions because the oxygenated fuels' molecules are carried to the combustion chamber's pyrolysis zone to make up for the oxygen deficiencies (Wei et al., 2018). Figure 4.5 represents the smoke vs. injection timing at different speed ranges.

From Figure 4.5, it is clear that at all injection timings, the smoke emissions reduce with an increase in engine speed. In terms of injection timing, the advanced injection timing had lower smoke than the retarded injection timing. The advanced injection timing paves the way for a longer resident duration and a higher cylinder temperature so that enough duration was provided for the soot oxidation process, and the ternary blends further enriched the oxygen to reduce the smoke emissions. In comparison to the default map condition, the advanced timing resulted in a 10.18%, 11.07%, and 14.74% reduction in smoke at 1500 rpm for diesel, B30M20D50, and B40M10D50, respectively. Diesel, B30M20D50, and B40M10D50 all reduced smoke by 12.99%, 13.99%, and 12.98% compared to the default map condition at 1250 rpm, respectively. In comparison to the default map condition, diesel, B30M20D50, and B40M10D50 produced 8.35%, 5.68%, and 9.62% less smoke at 1000 rpm, respectively.

The diesel, B30M20D50, and B40M10D50 blends emitted 16.66%, 3.59%, and 20.73% more smoke, respectively, when operated at 1500 rpm with the retarded injection timing compared to the conventional injection speed. When compared to the usual injection time of 1250 rpm, diesel, B30M20D50, and B40M10D50 generated smoke that was 12.31%, 4.42%, and 9.42% greater, respectively. Similarly, diesel, B30M20D50, and B40M10D50 all generated smoke that was 7.53%, 2.65%, and 9.01% greater than the default map condition when operated at 1000 rpm. In the literature, it is also reported that the increase in methanol fuel reduced smoke emissions, and advanced injection timing also contributed to the lesser smoke emission (Sayin et al., 2009).

4.3.5 CO_2 Emissions

For the complete combustion process of the hydrocarbon-rich fuel, the carbon atom and oxygen atom of the fuel will react to give carbon dioxide and water vapour. In incomplete combustion, CO will be formed instead of CO_2 due to oxygen deficiency. For oxygen-rich fuels such as methanol and WCO biodiesel, the CO_2 emissions are higher than those of neat diesel. Thus, increasing the concentration of methanol in the blends increases CO_2 emissions (Verma et al., 2020). The advanced injection timing phase resulted in an increased CO_2 level. The synergistic effect of the latent heat of vaporization and the oxygen content of methanol boosted the oxidation process and increased the CO_2 emission compared to diesel. Figure 4.6 represents the CO_2 vs. injection timing at different speed ranges.

Advanced injection timing at 1500 rpm increased the generation of CO_2 by 1.03%, 1.25%, and 0.86% compared to default map condition for diesel, B30M20D50, and B40M10D50, respectively. Diesel, B30M20D50, and B40M10D50 all produced 1.74%, 2.44%, and 2.226% more CO_2 than the default map condition, respectively,

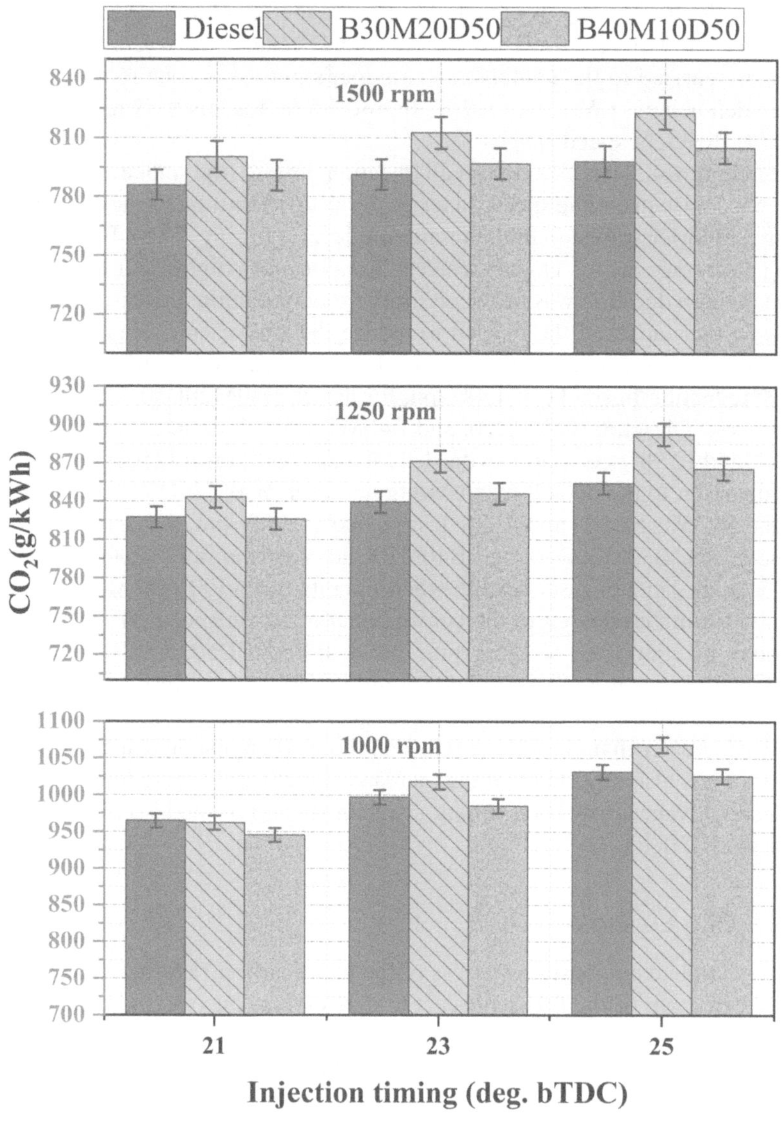

FIGURE 4.6 Variations of CO_2 with respect to injection timing at different speed ranges.

when operating at 1250 rpm. Compared with default map condition, diesel, B30M20D50, and B40M10D50 produced 3.43%, 4.88%, and 4.08% more CO_2 at 1000 rpm, respectively.

Compared with default map condition, diesel, B30M20D50, and B40M10D50 produced 0.76%, 1.53%, and 0.68% less CO_2 at 1500 rpm when the injection timing was retarded. Diesel, B30M20D50, and B40M10D50 all produced 2.40%, 1.42%, and 3.26% reductions in CO_2 emissions compared to default map condition at 1250 rpm, respectively. Diesel, B30M20D50, and B40M10D50 all reduced CO_2 emissions by 3.25%, 5.61%, and 4.09%, respectively, as compared to the default map condition at 1000 rpm.

4.3.6 Entropy-Weighted TOPSIS Method

By analysing several conflict factors, multicriteria decision-making (MCDM) offers a useful framework for comparison. Due to the widespread recognition that numerous criteria can adequately characterize numerous real-world issues, MCDM has rapidly expanded into a prominent topic of operational research. In most cases, a number of factors must be considered simultaneously during the decision-making process, necessitating the use of several industry-standard technologies. When faced with a situation requiring them to make a decision based on more than one criterion, decision-makers should exercise reason and evaluate each criterion separately or in combination to arrive at the best possible outcome (Chen, 2019). Among the several MCDM techniques, TOPSIS has become exceptionally well known. The TOPSIS technique analyses a set of alternatives by computing the geometric distance between each one and the ideal solution and normalizing scores for each criterion. It is a compensatory aggregation method. The idea is that, given a geometrical point and the optimal solution, the best alternative should be the one that is geometrically closest to the positive ideal solution and the furthest away from the negative ideal solution (Rahim et al., 2018).

To assign relative importance to various factors in a decision-making process, the entropy weight technique is employed. The term entropy is the thermodynamic perception, which measures the degree of data uncertainty or unpredictability and is computed for each criterion. Then, the entropy values are used to allocate weights; a greater entropy value means the criterion is given a higher weight (Chen, 2021). When decision-makers need to take into account the relative relevance of criteria (as measured by entropy weights) and the relative performance of alternatives (as measured by TOPSIS ranking), one can implement a combination of the two methods. The decision-making process is made more thorough and equitable as a result of this integration (Huang, 2008). To summarize, the entropy-weighted TOPSIS technique offers a strong foundation for decision-making through the use of entropy weights to establish the significance of criteria and the TOPSIS method to rank alternatives according to their distance from the optimal answer. This method improves decision-making by simultaneously considering the importance of criteria and the performance of alternatives. The literature states that the entropy-weighted TOPSIS significantly improved the recognition of the parameters (Chen, 2021). Figure 4.7 (a & b) represents the flowchart of the entropy-weighted TOPSIS and the methodology involved in the present study.

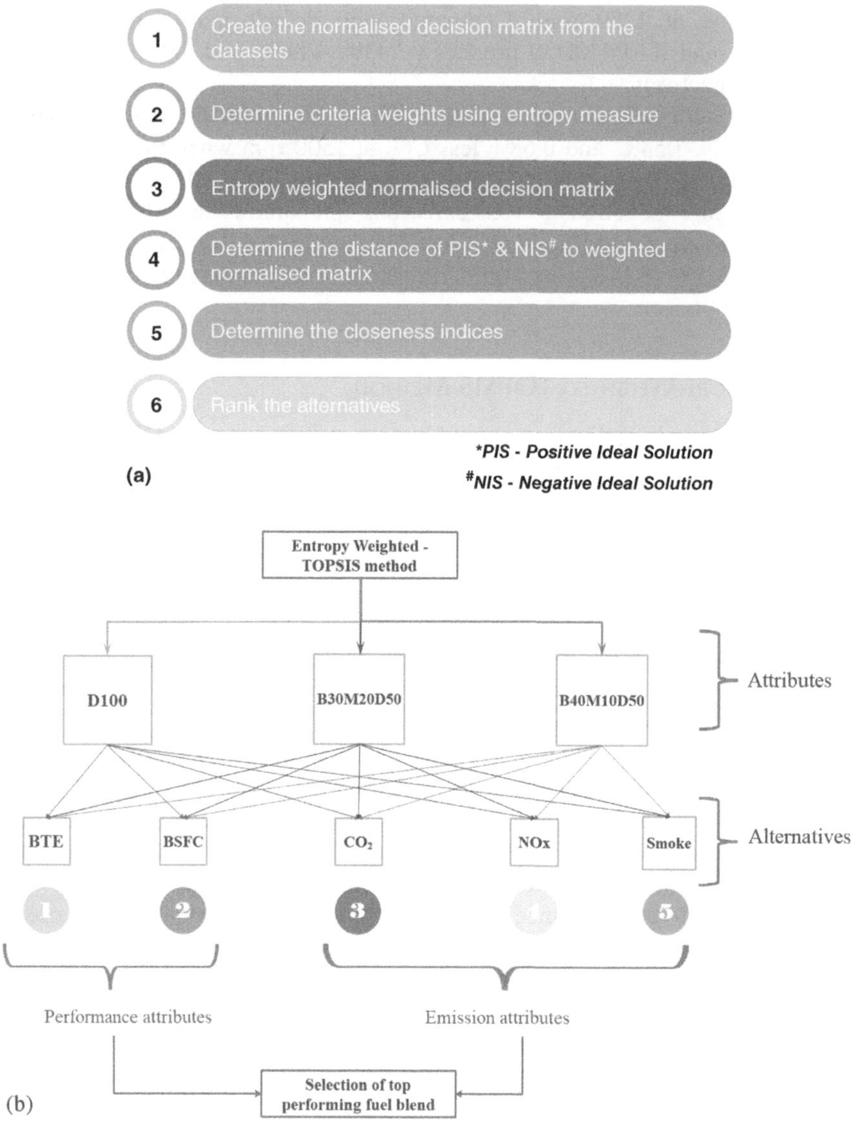

FIGURE 4.7 (a) Steps involved in entropy-weighted TOPSIS method; (b) methodology of TOPSIS in the current study.

Step 1: Decision matrix formulation

$$D_m = \begin{matrix} a_1 \\ a_2 \\ \vdots \\ a_q \end{matrix} \begin{bmatrix} x_{11} & x_{12} & x_{13} & \cdots & x_{1j} \\ x_{21} & x_{22} & x_{23} & \cdots & x_{2j} \\ \vdots & \vdots & \vdots & \vdots & \vdots \\ x_{i1} & x_{i2} & x_{i3} & \cdots & x_{ij} \end{bmatrix} \tag{4.3}$$

where $i = 1, 2, 3\ldots$ and $j = 1, 2, 3\ldots$. "a" is the *i*th study case. The columns inside the matrix represent the evaluation criteria, such as BTE, SFC, CO_2, NO_x, and smoke.

Step 2: Normalized decision matrix

Due to the multi-dimensional nature of the data in matrix D_m, normalization is required to convert the dimensions of the criteria into a non-dimensional form that can be compared with one another:

$$R = \left[r_{ij}\right] = \frac{x_{ij}}{\sqrt{\sum_{i=1}^{q} x_{ij}^{2}}} \tag{4.4}$$

where x_{ij} is the value in decision matrix and r_{ij} is the normalized value for *j*th criterion for *i*th alternative. Table 4.4 represents the normalized decision matrix of the experimental data, which was calculated using Equation 4.4.

TABLE 4.4
Normalized Decision Matrix of Experimental Data

Fuel	Inj. Timing	Speed	BTE	BSFC	CO	NO_x	Smoke
Diesel	21	1000	0.17726	0.18998	0.208412	0.16593	0.3196
Diesel	21	1250	0.20222	0.16653	0.18269	0.16529	0.17787
Diesel	21	1500	0.21306	0.15806	0.173397	0.14499	0.12081
Diesel	23	1000	0.16758	0.20096	0.220452	0.18225	0.2964
Diesel	23	1250	0.19573	0.17206	0.18875	0.18332	0.15724
Diesel	23	1500	0.20984	0.16049	0.176062	0.1658	0.10223
Diesel	25	1000	0.1596	0.21101	0.23148	0.19402	0.27263
Diesel	25	1250	0.19102	0.1763	0.193404	0.20238	0.13806
Diesel	25	1500	0.20723	0.16251	0.178275	0.18256	0.09233
B30M20D50	21	1000	0.17246	0.22366	0.208982	0.2017	0.27582
B30M20D50	21	1250	0.20104	0.19187	0.179275	0.19983	0.13428
B30M20D50	21	1500	0.21166	0.18224	0.170278	0.1698	0.08457
B30M20D50	23	1000	0.16695	0.23105	0.215887	0.20918	0.26861
B30M20D50	23	1250	0.1982	0.19461	0.181841	0.20979	0.12848
B30M20D50	23	1500	0.21022	0.18349	0.171445	0.1771	0.08158
B30M20D50	25	1000	0.16133	0.23911	0.22342	0.21256	0.25376
B30M20D50	25	1250	0.19479	0.19803	0.185032	0.22234	0.11168
B30M20D50	25	1500	0.20843	0.18507	0.172921	0.18713	0.07302
B40M10D50	21	1000	0.1762	0.20807	0.204774	0.19207	0.31601
B40M10D50	21	1250	0.20162	0.18183	0.178953	0.18999	0.16658
B40M10D50	21	1500	0.21058	0.1741	0.171341	0.17166	0.11294
B40M10D50	23	1000	0.16913	0.21676	0.213332	0.20415	0.28876
B40M10D50	23	1250	0.19685	0.18624	0.183292	0.20537	0.1516
B40M10D50	23	1500	0.20898	0.17543	0.172654	0.19156	0.09173
B40M10D50	25	1000	0.16235	0.22581	0.222228	0.21242	0.26227
B40M10D50	25	1250	0.19244	0.19051	0.187491	0.22069	0.13312
B40M10D50	25	1500	0.20684	0.17725	0.174444	0.20696	0.07914

Step 3: Entropy calculation for the outcomes

$$E_j = -k\sum_{i=1}^{m} p_{ij} \ln\left(p_{ij}\right) \tag{4.5}$$

where

$$k = \frac{1}{\ln(m)} \tag{4.6}$$

Step 4: Weight determination using the entropy values

The researcher selects the different alternatives based on a variety of criteria. So, it is necessary to give relative importance to each criterion. The relative importance value is called "weight." There are different methods to determine the weights, such as the AHP method, the eigen vector method, and the weighted least squares method. Out of all the methods, the entropy-weighted method is the most reliable to determine the weights for the criterion. The cumulative sum of the entropy weights for all the alternative sums should be exactly 1. Table 4.5 represents the entropy weights for this study, which were calculated using Equations 4.5–4.7:

$$EW_j = \frac{1 - E_j}{\sum_{j=1}^{n}\left(1 - E_j\right)} \tag{4.7}$$

Step 5: Weighted normalized decision matrix

$$V = EW_j * R \tag{4.8}$$

Table 4.6 represents the weighted normalized decision matrix, which was calculated from Equation 4.8.

Step 6: Determination of positive (S^+) and negative (S^-) ideal solution

$$\begin{aligned} S^+ &= \left\{v_1^+, v_2^+ .....v_n^+\right\}, \left\{\left(\max v_{ij} \mid i \in I\right), \left(\min v_{ij} \mid i \in I\right)\right\} \\ S^- &= \left\{v_1^-, v_2^- .....v_n^-\right\}, \left\{\left(\max v_{ij} \mid i \in I\right), \left(\min v_{ij} \mid i \in I\right)\right\} \end{aligned} \tag{4.9}$$

where *i* is beneficial criteria and *j* is non-beneficial criteria.

TABLE 4.5
Weights Obtained from Entropy Method

	BTE	BSFC	CO	NO_x	Smoke
Condition	Higher is better	Lesser is better	Lesser is better	Lesser is better	Lesser is better
Weights	0.033889	0.048225	0.037121	0.03803	0.842734

TABLE 4.6
Weighted Normalized Decision Matrix of Experimental Data

Fuel	Inj. Timing	Speed	BTE	BSFC	CO	NO_x	Smoke
Diesel	21	1000	0.00601	0.00916	0.007737	0.00631	0.26934
Diesel	21	1250	0.00685	0.00803	0.006782	0.00629	0.1499
Diesel	21	1500	0.00722	0.00762	0.006437	0.00551	0.10181
Diesel	23	1000	0.00568	0.00969	0.008183	0.00693	0.24978
Diesel	23	1250	0.00663	0.0083	0.007007	0.00697	0.13251
Diesel	23	1500	0.00711	0.00774	0.006536	0.00631	0.08615
Diesel	25	1000	0.00541	0.01018	0.008593	0.00738	0.22976
Diesel	25	1250	0.00647	0.0085	0.007179	0.0077	0.11635
Diesel	25	1500	0.00702	0.00784	0.006618	0.00694	0.07781
B30M20D50	21	1000	0.00584	0.01079	0.007758	0.00767	0.23244
B30M20D50	21	1250	0.00681	0.00925	0.006655	0.0076	0.11317
B30M20D50	21	1500	0.00717	0.00879	0.006321	0.00646	0.07127
B30M20D50	23	1000	0.00566	0.01114	0.008014	0.00796	0.22636
B30M20D50	23	1250	0.00672	0.00939	0.00675	0.00798	0.10827
B30M20D50	23	1500	0.00712	0.00885	0.006364	0.00674	0.06875
B30M20D50	25	1000	0.00547	0.01153	0.008294	0.00808	0.21385
B30M20D50	25	1250	0.0066	0.00955	0.006869	0.00846	0.09412
B30M20D50	25	1500	0.00706	0.00892	0.006419	0.00712	0.06154
B40M10D50	21	1000	0.00597	0.01003	0.007602	0.0073	0.26631
B40M10D50	21	1250	0.00683	0.00877	0.006643	0.00723	0.14038
B40M10D50	21	1500	0.00714	0.0084	0.00636	0.00653	0.09518
B40M10D50	23	1000	0.00573	0.01045	0.007919	0.00776	0.24335
B40M10D50	23	1250	0.00667	0.00898	0.006804	0.00781	0.12776
B40M10D50	23	1500	0.00708	0.00846	0.006409	0.00729	0.0773
B40M10D50	25	1000	0.0055	0.01089	0.008249	0.00808	0.22102
B40M10D50	25	1250	0.00652	0.00919	0.00696	0.00839	0.11218
B40M10D50	25	1500	0.00701	0.00855	0.006476	0.00787	0.06669
Ideal best solution		0.00722	0.007622	0.006321	0.005514	0.061537	0.00722
Ideal worst solution		0.005408	0.011531	0.008593	0.008456	0.269339	0.005408

Step 7: Determination of Euclidian distance between S^+ and S^-

$$A_i^+ = \sqrt{\sum_{j=1}^{n} \left(V - S^+\right)^2}$$
$$A_i^- = \sqrt{\sum_{j=1}^{n} \left(V - S^-\right)^2} \tag{4.10}$$

where $i = 1, 2, 3\ldots\ldots n$.

Step 8: Evaluating closeness coefficient

$$C_i = \frac{A_i^-}{A_i^+ + A_i^-} \tag{4.11}$$

where C_i is the ranking index or closeness coefficient, which is seen as an overall scoring factor. The range of the overall score lies between 0 and 1. The alternative with the highest value is preferred.

Step 9: Ranking of alternatives

According to the decreasing values of the closeness coefficients, the alternatives are ranked from most preferred to least preferred. Based on the closeness coefficient (C_i), the option that is considered to be the closest is selected. Equations 4.10 and 4.11 calculated the Euclidian distance and closeness coefficient, as shown in Table 4.7. The final ranking is obtained from the maximum values of the closeness coefficient.

TABLE 4.7
Ranking of Experimental Alternatives

Fuel	Inj. timing	Speed	S_i^+	S_i^-	C_i	Rank
Diesel	21	1000	0.207817	0.003363	0.015923	27
Diesel	21	1250	0.088365	0.119536	0.574967	18
Diesel	21	1500	0.04027	0.167627	0.806299	10
Diesel	23	1000	0.188277	0.019708	0.094759	25
Diesel	23	1250	0.071001	0.136886	0.658463	16
Diesel	23	1500	0.024629	0.183258	0.881527	7
Diesel	25	1000	0.168274	0.03962	0.190578	22
Diesel	25	1250	0.054876	0.15303	0.736055	14
Diesel	25	1500	0.016337	0.191592	0.921432	6
B30M20D50	21	1000	0.170958	0.036926	0.177629	23
B30M20D50	21	1250	0.051698	0.156211	0.751343	13
B30M20D50	21	1500	0.009845	0.198122	0.952662	4
B30M20D50	23	1000	0.164898	0.042985	0.206775	21
B30M20D50	23	1250	0.046838	0.161098	0.774748	11
B30M20D50	23	1500	0.007418	0.200634	0.964343	3
B30M20D50	25	1000	0.152412	0.055487	0.266894	19
B30M20D50	25	1250	0.032779	0.175246	0.842427	8
B30M20D50	25	1500	0.002073	0.20784	0.990123	1
B40M10D50	21	1000	0.204806	0.003744	0.017953	26
B40M10D50	21	1250	0.07887	0.129019	0.620614	17
B40M10D50	21	1500	0.033668	0.17422	0.838048	9
B40M10D50	23	1000	0.181862	0.026031	0.125214	24
B40M10D50	23	1250	0.066276	0.141625	0.681213	15
B40M10D50	23	1500	0.015889	0.192083	0.923601	5
B40M10D50	25	1000	0.15956	0.048324	0.232455	20
B40M10D50	25	1250	0.050759	0.157188	0.755906	12
B40M10D50	25	1500	0.005747	0.202689	0.972426	2

4.3.7 Optimized Results Obtained from Entropy Weighted TOPSIS Approach

In the current work, engine data of neat diesel is used to compare with the two ternary fuel blends named B30M20D50 and B40M10D50. The engine output parameters, such as BTE, BSFC, NO_x, smoke, and CO_2, are considered the criteria for determining the optimized conditions and best-performing fuel blend. The fuel and its blends are considered alternatives, and the engine data is considered an attribute. The ternary blend B30M20D50 ranked first with a closeness coefficient of 0.990123 with the optimized input conditions of 25 deg. bTDC and 1500 rpm. With similar optimized input conditions, another ternary blend, B40M10D50, ranked second with a closeness coefficient of 0.972426. On comparing with diesel, B30M20D50 outperformed well at the injection timing of 25 deg. bTDC in terms of BTE and smoke emissions. Also, the CO_2 was higher in the same conditions, which indicates that the combustion was complete. In terms of NO_x and BSFC, the B30M20D50 underperformed the diesel. Considering overall performance and emissions, B30M20D50 performed better than diesel.

4.4 CONCLUSION

In the present study, the engine performance and emission characteristics of diesel, B30M20D50, and B40M10D50 blends are investigated at three different injection timings and speeds in a single-cylinder diesel engine. Also, the study is optimized by determining the best-performing fuel blend along with the input parameters using the entropy-weighted TOPSIS method. The following conclusions are derived from the current investigations:

- When compared to the default map condition, the BTE was enhanced when the injection was delayed by 21 deg. bTDC. At retarded injection timings of 1000, 1250, and 1500 rpm, B30M20D50 produced 3.25%, 1.42%, and 0.68% higher BTE compared to the default map condition of 23 deg. bTDC.
- The BSFC increased when the injection timing was advanced, whereas it decreased when the injection timing was retarded. Out of all the input conditions, B30M20D50 resulted in higher BSFC due to the lesser calorific value of the fuel.
- Since B30M20D50 has a higher concentration of methanol than other blends, the NO_x emissions are higher due to the increased oxygen content of WCO biodiesel and methanol. B30M20D50 produced less NO_x at retarded injection timing. It emitted 4.21%, 4.86%, and 3.64% less NO_x at retarded injection timing than at default map condition.
- Both the ternary fuel blends emitted lower smoke emissions than the diesel at all input conditions. The smoke emissions decreased with increasing the speed range and advancing the injection timing. B30M20D50 emitted the fewest smoke emissions among all due to the higher methanol concentrations. At advanced injection timing, it produced 11.07%, 13.99%, and 5.68% less smoke than the default map condition.

- The entropy-weighted TOPSIS determined B30M20D50 as the overall best-performing blend with a closeness coefficient of 0.990123. The optimized input conditions are 25 deg. bTDC injection timing with 1500 rpm. The experimental conditions also validated the statement by performing well in terms of both engine performance and emissions.

REFERENCES

Adhikesavan, C., Ganesh, D., & Charles Augustin, V. (2022). Effect of quality of waste cooking oil on the properties of biodiesel, engine performance and emissions. *Cleaner Chemical Engineering*, *4*, 100070. https://doi.org/10.1016/j.clce.2022.100070

Ağbulut, Ü, Karagöz, M., Sarıdemir, S., & Öztürk, A. (2020). Impact of various metal-oxide based nanoparticles and biodiesel blends on the combustion, performance, emission, vibration and noise characteristics of a CI engine. *Fuel*, *270*, 117521. https://doi.org/10.1016/j.fuel.2020.117521

Ananda Murugan, M., & Nataraj, G. (2024). Exploring the multifarious blend ratios of waste fried edible oil biodiesel/diesel/low carbon methanol in an automotive engine: An approach towards fuel characterization, experimental, and multicriteria decision making method. *Environmental Progress & Sustainable Energy*. https://doi.org/10.1002/ep.14394

Canakci, M. (2007). Combustion characteristics of a turbocharged DI compression ignition engine fueled with petroleum diesel fuels and biodiesel. *Bioresource Technology*, *98*(6), 1167–1175. https://doi.org/10.1016/j.biortech.2006.05.024

Chen, C.-H. (2019). A new multi-criteria assessment model combining GRA techniques with intuitionistic fuzzy entropy-based TOPSIS method for sustainable building materials supplier selection. *Sustainability*, *11*(8), 2265. https://doi.org/10.3390/su11082265

Chen, P. (2021). Effects of the entropy weight on TOPSIS. *Expert Systems with Applications*, *168*, 114186. https://doi.org/10.1016/j.eswa.2020.114186

Cheng, C. H., Cheung, C. S., Chan, T. L., Lee, S. C., Yao, C. D., & Tsang, K. S. (2008). Comparison of emissions of a direct injection diesel engine operating on biodiesel with emulsified and fumigated methanol. *Fuel*, *87*(10–11), 1870–1879. https://doi.org/10.1016/j.fuel.2008.01.002

Coughlin, B., & Hoxie, A. (2017). Combustion characteristics of ternary fuel blends: Pentanol, butanol and vegetable oil. *Fuel*, *196*, 488–496. https://doi.org/10.1016/j.fuel.2017.01.104

E, J., Pham, M., Zhao, D., Deng, Y., Le, D., Zuo, W., Zhu, H., Liu, T., Peng, Q., & Zhang, Z. (2017). Effect of different technologies on combustion and emissions of the diesel engine fueled with biodiesel: A review. *Renewable and Sustainable Energy Reviews*, *80*, 620–647. https://doi.org/10.1016/j.rser.2017.05.250

El-Seesy, A. I., Waly, M. S., He, Z., El-Batsh, H. M., Nasser, A., & El-Zoheiry, R. M. (2021). Influence of quaternary combinations of biodiesel/methanol/n-octanol/diethyl ether from waste cooking oil on combustion, emission, and stability aspects of a diesel engine. *Energy Conversion and Management*, *240*, 114268. https://doi.org/10.1016/j.enconman.2021.114268

Emiroğlu, A. O., & Şen, M. (2018). Combustion, performance and exhaust emission characterizations of a diesel engine operating with a ternary blend (alcohol-biodiesel-diesel fuel). *Applied Thermal Engineering*, *133*, 371–380. https://doi.org/10.1016/j.applthermaleng.2018.01.069

Fayad, M. A. (2019). Effect of fuel injection strategy on combustion performance and NOx/smoke trade-off under a range of operating conditions for a heavy-duty DI diesel engine. *SN Applied Sciences*, *1*(9), 1088. https://doi.org/10.1007/s42452-019-1083-2

Feng, Q., Jalali, A., Fincham, A. M., Wang, Y. L., Tsotsis, T. T., & Egolfopoulos, F. N. (2012). Soot formation in flames of model biodiesel fuels. *Combustion and Flame*, *159*(5), 1876–1893. https://doi.org/10.1016/j.combustflame.2012.01.003

Ge, J. C., Wu, G., Yoo, B.-O., & Choi, N. J. (2022). Effect of injection timing on combustion, emission and particle morphology of an old diesel engine fueled with ternary blends at low idling operations. *Energy*, *253*, 124150. https://doi.org/10.1016/j.energy.2022.124150

Gülüm, M. (2023). Performance, combustion and emission characteristics of a diesel engine fuelled with diesel fuel + corn oil + alcohol ternary blends. *Environmental Science and Pollution Research*, *30*(18), 53767–53777. https://doi.org/10.1007/s11356-023-26053-x

Hasan, A. O., Osman, A. I., Al-Muhtaseb, A. H., Al-Rawashdeh, H., Abu-jrai, A., Ahmad, R., Gomaa, M. R., Deka, T. J., & Rooney, D. W. (2021). An experimental study of engine characteristics and tailpipe emissions from modern DI diesel engine fuelled with methanol/diesel blends. *Fuel Processing Technology*, *220*, 106901. https://doi.org/10.1016/j.fuproc.2021.106901

Huang, J. (2008). Combining entropy weight and TOPSIS method For information system selection. *2008 IEEE International Conference on Automation and Logistics*, 1965–1968. https://doi.org/10.1109/ICAL.2008.4636483

Jindal, S. (2011). Effect of injection timing on combustion and performance of a direct injection diesel engine running on Jatropha methyl ester. *International Journal of Energy, Environment and Economics*, *2*(1). Online. www.IJEE.IEEFoundation.org

Krishnamoorthy, V., Dhanasekaran, R., Rana, D., Saravanan, S., & Rajesh Kumar, B. (2018). A comparative assessment of ternary blends of three bio-alcohols with waste cooking oil and diesel for optimum emissions and performance in a CI engine using response surface methodology. *Energy Conversion and Management*, *156*, 337–357. https://doi.org/10.1016/j.enconman.2017.10.087

Kulkarni, M. G., & Dalai, A. K. (2006). Waste cooking OilAn economical source for biodiesel: A review. *Industrial & Engineering Chemistry Research*, *45*(9), 2901–2913. https://doi.org/10.1021/ie0510526

Kumar, A. N., Ashok, B., Nanthagopal, K., Ong, H. C., Geca, M. J., Victor, J., Vignesh, R., Jeevanantham, A. K., Kannan, C., & Kishore, P. S. (2022). Experimental analysis of higher alcohol–based ternary biodiesel blends in CI engine parameters through multivariate and desirability approaches. *Biomass Conversion and Biorefinery*, *12*(5), 1525–1540. https://doi.org/10.1007/s13399-020-01134-w

Kumar, S., & Dinesha, P. (2021). Combined influence of fuel injection strategy and nanoparticle additives on the performance and emission characteristics of a biodiesel fueled engine. *Cogent Engineering*, *8*(1). https://doi.org/10.1080/23311916.2021.1939515

Li, R., Wang, Z., Ni, P., Zhao, Y., Li, M., & Li, L. (2014). Effects of cetane number improvers on the performance of diesel engine fuelled with methanol/biodiesel blend. *Fuel*, *128*, 180–187. https://doi.org/10.1016/j.fuel.2014.03.011

Lozhkina, O. V., & Lozhkin, V. N. (2016). Estimation of nitrogen oxides emissions from petrol and diesel passenger cars by means of on-board monitoring: Effect of vehicle speed, vehicle technology, engine type on emission rates. *Transportation Research Part D: Transport and Environment*, *47*, 251–264. https://doi.org/10.1016/j.trd.2016.06.008

Mat Yasin, M. H., Yusaf, T., Mamat, R., & Fitri Yusop, A. (2014). Characterization of a diesel engine operating with a small proportion of methanol as a fuel additive in biodiesel blend. *Applied Energy*, *114*, 865–873. https://doi.org/10.1016/j.apenergy.2013.06.012

Mohammed, A. S., Atnaw, S. M., Ramaya, A. V., & Alemayehu, G. (2023). A comprehensive review on the effect of ethers, antioxidants, and cetane improver additives on biodiesel-diesel blend in CI engine performance and emission characteristics. *Journal of the Energy Institute*, *108*, 101227. https://doi.org/10.1016/j.joei.2023.101227

Qi, D. H., Chen, H., Geng, L. M., & Bian, Y. Z. (2011). Effect of diethyl ether and ethanol additives on the combustion and emission characteristics of biodiesel-diesel blended fuel engine. *Renewable Energy*, *36*(4), 1252–1258. https://doi.org/10.1016/j.renene.2010.09.021

Rahim, R., Supiyandi, S., Siahaan, A. P. U., Listyorini, T., Utomo, A. P., Triyanto, W. A., Irawan, Y., Aisyah, S., Khairani, M., Sundari, S., & Khairunnisa, K. (2018). TOPSIS method application for decision support system in internal control for selecting best employees. *Journal of Physics: Conference Series*, *1028*, 012052. https://doi.org/10.1088/1742-6596/1028/1/012052

Rajak, U., Nashine, P., Chaurasiya, P. K., Verma, T. N., Dasore, A., Pathak, K. K., Dwivedi, G., Shukla, A. K., & Saini, G. (2022). The effects on performance and emission characteristics of DI engine fuelled with CeO_2 nanoparticles addition in diesel/tyre pyrolysis oil blends. *Environment, Development and Sustainability*. https://doi.org/10.1007/s10668-022-02358-8

Sayin, C., Ilhan, M., Canakci, M., & Gumus, M. (2009). Effect of injection timing on the exhaust emissions of a diesel engine using diesel–methanol blends. *Renewable Energy*, *34*(5), 1261–1269. https://doi.org/10.1016/j.renene.2008.10.010

Swarna, S., Swamy, M. T., Divakara, T. R., Krishnamurthy, K. N., & Shashidhar, S. (2022). Experimental assessment of ternary fuel blends of diesel, hybrid biodiesel and alcohol in naturally aspirated CI engine. *International Journal of Environmental Science and Technology*, *19*(9), 8523–8554. https://doi.org/10.1007/s13762-021-03586-7

Tuan Hoang, A., Nižetić, S., Chyuan Ong, H., Tarelko, W., Viet Pham, V., Hieu Le, T., Quang Chau, M., & Phuong Nguyen, X. (2021). A review on application of artificial neural network (ANN) for performance and emission characteristics of diesel engine fueled with biodiesel-based fuels. *Sustainable Energy Technologies and Assessments*, *47*, 101416. https://doi.org/10.1016/j.seta.2021.101416

Verma, T. N., Nashine, P., Chaurasiya, P. K., Rajak, U., Afzal, A., Kumar, S., Singh, D. V., & Azad, A. K. (2020). The effect of ethanol-methanol-diesel-microalgae blends on performance, combustion and emissions of a direct injection diesel engine. *Sustainable Energy Technologies and Assessments*, *42*, 100851. https://doi.org/10.1016/j.seta.2020.100851

Verma, T. N., Shrivastava, P., Rajak, U., Dwivedi, G., Jain, S., Zare, A., Shukla, A. K., & Verma, P. (2021). A comprehensive review of the influence of physicochemical properties of biodiesel on combustion characteristics, engine performance and emissions. *Journal of Traffic and Transportation Engineering (English Edition)*, *8*(4), 510–533. https://doi.org/10.1016/j.jtte.2021.04.006

Wei, J., Yin, Z., Wang, C., Lv, G., Zhuang, Y., Li, X., & Wu, H. (2021). Impact of aluminium oxide nanoparticles as an additive in diesel-methanol blends on a modern DI diesel engine. *Applied Thermal Engineering*, *185*, 116372. https://doi.org/10.1016/j.applthermaleng.2020.116372

Wei, L., Cheung, C. S., & Ning, Z. (2018). Effects of biodiesel-ethanol and biodiesel-butanol blends on the combustion, performance and emissions of a diesel engine. *Energy*, *155*, 957–970. https://doi.org/10.1016/j.energy.2018.05.049

Zhang, Y. (2003). Biodiesel production from waste cooking oil: 1. Process design and technological assessment. *Bioresource Technology*, *89*(1), 1–16. https://doi.org/10.1016/S0960-8524(03)00040-3

Zhu, L., Cheung, C. S., Zhang, W. G., & Huang, Z. (2010). Emissions characteristics of a diesel engine operating on biodiesel and biodiesel blended with ethanol and methanol. *Science of The Total Environment*, *408*(4), 914–921. https://doi.org/10.1016/j.scitotenv.2009.10.078

5 Combustion Stability Analysis for Advanced Combustion-Mode Engines

Prem Kumar, Satsimran Singh, Sarbjot Singh Sandhu, Sumit Kanchan, and Dipak Kumar Mandal

5.1 INTRODUCTION

Optimizing engine performance is a perpetual pursuit in the automotive industry, with combustion stability playing a crucial role, especially in advanced combustion-mode engines. In this comprehensive exploration, cycle-to-cycle combustion variability is examined, and its diverse implications are explored. Cycle-to-cycle combustion variability can be observed under specific engine operating conditions, including low load, high exhaust gas recirculation (EGR) idle, and diluted mixtures. Following engine combustion cycles produces unique fingerprints that impact torque development, engine speed, and other performance parameters. Combustion variability is characterized by engine roughness, compromised torque, high emissions, reduced efficiency, and lower fuel usage (Kumar et al., 2022).

Combustion variability impacts more than just power and torque. The complexity of combustion causes conditions where some cycles exhibit faster combustion, leading to a tendency for knocking (Kumar & Sandhu, 2022b). Conversely, slower combustion cycles may not be completed before the exhaust valve opens, resulting in high unburned hydrocarbon emissions and increased fuel consumption. There is a significant impact of combustion variability in all engine operating conditions, whether it is EGR, throttled conditions, or lean mixtures. Variations in engine speed and torque, as well as a lack of brake power, are some of the negative effects of cycle to cycle variations (CCVs) on engine performance. Also, these variations reduce drivability, particularly in transmissions and engines with CCVs. CCVs in combustion also contribute to engine noise. Fluctuations in in-cylinder pressure led to periodic oscillations in engine speed, influencing the acoustic signature. Reduction in cyclic variations in combustion has the potential to mitigate engine noise, enhancing the overall driving experience. Understanding the origins of cyclic combustion variations is crucial for developing effective control strategies. Factors contributing to combustion variability include cyclic variations in gas motion, variations in the

DOI: 10.1201/9781003482277-5

amount of fuel, air, and residual gases inducted in each engine cycle, gas charge motion, mixing homogeneity, fuel preparation strategies, long combustion duration, disproportionate dilution (by EGR or air), and low ignition energy (Saxena & Maurya, 2018b). These factors are categorized into prior-cycle effects (residual gases, misfires) and same-cycle effects (in-cylinder flow).

The compression ignition (CI) engine often spontaneously ignites diesel fuel, resulting in low cyclic combustion variations due to non-premixed combustion that is primarily controlled by fuel-air mixing. It is possible to minimize variations by controlling pulse profiles, injecting multiple pilots, and optimizing glow plug temperatures (Sharma et al., 2023). The cyclic variation input parameters for low-temperature combustion (LTC) engines, including RCCI engines, are more sensitive than those for conventional diesel engines (Abdalla et al., 2019). It is crucial that LTC engines deliver fuel accurately and consistently, as well as manage temperature and EGR sensitivity, to minimize cyclic variations. A variety of factors must be precisely controlled and managed in order to mitigate cyclic variations and optimize engine performance.

LTC engines, such as reactivity controlled compression Ignition (RCCI), homogeneous charge compression ignition (HCCI), and partially premixed compression ignition (PPCI), operate under conditions that promote LTC, resulting in reduced emissions and improved fuel efficiency (Pan et al., 2020). However, these engines are more susceptible to CCVs in combustion parameters compared to conventional CI engines. To minimize cyclic variations in LTC engines, accurate and consistent fuel delivery is crucial. Cyclic variations result from fluctuations in combustion processes between cycles, impacting performance and emissions. The fuel injection process plays an important role in the combustion process, and any inconsistencies in fuel delivery can impact ignition timing, combustion rates, and overall performance (Rezaei et al., 2013). EGR, or reintroducing exhaust gases into the combustion chamber, alters combustion temperatures and oxygen concentration. Minimizing cyclic variations requires effective management of these parameters.

PPCI engines, another category closely related to LTC engines, also play a significant role in the context of cyclic variations (Kumar & Sandhu, 2022a). PPCI engines achieve combustion by partially premixing the fuel and air before ignition. This approach aims to combine the benefits of conventional diesel combustion and premixed combustion, offering a compromise between efficiency and emissions. PPCI engines face challenges similar to those encountered in LTC engines, particularly in managing cyclic variations (Cheng et al., 2016; Oliver, 2013; Zhang et al., 2012).

The LTC strategy can be implemented by modifying existing diesel engines with sensors and an open engine control unit (ECU) system. These modifications allow for precise control of the combustion process, enhancing stability by minimizing cycle-to-cycle variations. By optimizing combustion parameters using real-time data from the sensors, the LTC strategy improves fuel efficiency and reduces emissions, providing a more sustainable and efficient solution for diesel engines. This stability ensures consistent engine performance, reduced emissions, and improved efficiency. To enhance combustion stability in these advanced engines, advanced control algorithms, sophisticated fuel injection systems, temperature management, and EGR management are continuously researched and implemented. As the automotive industry moves toward more sustainable and efficient technologies, ongoing research is focused on understanding and mitigating cyclic variations in LTC, PPCI, and similar advanced combustion engines. By refining these engines, advances in sensor technologies, control strategies, and combustion modeling enable them to perform optimally and comply with stringent emission regulations.

5.2 LOW-TEMPERATURE COMBUSTION

LTC represents an innovative combustion strategy devised by researchers to circumvent diffusion combustion, aiming to achieve optimal premixing of fuel and air. In LTC mode, the heat release rate (HRR) exhibits distinctive characteristics, notably a two-stage heat release, which differs from the traditional diesel HRR (He et al., 2014; Maurya & Agarwal, 2011; Noh & No, 2017). Notably, LTC mode operation with higher research octane number (RON) fuels demonstrates promising results in addressing exhaust emission challenges inherent in traditional CI engines (Kumar et al., 2021). Moreover, the utilization of fuels with high lower heating value (LHV) contributes to reduced compression work, thereby enhancing engine efficiency (Shamun et al., 2017). Despite the significant potential for emission reduction, achieving combustion controllability remains a crucial concern in LTC engines. Consequently, these combustion strategies are typically applicable only within specific ranges of the engine load and speed map (Imtenan et al., 2014).

However, among the array of LTC strategies, the partially premixed combustion (PPC) strategy exhibits superior combustion controllability compared to HCCI. In HCCI engines, spontaneous auto-ignition predominantly governed by the chemical kinetics of the homogeneous charge leads to excessive pressure rise rate (PRR) at high loads, contributing to combustion instability (Dempsey et al., 2014). To mitigate this instability, it's imperative to separate the combustion event from the fuel injection event. Fuel injection timings in the PPC strategy are deliberately delayed compared to those in the HCCI strategy, resulting in partially premixed air-fuel mixtures rather than fully premixed ones. This partial premixing occurs because the injection is decoupled from the start of combustion (SOC). By using this approach, it becomes possible to better control the combustion phasing of the Premixed Charge Compression Ignition (PPCI) engine, leading to reduced cyclic variability. It has been observed that when the combustion phasing range is closer to top dead center (TDC), CCVs can be further improved. Moreover, the PPC strategy yields a more moderate HRR, enabling it to overcome the high load limitations encountered by HCCI engines (Manente et al., 2010). Various pressure-related parameters extracted from the pressure history provide insights into the CCVs (Kumar et al., 2021).

Few important parameters of pressure are as follows:

1. In-cylinder peak pressure
2. Individual cycles indicated mean effective pressure (IMEP)
3. Maximum rate of pressure rise

A prevalent technique employed to assess cyclic variation involves a statistical method such as the coefficient of variation (COV) of IMEPn. IMEPn can be computed using the measured in-cylinder pressure, as per Equation (5.1). Combustion stability can then be evaluated using the COV of IMEPn, denoted as COV_{IMEPn}, calculated through Equations (5.2)–(5.3). IMEPn is derived from in-cylinder pressure measurements.

$$\mathrm{IMEP_n} = \frac{1}{\mathrm{V_d}} \int_{0}^{720} p\,dv \tag{5.1}$$

$$\bar{\mathrm{X}} = \frac{1}{N} \times \sum_{1}^{N} X_i \tag{5.2}$$

$$\mathrm{COVIMEPn} = \left(\sqrt{\sum_{i=1}^{N} \frac{\left(X_i - \bar{X}\right)^2}{N}} \Big/ \bar{X} \right) 100\% \tag{5.3}$$

where N represents consecutive sampled cycles and X_i denotes the IMEPn of a particular combustion cycle. As outlined in the report by Przybyla et al., stable combustion is typically characterized by a COV_{IMEPn} upper limit of 5%. Beyond this threshold, the engine tends to exhibit unacceptably high CCVs (Przybyła et al., 2016). Furthermore, if the upper limit of COV_{IMEPn} surpasses 10%, the engine is susceptible to misfiring (Heywood, 2011).

5.3 WAVELET TRANSFORM ANALYSIS

In internal combustion engines (ICE), it has proven to be an effective tool for examining signal singularities, intermittent combustion, and complex non-stationary time series. The wavelet transform (WT) algorithm finds applications in a wide range of research domains and sectors, demonstrating its efficiency in signal analysis (Kumar & Sandhu, 2022a, 2022b; Kumar et al., 2022). Moreover, using WT, one can calculate the amplitude of cyclic variability, determine its periodicity, and determine its temporal extent. This information can then be utilized to modify and control systems.

In the time-frequency plane, continuous WTs (CWTs) use a variable-size window to enable adaptive changes in frequency and time resolutions. As illustrated in Equation 5.4, the CWT formulation for a mother wavelet ψ (t) (Torrence & Compo, 1998).

$$\mathrm{CWT_n} = \left(\frac{\delta \mathrm{t}}{\mathrm{a}}\right)^{\frac{1}{2}} \sum_{\mathrm{n'}=0}^{\mathrm{N}-1} \mathrm{Xn'} \psi^{*} \left[\frac{(\mathrm{n'}-\mathrm{n})\delta \mathrm{t}}{\mathrm{a}}\right] \tag{5.4}$$

A wavelet power spectrum (WPS) is calculated by using Equation (5.5) in order to calculate the square modulus of a CWT (Christopher Torrence & Compo, 1998).

$$WPS = \left|CWT_n(a)\right|^2 \tag{5.5}$$

With this method, both maximum pressure (Pmax) time-series data and IMEP data can be analyzed for cyclic variations across a time and frequency domain (number of combustion cycles). Therefore, the WPS patterns can be projected onto a two-dimensional plane for representation in frequency-time. The global wavelet spectrum (GWS), calculated by Equation (5.6) (Torrence & Compo, 1998), is an estimate of average time-series characteristics of the WPS. Peaks in the GWS are indicative of predominant periodic patterns.

$$GWS = \frac{1}{N}\sum_{n-1}^{N}\left|CWT_n(a)\right|^2 \tag{5.6}$$

5.4 TEST METHOD

Using neat n-butanol to achieve the PPC strategy requires higher intake air temperatures, which can decrease volumetric efficiency. To address this, blending n-butanol with diesel fuel is more suitable for CI engines. This blend optimizes the PPC strategy by balancing the benefits of higher air temperatures and maintaining efficient volumetric performance, thus enhancing the overall combustion process in CI engines. Since the density of DB20 fuel is close to that of diesel (DB00) fuel, it mixes readily with diesel. Diesel/n-butanol blends with n-butanol volume ratios of 0% (DB00), 20% (DB20), and 60% (DB40) were tested on the CI engine. The experiments were conducted under torque loads of 15% and 45% of the maximum torque output, corresponding to 2 bar and 6 bar BMEP. Injection timings of CA before top dead center (BTDC) at 4° and 8° were considered delayed, while injection timings at 12° and 16° were considered advanced.

5.5 TEST ENGINE AND MEASURE EQUIPMENT

Fuel injection system (FIS) modifications and possibly intake system modifications are required for CI engines to operate in the n-butanol/diesel PPCI combustion mode. The FIS may need to be adapted to inject the diesel/n-butanol blend in a controlled manner, ensuring correct timing and precise amounts.

The FIS in this setup operates as an open-loop system and includes software, a communication interface, and an electronic control unit (ECU). Figure 5.1 presents a schematic diagram of the experimental setup, showcasing various actuators and sensors. In this experiment, the BOSCH-MCP4-i7 ECU individually controlled the injection mass, boost pressure, injection timing, injection pressure, and number of injections.

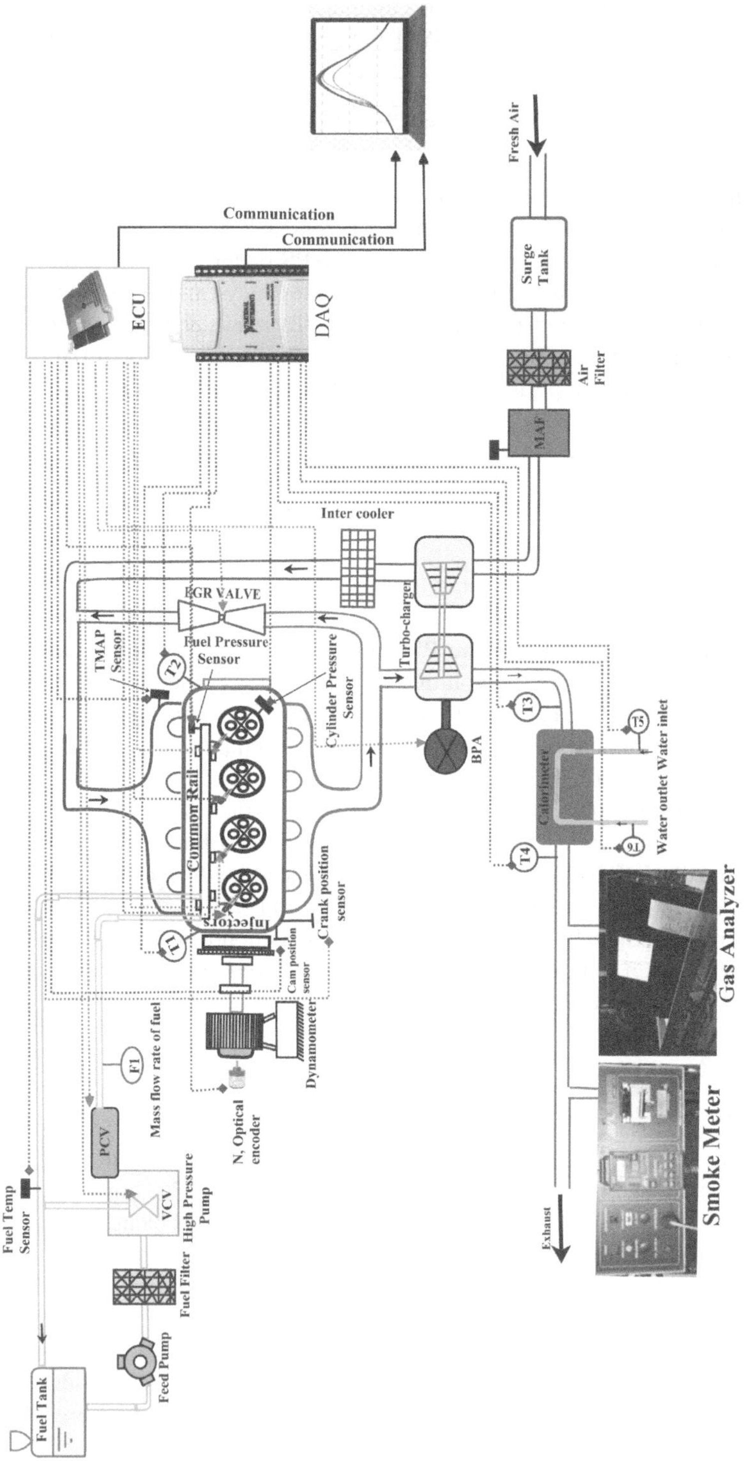

FIGURE 5.1 Schematic diagram of the experimental setup.

5.6 COMBUSTION STABILITY ANALYSIS

A variation in IMEP and maximum pressure (Pmax) from cycle to cycle can significantly impact the combustion characteristics of the engine, making it difficult to drive and causing it to consume more fuel.

5.6.1 Statistical Analysis of Combustion Stability

In PPCI mode engines, the start of injection (SOI) of fuel, especially in combination with diesel/n-butanol fuel blends, plays an important role in controlling combustion parameters since it is a critical control variable. Considering this, a meticulous analysis of their effects on CCVs of both maximum pressure and IMEP is performed over 200 consecutive combustion cycles. Under all tested conditions, this study uses a variety of statistical methods, including standard deviation (STD) calculation in time-series data for Pmax and IMEP, normal probability assessment of IMEP, and variance calculation.

Based on Figures 5.2(a, b), it is demonstrated that SOI of 12° CA BTDC leads to a reduction in STD in the IMEP time series across all fuel variants (Kumar & Sandhu, 2022b). Nevertheless, Figures 5.2(c, d) demonstrate a departure from the observed trend in the STD of Pmax time series for DB40 fuel compared to IMEP time series. Moreover, as illustrated in Figures 5.2(c, d), adding n-butanol to DB00 fuel results in an increase in the STD of Pmax time series.

Therefore, we selected the IMEP time series to analyze CCVs fluctuations spectral. A reduction in STD is also observed in Figures 5.2(a–d) when the early SOI timing is used as an alternative to other experimental conditions in the IMEP time series for DB20. As a result of the oxidation process that precedes combustion during early SOI, which results in higher combustion temperatures and a lower STD, this reduction can be attributed.

COV_{IMEP} and COV_{Pmax} for diesel/n-butanol fuel blends at early and late SOI timing are illustrated in Figure 5.3.

The study indicates that with an early SOI timing of 12° CA BTDC, both COV_{IMEP} and COV_{Pmax} decrease under 2 bars of BMEP. Early SOI timing can be attributed to a higher mean combustion temperature and higher combustion angle at 50% mass fraction (CA50) (Kumar & Sandhu, 2022b). Kumar et al. stated that the peaks of the SOC and HRR are shifted earlier when the SOI timing is advanced from 4° CA BTDC to 12° CA BTDC (Kumar & Sandhu, 2022b). The increased combustion temperatures and decreased piston speed result from this advancement. Thus, both the fresh charge and residual gas from the previous cycle are less sensitive to pressure and temperature fluctuations, resulting in enhanced combustion stability. When n-butanol fuel is introduced into diesel, the COV_{Pmax} and COV_{IMEP} are exacerbated (indicating higher cyclic variability) at 2 bar BMEP load and 4° CA BTDC compared to DB00 fuel. In DB40 fuel, the heating value is lower (LHV) and the heating efficiency is higher (LHE), resulting in a delayed combustion phasing due to the slower end of injection than in D00 and DB20 fuel. In both early and late SOI timing configurations, DB40 exhibits significant cyclic variability. A number of factors influence the CCVs of CI engines, including SOI timing and the use of

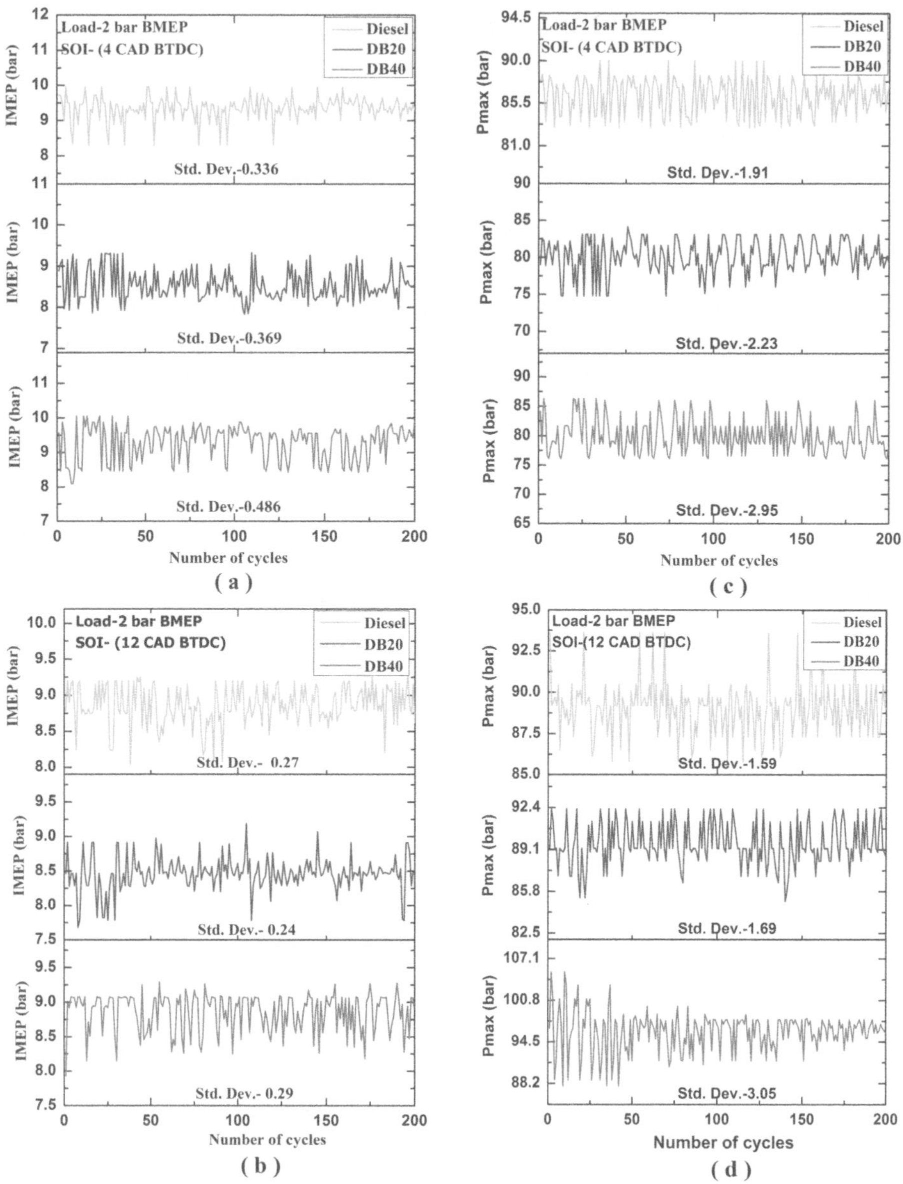

FIGURE 5.2 (a–d) N-butanol/diesel blended fuels and early and late SOI timing on time series of Pmax and IMEP. (Kumar & Sandhu, 2022b)

blended fuels containing n-butanol and diesel. Due to the increased CCVs, excessively delayed injection timings or high n-butanol ratios cannot be used.

A PPCI engine may exhibit CCVs in a variety of patterns due to changes in air/fuel ratios and underlying physics within the combustion chamber (Shahbakhti & Koch, 2008). A further analysis of the ensemble IMEP derived from 200 consecutive

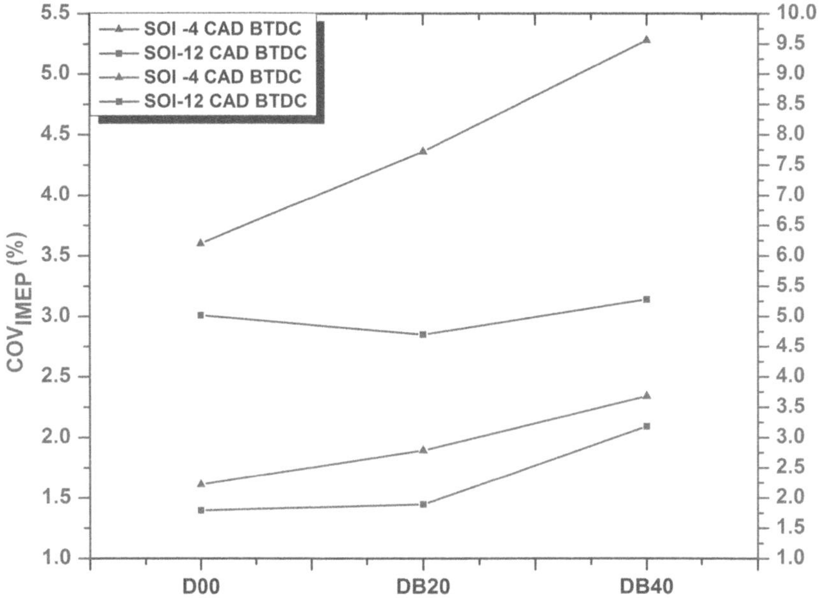

FIGURE 5.3 COV_{IMEP} and COV_{Pmax} of an n-butanol and diesel fuel blends at early and late SOI timing.

combustion cycles will be conducted to understand the trends of CCVs in PPCI engines. The ensemble IMEP is shown to exhibit a variety of distributions across different engine operating conditions. Understanding these IMEP distributions will aid in pinpointing areas of high CCVs in PPCI engines. An effective way to determine if IMEP time series adhere to a normal distribution is to use a normal probability plot. A normal distribution is indicated by a straight line, and deviations from this line indicate deviations from normality.

IMEP plots for n-butanol and diesel blended fuels at a 4° CA BTDC late IT are presented in Figures 5.4(a–c) (Kumar & Sandhu, 2022b).

Figure 5.4(c) illustrates that the DB40 blended fuel departs significantly from the normal distribution. On the other hand, the DB00 fuel distribution is much closer to normal. In the IMEP time-series data, deviations from the normal distribution indicate deterministic components, which indicate predictable patterns that can be modeled and predicted to control the data. Although statistical approaches are useful for analyzing the temporal variation in the maximum pressure (Pmax) and IMEP time series, they cannot characterize the spectral fluctuations or internal evolution of the series.

5.6.2 Wavelet Transforms Approach to Combustion Stability

To explore the spectrum fluctuations associated with PPCI engine combustion parameters, the WT is utilized. By analyzing time-series data encompassing 200 combustion cycles in IMEP, cyclic variations can be examined using both GWS and the WPS. In the WPS, the x-axis represents the number of combustion cycles,

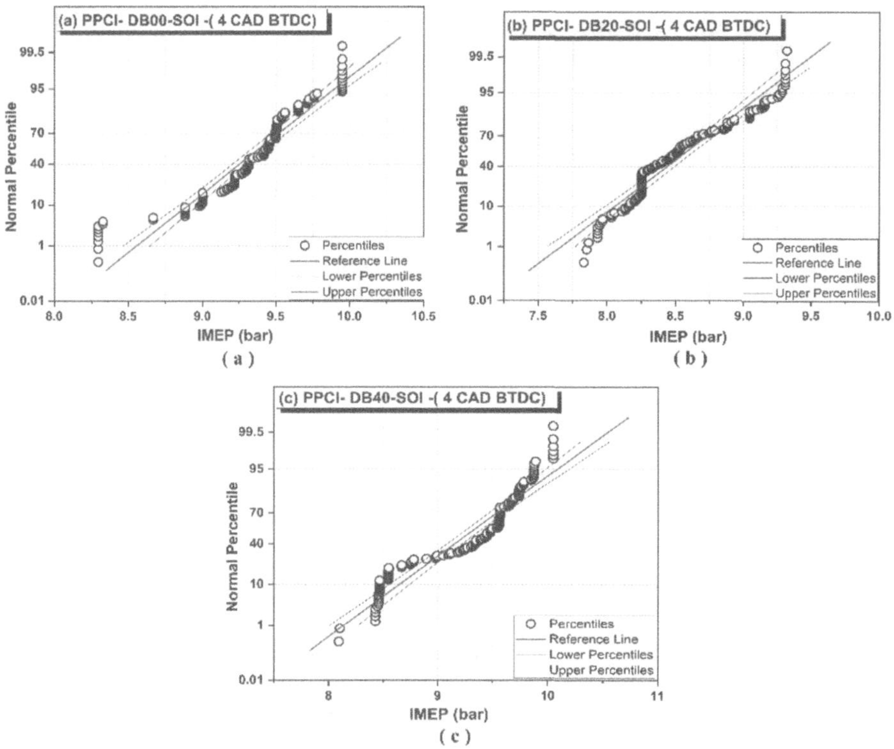

FIGURE 5.4 (a–c) Normal probability plots of IMEP for n-butanol/diesel blended fuels at late injection timings.

while the y-axis represents the frequency of the combustion parameters. For the GWS, the power axis is on the x-axis, and the period axis is on the vertical axis. In spectrogram analysis, red indicates signals at their highest energy, and blue represents signals at their lowest (Kumar et al., 2022). The colors in the WPS reflect the signal's energy contribution and the amplitude of the WT. Sen et al., limited their analysis to the areas above the cone of influence (COI), which are delineated by U-shaped traces marking the boundaries of the COI (Sen et al., 2008).

IMEP time-series data are investigated using wavelet analysis for periodicity and amplitude fluctuations. For analyzing cyclic variability using the WT approach, spectrograms are commonly employed to assess the energy contribution and temporal fluctuations of each frequency band (Qi, 2000; Saxena & Maurya, 2018a; Sen et al., 2010; Tily & Brace, 2011). In PPCI engines, IMEP fluctuates across multiple time scales. According to Figure 5.5(a), a periodic band with strong intensity within the 4–6 cycle range exhibits pronounced fluctuations within cycles spanning 5–34 and 116–128, respectively. The presence of intense periodicity at specific combustion cycles indicates heightened cyclic variability in combustion parameters, as evidenced by the increased variation within the highlighted periodic bands (Kumar & Sandhu, 2022a). Meanwhile, lower-intensity periodicity suggests reduced CCVs. In Figure 5.5(a), a nearly continuous periodic band is observed around the 48–64 cycle range, spanning cycles 66–131.

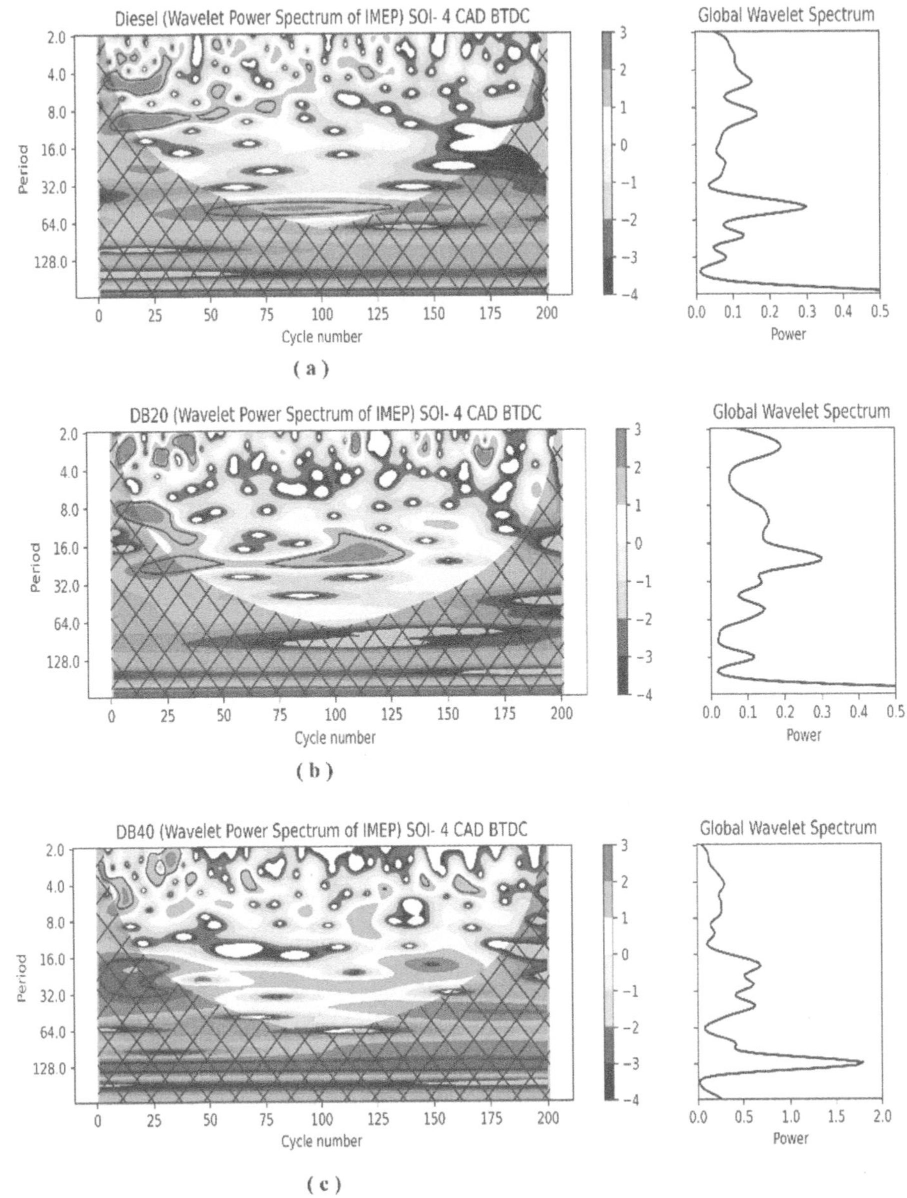

FIGURE 5.5 (a–c) GWS and WPS of IMEP for various diesel/n-butanol blends at SOI – 4° CA BTDC (Kumar & Sandhu, 2022b).

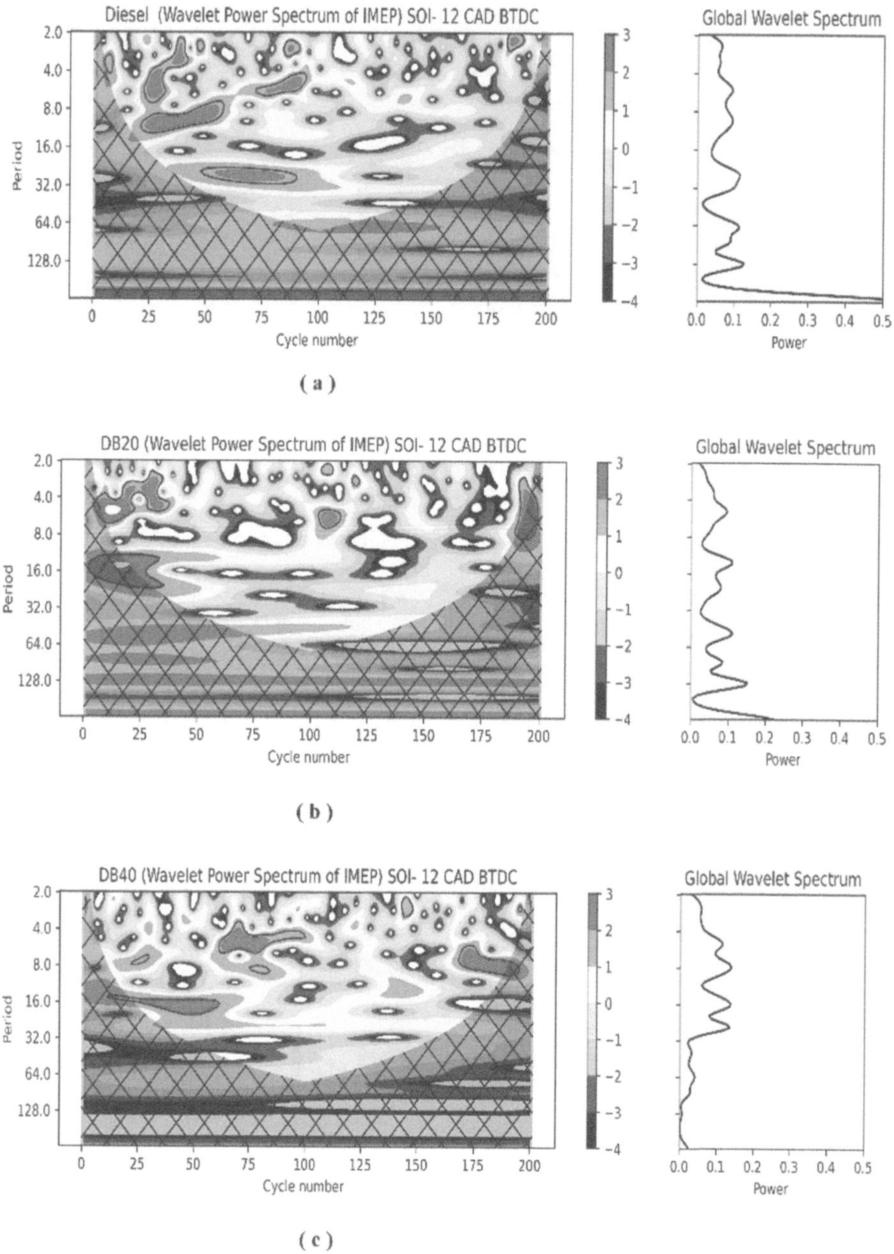

FIGURE 5.6 (a–c) GWS and WPS of IMEP for various diesel/n-butanol blends at SOI – 12° CAD BTDC (Kumar & Sandhu, 2022b)

DB00 fuel exhibits significant cyclic variability with a lower frequency at an SOI timing of 4° CA BTDC, as evidenced by the presence of extended periodic bands over numerous cycles. Additionally, Figure 5.5(b, c) shows that adding n-butanol to fuel amplifies IMEP cyclic variability at a 4° CA BTDC IT. Specifically, Figure 5.5(b) indicates that the GWS of IMEP for DB20 fuel increases by 0.1 compared to DB00 fuel, suggesting a notable rise in IMEP cyclic variation. The figure also demonstrates scattered strong intensity periodic bands throughout the WPS for DB20 fuel, with intense bands spanning cycles 3–11, 20–26, and 28–38. Moreover, lower frequency regions between periods 14–24 and 16–20 show strong intensity periodic bands spanning cycles 70–131 and 24–50, respectively. With an increased n-butanol blend, the maximum GWS power significantly rises to 0.8 for DB40 fuel, indicating severe CCVs in IMEP at a 4° CA BTDC IT. The increased cycle-to-cycle variability of DB20 and DB40 fuels may be attributed to their differing combustion characteristics.

Furthermore, Figure 5.6(a) indicates a higher degree of cyclic variation for DB00 fuel, highlighted by two strong intensity periodic bands of 8–14 and 24–32 over numerous cycles, ranging from 21–58 to 48–90, respectively. Compared to diesel fuel, the peak GWS (0.12 at period 16) for DB20 fuel at 12° CA BTDC shows only a minimal difference in GWS power between diesel and DB20 fuel under a 2 bar BMEP load, as shown in Figures 5.6(a, b). As illustrated in Figure 5.6(b), high-frequency fluctuations in IMEP are evident for DB20 fuel, particularly between cycles 3–8 and 10–20, extending over more than 20 engine cycles from cycle 16 to 36. Increasing the n-butanol ratio to 40% in DB00 fuel results in strong intermittent oscillations, with a prominent periodic band observed between cycles 60–98 and 167–189. The peak GWS power value (GWS peak=0.13 at period=16) for DB40 fuel at 12° CA BTDC rises, as a nearly continuous periodic band spans cycles 21–60 around periods 16–24, as depicted in Figure 5.6(c). Consequently, DB40 fuel exhibits higher cyclic variability in IMEP compared to DB20 and DB00 fuels, indicating that the impact of n-butanol fuel and ignition timing on IMEP is significant.

5.6.3 The Influence of Blending Ratios on Cycle-to-Cycle Variation in Combustion Parameters through Wavelet Analysis

Three panels of Figures 5.7(a–c) show the WPS and the GWS of the IMEP time series of the n-butanol/diesel blends fuels (DB00, DB20, and DB40) at an injection start timing of 8° CA BTDC under 2 bar BMEP. IMEP time series consists of 200 engine cycles and limits the periodicities to less than 64 cycles. At an 8° CA BTDC injection timing, WPS analysis reveals distinct patterns for various diesel/n-butanol blend fuels.

In DB00 fuel, WPS is restricted with strong intensity occurring intermittently between 2 and 6 periods (higher frequency), spanning cycles 10–19, 31–41, and 133–150. There is also evidence of a strong intensity band spanning periods 6–16, occurring in cycles 10–48, along with another period of intensity spanning periods 16–30, which persists for 13 cycles from cycle 25 to 38. Within the COI, these periodic bands result in a GWS maximum power of 0.16 (at period 12). In contrast, DB20 fuel exhibits diminished cyclic variations, as indicated by a lower GWS peak power (GWSpeak = 0.11, at period 3). As a result of the higher oxygen

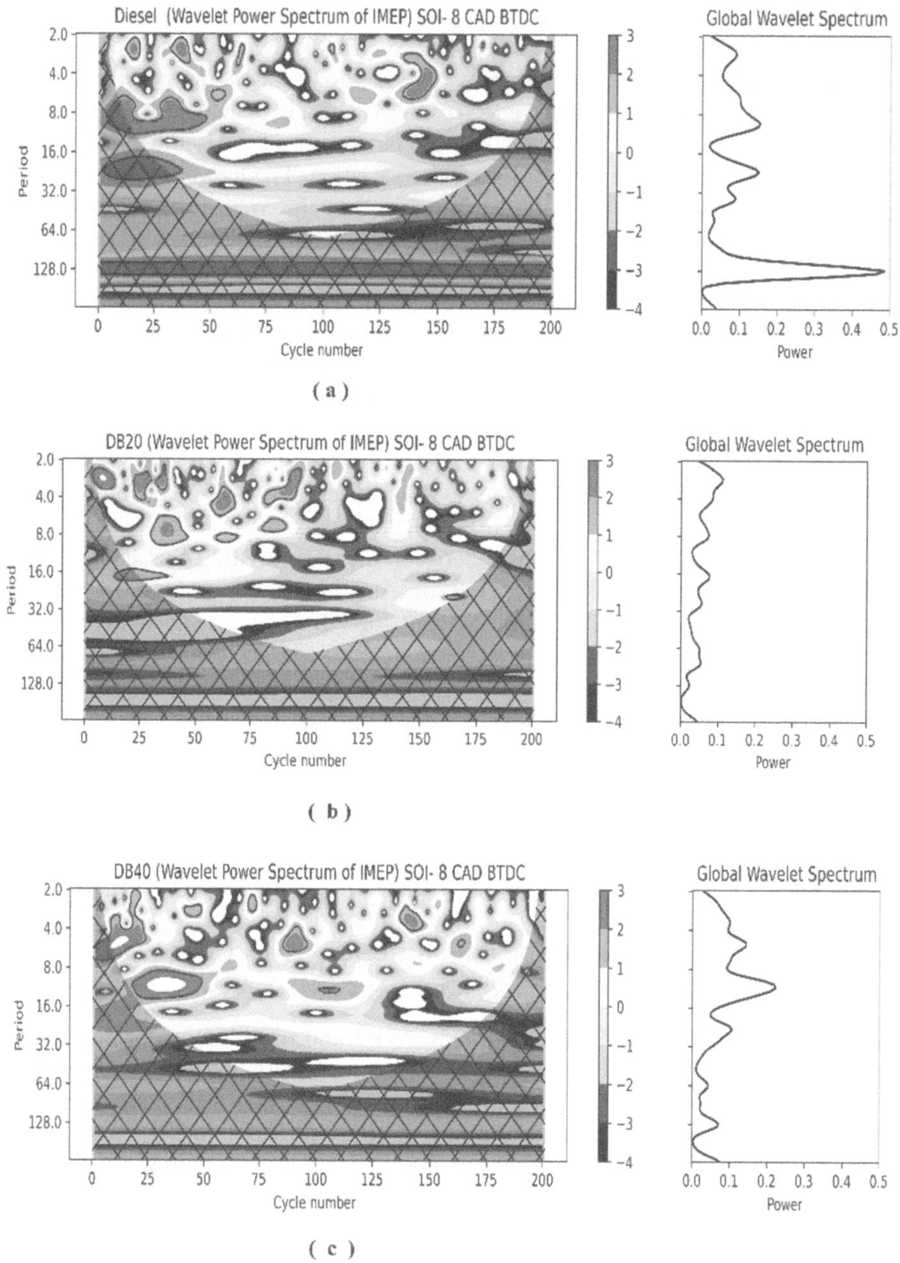

FIGURE 5.7 (a–c) WPS and GWS of IMEP for n-butanol/diesel blends (a) D00, (b) DB20, and (c) DB40 at SOI – 8° CAD BTDC (Kumar & Sandhu, 2022a).

concentration in DB20 fuel and the phasing of combustion near TDC, this reduces the rate of combustion instability (Kumar & Sandhu, 2022a). Variations are mostly concentrated in cycles 2–10, with occasional occurrences in cycles 2–14, 21–30, 29–46, 31–34, 55–67, and 81–97. On the other hand, DB40 fuels exhibit greater cyclic variation, as evidenced by their higher GWS peak power (GWSpeak = 0.22, at period 10) than their counterparts DB20 and DB00. A strong intensity band is evident in DB40, with intermittent fluctuations occurring in periods 3–7, spanning cycles 7–24, 84–96, and 136–148.

Figures 5.6(a–c) present GWS and WPS for various diesel/n-butanol blends under 12° CA BTDC injection timings. In Figure 5.6(a), both GWS and WPS are shown for the IMEP time-series data of DB00 fuel. The WPS analysis reveals a relatively strong intensity band encompassing periods 3–8, with fewer cycles in the high-frequency region (lower period), including periods 22–42, 84–91, 71–94, and 181–189. A period band of 8–12 periods is also prominent in cycles 18–50 and 87–121. Overall, the wavelet transform provides valuable insights into the spectrum behavior of IMEP time series by explaining how diesel/n-butanol blend fuels affect PPC parameters at multiple scales. By utilizing in-cylinder pressure feedback as a fuel injection control strategy, these findings can be utilized.

ABBREVIATIONS

BMEP	Brake mean effective pressure
CCVs	Cycle to cycle variations
CRDI	Common rail direct injection
DB20	20% N-butanol biofuel blended with 80% diesel
ECU	Electronic control unit
FIT	Fuel Injection Timing
GWS	Global Wavelet Spectrum
HRR	Heat release rate
IMEP	Indicated mean effective pressure
Pmax	Maximum cylinder pressure
PPCI	Partially Premixed Compression Ignition
SOC	Start of combustion
SOI	Start of Injection
STD	Standard deviation
WPS	Wavelet power spectrum

5.7 CONCLUSIONS

This study examined the impact of early and late starting of injection (SOI) timings, along with diesel/n-butanol blends, on CCVs for a 4-cylinder turbocharged CRDI diesel engine at maximum pressure (Pmax) and IMEP. As a summary of this research, the following findings are important:

1. Opting for an advanced fuel injection strategy across all examined fuels leads to increased combustion pressure (CP) and diminished premixed

combustion stage compared to adopting a delayed fuel injection approach. Furthermore, the inclusion of n-butanol into diesel fuel results in heightened premixed combustion HRR and in-cylinder pressure within the parameters of the conducted experiments.

2. Cycle-to-cycle variabilities (CCVs) in IMEP and maximum pressure (Pmax) are investigated in the 4-cylinder CRDI engine through statistical analysis and wavelet analysis. Observations reveal that an early SOI timing results in a reduced STD in the time series of IMEP data compared to a late injection timing. Additionally, the COV of IMEP (COV_{IMEP}) and the COV of Pmax (COV_{Pmax}) values decrease with the advancement of SOI timing across all tested fuels. Experimental findings underscore that DB20 demonstrates lower CCVs in IMEP and Pmax when employing an early SOI timing at 12° CA BTDC.
3. Incorporating n-butanol into DB00 fuel results in heightened deviation from a normal distribution at a late IT of 4° CA BTDC. The introduction of DB40 fuel diverges from a normal distribution primarily because its combustion angle at 50% mass fraction (CA50) is distant from TDC.

The study analyzes 200 consecutive combustion cycles of IMEP time-series data and reveals that when using a 12° CA BTDC SOI under 2 bar BMEP, DB20 shows a reduced cyclic variability, as indicated by its lower GWS power. Also, the DB40 fuel exhibits the highest GWS power at a 4° CA BTDC SOI timing among all tested fuels, with the highest cycle-to-cycle variability (CCVs) observed in IMEP within a periodic interval of 16–32 cycles. By analyzing cyclic variability from a spectral perspective, we gain valuable insights that can be applied to the development of an effective engine control strategy.

REFERENCES

Abdalla, A. N., Ali, O. M., Awad, O. I., & Tao, H. (2019). Wavelet analysis of an SI engine cycle-to-cycle variations fuelled with the blending of gasoline-fusel oil at a various water content. *Energy Conversion and Management, 183*(January), 746–752. https://doi.org/10.1016/j.enconman.2019.01.034

Cheng, X., Li, S., Yang, J., & Liu, B. (2016). Investigation into partially premixed combustion fueled with n-butanol-diesel blends. *Renewable Energy, 86*(X), 723–732. https://doi.org/10.1016/j.renene.2015.08.067

Dempsey, A. B., Walker, N. R., Gingrich, E., & Reitz, R. D. (2014). Comparison of low temperature combustion strategies for advanced compression ignition engines with a focus on controllability. *Combustion Science and Technology, 186*(2), 210–241. https://doi.org/10.1080/00102202.2013.858137

He, B., Yuan, J., Liu, M., & Zhao, H. (2014). Combustion and emission characteristics of a n-butanol HCCI engine. *Fuel, 115*, 758–764. https://doi.org/10.1016/j.fuel.2013.07.089

Heywood, J. B. (2011). *Internal combustion engine fundamentals* (New York, McGraw Hill Education, 1988), ISBN:978-0070286375.

Imtenan, S., Varman, M., Masjuki, H. H., Kalam, M. A., Sajjad, H., Arbab, M. I., & Rizwanul Fattah, I. M. (2014). Impact of low temperature combustion attaining strategies on diesel engine emissions for diesel and biodiesels: A review. *Energy Conversion and Management, 80*(x), 329–356. https://doi.org/10.1016/j.enconman.2014.01.020

Kumar, P., & Sandhu, S. S. (2022a). An attempt to implement partially premixed combustion strategy in a multi-cylinder CRDI engine: A detailed experimental and wavelet transform analysis. *Fuel, 323*(February), 124372. https://doi.org/10.1016/j.fuel.2022.124372

Kumar, P., & Sandhu, S. S. (2022b). An experimental investigation into the combustion stability and emissions of an n-butanol/diesel blended fueled partially premixed compression ignition (PPCI) engine. *International Journal of Engine Research, 24*, 2451–2464.

Kumar, P., Sandhu, S. S., Singh, M., & Deep, A. (2021). Potential assessment of methanol to reduce the emission in LTC mode diesel engine. In Agarwal A.K., Valera H., Pexa M., Čedík J. (eds) *Methanol. Energy, environment, and sustainability.* Springer, Singapore. https://doi.org/10.1007/978-981-16-1280-0_11

Kumar, P., Singh, M., & Sandhu, S. S. (2022). Wavelet analysis for cyclic combustion dynamics of a multi-cylinder CRDI diesel engine fuelled with a blending of argemone biodiesel–diesel oil. *Chaos: An Interdisciplinary Journal of Nonlinear Science, 32*(4), 043107. https://doi.org/10.1063/5.0080910

Manente, V., Tunestal, P., Johansson, B., & Cannella, W. J. (2010). Effects of ethanol and different type of gasoline fuels on partially premixed combustion from low to high load. *SAE Technical Papers.* https://doi.org/10.4271/2010-01-0871

Maurya, R. K., & Agarwal, A. K. (2011). Experimental study of combustion and emission characteristics of ethanol fuelled port injected homogeneous charge compression ignition (HCCI) combustion engine. *Applied Energy, 88*(4), 1169–1180. https://doi.org/10.1016/j.apenergy.2010.09.015

Noh, H. K., & No, S. Y. (2017). Effect of bioethanol on combustion and emissions in advanced CI engines: HCCI, PPC and GCI mode – A review. *Applied Energy, 208*(August), 782–802. https://doi.org/10.1016/j.apenergy.2017.09.071

Oliver, J. (2013). Combustion and emissions characteristics for DI diesel engine run by partially-premixed (PPCI) low temperature combustion (LTC) mode. *Journal of Chemical Information and Modeling, 53*(9), 1689–1699. https://doi.org/10.1017/CBO9781107415324.004

Pan, S., Liu, X., Cai, K., Li, X., Han, W., & Li, B. (2020). Experimental study on combustion and emission characteristics of iso-butanol/diesel and gasoline/diesel RCCI in a heavy-duty engine under low loads. *Fuel, 261*(2019), 116434.

Przybyła, G., Postrzednik, S., & Żmudka, Z. (2016). The impact of air-fuel mixture composition on SI engine performance during natural gas and producer gas combustion. *IOP Conference Series: Materials Science and Engineering, 148*(1), 12082.

Qi, G. (2000). Wavelet-based AE characterization of composite materials. *NDT and E International, 33*(3), 133–144. https://doi.org/10.1016/S0963-8695(99)00037-7

Rezaei, S. Z., Zhang, F., Xu, H., Ghafourian, A., Herreros, J. M., & Shuai, S. (2013). Investigation of two-stage split-injection strategies for a dieseline fuelled PPCI engine. *Fuel, 107*(x), 299–308. https://doi.org/10.1016/j.fuel.2012.11.048

Saxena, M. R., & Maurya, R. K. (2018a). *Characterization of Cycle-to-Cycle Variations in Conventional Diesel Engine Using Wavelets.* 135–155. https://doi.org/10.1007/978-981-10-7575-9_7

Saxena, M. R., & Maurya, R. K. (2018b). Effect of butanol addition on performance, combustion stability and nano-particle emissions of a conventional diesel engine. *SAE Technical Papers, 2018-Septe*, 1–13. https://doi.org/10.4271/2018-01-1795

Sen, A. K., Litak, G., Finney, C. E. A., Daw, C. S., & Wagner, R. M. (2010). Analysis of heat release dynamics in an internal combustion engine using multifractals and wavelets. *Applied Energy, 87*(5), 1736–1743. https://doi.org/10.1016/j.apenergy.2009.11.009

Sen, A. K., Litak, G., Taccani, R., & Radu, R. (2008). Wavelet analysis of cycle-to-cycle pressure variations in an internal combustion engine. *Chaos, Solitons and Fractals, 38*(3), 886–893. https://doi.org/10.1016/j.chaos.2007.01.041

Shahbakhti, M., & Koch, C. R. (2008). Characterizing the cyclic variability of ignition timing in a homogeneous charge compression ignition engine fuelled with n-heptane/iso-octane blend fuels. *International Journal of Engine Research*, *9*(5), 361–397. https://doi.org/10.1243/14680874JER01408

Shamun, S., Haşimoğlu, C., Murcak, A., Andersson, Ö, Tunér, M., & Tunestål, P. (2017). Experimental investigation of methanol compression ignition in a high compression ratio HD engine using a Box-Behnken design. *Fuel*, *209*, 624–633.

Sharma, A., Kumar, P., Sandhu, S. S., & Singh, M. (2023). *Experimental investigation into the effects of Argemone biodiesel/diesel blends on cyclic variations in a multi-cylinder CRDI engine*. 7(2), 363–374.

Tily, R., & Brace, C. J. (2011). Analysis of cyclic variability in combustion in internal combustion engines using wavelets. *Proceedings of the Institution of Mechanical Engineers, Part D: Journal of Automobile Engineering*, *225*(3), 341–353. https://doi.org/10.1177/09544070JAUTO1723

Torrence, C., & Compo, G. P. (1998). A practical guide to wavelet analysis. *Bulletin of the American Meteorological Society*, *79*(1), 61–78. https://doi.org/10.4324/9780429311369-6

Zhang, F., Xu, H., Zeraati Rezaei, S., Kalghatgi, G., & Shuai, S. J. (2012). Combustion and emission characteristics of a PPCI engine fuelled with dieseline. *SAE Technical Papers*. https://doi.org/10.4271/2012-01-1138

6 Exploration of Fuel Property Effects on Advanced Combustion-Mode Engines

Giovani Dambros Telli, Josimar Souza Rosa, Paulo Roberto Wander, and Luiz Alberto Oliveira Rocha

6.1 INTRODUCTION

In compression ignition (CI) engines, alcohols are primarily used in advanced combustion modes such as reactivity controlled compression ignition (RCCI), premix charge compression ignition (PCCI), and homogeneous charge compression ignition (HCCI), among others. The performance of these combustion modes is highly sensitive to the properties of the fuels used, making it important to thoroughly understand these properties in detail (Y. Li et al., 2016).

The literature broadly classifies fuels used in advanced combustion modes into two categories: low-reactivity fuels (LRFs) and high-reactivity fuels (HRFs). Conventional LRFs, characterized by a high-octane rating, include gasoline, natural gas, and ethanol (or hydrous ethanol/wet ethanol (Telli et al., 2023)), while HRFs, marked by a high cetane rating, encompass diesel, biodiesel, and other oils. The ratio between LRF and HRF varies depending on the load applied to the engine. At higher loads, a greater amount of LRF is required to prevent premature combustion, while at lower loads, a higher proportion of HRF is needed (Rosa et al., 2020).

Binary mixtures used in advanced combustion modes consist of two distinct fuels, combined to enhance specific characteristics (Ashok et al., 2019; Ge et al., 2022; Kandasamy et al., 2019). Similarly, ternary mixtures, composed of three different fuels, can be prepared for the same purpose (Ghadikolaei et al., 2018; Kukana & Jakhar, 2022; Zuo et al., 2022). For example, biodiesel has certain unfavorable physical properties that can deteriorate engine performance, such as higher viscosity and a higher pour point compared to conventional diesel. To address these issues, biodiesel can be blended with alcohol to form a binary mixture or combined with both alcohol and diesel to create a ternary mixture (Ashour & Elwardany, 2020). While these mixtures can improve viscosity and pour point, they may also lower the overall calorific value (because of the presence of oxygen in its molecule), which can

DOI: 10.1201/9781003482277-6

negatively impact engine efficiency and power output. Furthermore, the high latent heat of vaporization (HoV) of alcohols also impacts engine performance, which will be discussed in more detail in the following sections. Consequently, a significant number of studies have focused on this topic in recent years.

Power output also tends to decrease when using different fuels, because the engine was not designed to operate with them. Obviously, fuels with lower calorific value tend to reduce power output, especially if strategies to correct the operation, such as adjusting the fuel injection angles, are not provided.

In the realm of advanced combustion modes, various strategies are employed for the simultaneous use of two or three fuels, extending beyond the simple premixing of liquid fuels. For instance, these strategies may involve the combined use of direct injection (DI) and manifold injection. Typically, the HRF is delivered to the engine via DI into the combustion chamber, where it facilitates ignition. Meanwhile, the LRF can be supplied to the manifold (Rosa et al., 2021; Rosa, Lorenzini, et al., 2019; Telli et al., 2018a, 2018b, 2021) or alternatively injected into the combustion chamber during the intake stroke or even during the compression.

In any fuel supply strategy, the properties of the fuel mixture differ from those of the individual fuels. Furthermore, the effects on combustion vary depending on whether the two fuels are supplied as a premixed blend or through separate injection systems, as the formation of the fuel-air mixture differs in each case. Figure 6.1 provides a graphical summary of the fuel injection strategies commonly discussed in the literature.

This chapter explores the properties of fuels and how various strategies employed in advanced combustion modes (HCCI, RCCI, and PCCI) can impact engine performance and emissions. Initially, it is provided a brief review of fuel properties and their effects on engine operation, with a particular focus on alcohols and other HRFs- or LRFs used in advanced combustion modes. The second part of the chapter

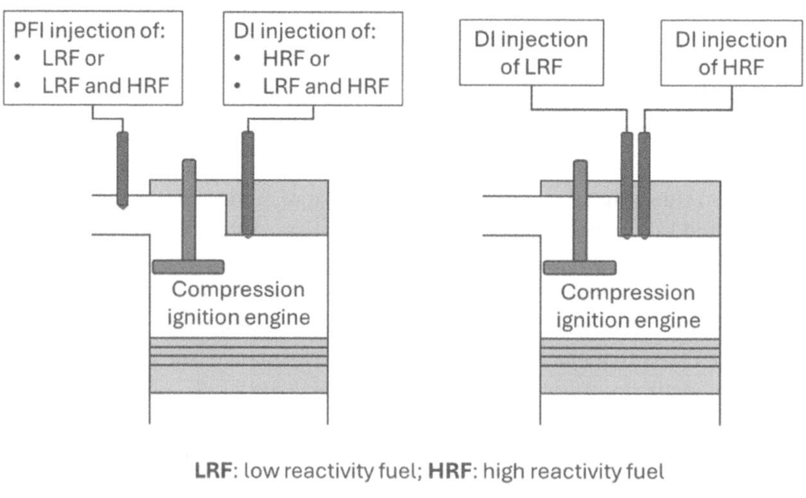

FIGURE 6.1 Injection strategies used in advanced combustion modes.

presents a literature review to highlight the results reported and discussed by various authors in recent years.

6.2 PROPERTIES OF ALCOHOLS AND OTHER FUELS OF INTEREST IN ADVANCED COMBUSTION MODES

The properties of fuels used in advanced combustion modes have a direct influence on air-fuel mixture formation, and consequently, on engine combustion, performance, and emissions characteristics. The following sections discuss the fuel properties most frequently cited in the literature as impactful in the combustion process for these advanced modes.

6.2.1 HEAT OF VAPORIZATION (HoV)

Alcohols generally have a higher HoV compared to petroleum-derived fuels (Kukana & Jakhar, 2022; Rosa, Altafini, et al., 2019). For example, the HoV of methanol is 3.7 times greater than that of gasoline (Jing et al., 2022). This property represents the energy required to vaporize the fuel, and fuels with a high HoV can significantly lower air temperature when injected, which in turn cools the air-fuel mixture. This temperature reduction directly influences the ignition process, especially in CI engines and it is called "cooling effect". The effect of HoV can vary depending on whether the engine uses separate injection systems for LRFs and HRFs, or if it operates with premixed fuel blends. These two methods of supplying fuels to the engine are illustrated in Figure 6.1. Figure 6.2 adapted from (Rosa et al., 2021) illustrates the pressure effect during compression in a CI engine using ethanol and diesel.

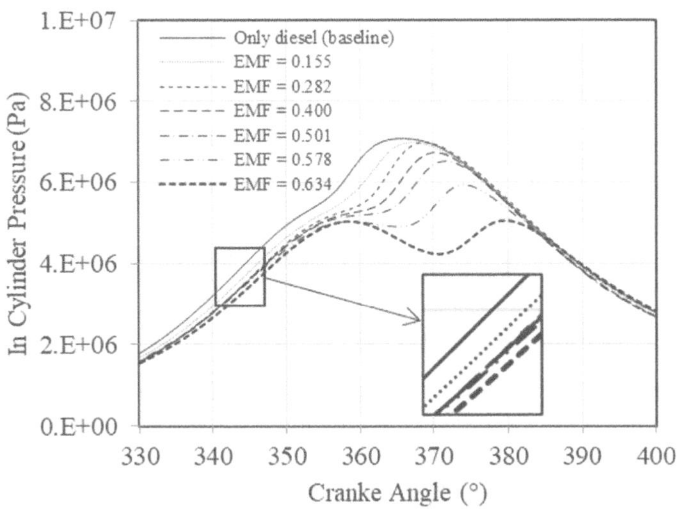

FIGURE 6.2 Pressure drop during compression with ethanol in a compression ignition engine. (Adapted from Rosa et al., 2021 with permission of Elsevier)

TABLE 6.1
Typical Values of HoV of Fuels for Advanced Combustion Modes

Fuel	HoV (kJ/kg)	Ref.
Methanol	1162.64	Ghadikolaei et al. (2018)
Ethanol	918.42	Ghadikolaei et al. (2018)
Butanol	581.4	Ghadikolaei et al. (2018)
Gasoline	290–420	Gawale and Naga Srinivasulu (2020)
Diesel	230–600	Rosa et al. (2020)
Biodiesel	300	Ghadikolaei et al. (2018)
PODE	359	J. Li et al. (2022)

The greater the ethanol mass fraction (EMF) (relative to diesel), known as the EMF, the lower the pressure during the compression stroke and, consequently, the temperature of the premixed charge. This temperature reduction is attributed in the literature to ethanol's high HoV (Goldsworthy, 2013; Morsy, 2015). The cooling effect of the intake charge serves as an effective means of controlling both the combustion phase and the in-cylinder temperature, helping to reduce NOx and particulate matter formation. This is one of the key reasons alcohols are commonly used in advanced combustion methods, especially when combined with HRFs.

When the engine employs fuel blending, the HoV of the mixture can be calculated as follows:

$$HoV = \sum_{i=1}^{j} x_i HV_i \tag{6.1}$$

- x_i: mass fraction of i component.
- HV_i: HoV of i component (kJ/kg).

Typical values of HoV are shown in Table 6.1.

Other fuels commonly used in advanced combustion modes, such as LRF, include dimethyl ether (DME), ammonia, and compressed natural gas (CNG). Their HoV values is not listed, as they are gases under standard ambient conditions.

6.2.2 Lower Heating Value (LHV)

Alcohols typically have a lower heating value (LHV) compared to other fuels. For example, methanol has an LHV that is 53% lower than diesel, ethanol is 37% lower, and butanol is 22% lower. Similar disparities are seen when compared to gasoline, which has an LHV of approximately 43.8 MJ/kg (Gawale & Naga Srinivasulu, 2020). Even when compared to other alternative fuels, such as biodiesel, alcohols exhibit lower LHV values (Zhao et al., 2021). Table 6.2 presents the LHV values for various fuels used in advanced combustion modes, including DME, ammonia, and CNG.

TABLE 6.2
Typical LHV of Fuels for Advanced Combustion Modes

Fuel	LHV (MJ/kg)	Ref.
Methanol	20.1	Rajesh Kumar and Saravanan (2016)
Ethanol	26.9	Rajesh Kumar and Saravanan (2016)
Butanol	33.2	Pan et al. (2017)
Gasoline	43.8	Gawale and Naga Srinivasulu (2020)
Diesel	42.5	Rajesh Kumar and Saravanan (2016)
Biodiesel	37.0	Zhao et al. (2021)
PODE	19.0	J. Li et al. (2022)
Dimethyl ether (DME)	28.4	Park et al. (2016)
Ammonia	18.6	Uddeen et al. (2024)
CNG	39.3	Garlet et al. (2024)

The relatively low LHV of alcohols compared to other fuels poses a challenge to their wider application. These lower calorific values can cause ignition and combustion issues, especially in systems not originally designed for alcohol fuels (Jing et al., 2022). The weighted LHV of two or more fuels used in advanced combustion modes, whether supplied as blends or separately, can be calculated using the following formula:

$$LHV_b = \sum_{i=1}^{j} x_i LHV_i \tag{6.2}$$

- LHV_b: low heating value of blend or combination of fuels (MJ/kg).
- x_i: mass fraction of i component.
- LHV_i: LHV of i component (MJ/kg).

The total energy rate ($\dot{Q}$) supplied by two or more fuels to the engine can be determined by

$$\dot{Q} = \sum_{i=1}^{j} \dot{m}_i LHV_i = \dot{m}_b LHV_b \tag{6.3}$$

- $\dot{m}_i$: mass flow of i component (fuel) (kg/s).
- $\dot{m}_b$: mass flow of blend or combination of fuels (kg/s).

6.2.3 Octane and Cetane Number

Evaluating a fuel's ignition behavior is critical to preventing engine damage, and fuel manufacturers must monitor the quality of the fuels they produce. As part of current fuel quality standards, the octane number is measured for spark-ignition engines, and the cetane number is measured for compression-ignition engines

TABLE 6.3
Typical Octane Number and Cetane Number of Fuels for Advanced Combustion Modes

Fuel	Octane Number	Ref.	Cetane Number	Ref.
Methanol	111	Paul et al. (2013)	3–5	Paul et al. (2013)
Ethanol	108	Gawale and Naga Srinivasulu (2020)	8	Ge et al. (2022)
Butanol	96	Zhao et al. (2021)	17–25	Zhao et al. (2021)
Gasoline	85–88	Gawale and Naga Srinivasulu (2020)	–	–
Diesel	20–30	Paul et al. (2013)	55.8	Ashok et al. (2019)
Biodiesel	–	–	53	Ge et al. (2022)
PODE	13.3	Wang et al. (2022)	78	J. Li et al. (2022)
Dimethyl ether (DME)	–	–	68	Park et al. (2016)
Ammonia	>130	Uddeen et al. (2024)	–	–
CNG	130	Telli et al. (2018a)	–	–

(Druzgalski et al., 2019). The standard method used in major industries to determine both octane and cetane numbers is in a Cooperative Fuel Research (CFR) engine, which has a variable compression ratio (CR). This allows for the comparison of fuels with standard mixtures under various thermodynamic conditions within the cylinder. Table 6.3 presents typical values reported in the literature.

The octane number essentially measures a fuel's resistance to premature detonation. A higher octane number indicates better anti-knock performance and enables the engine to operate at a higher CR (Y. L. Li et al., 2015). On the other hand, the cetane number reflects a fuel's ignition quality, with a higher cetane number corresponding to a shorter ignition delay, a desirable property for fuels used in conventional compression-ignition engines.

When mixtures of two fuels with different octane numbers or different cetane numbers are used, which is typical in dual-fuel HCCI combustion mode, the value of these properties for the mixture can be calculated by (Y. L. Li et al., 2015)

$$ON_b = \sum_{i=1}^{j} \frac{V_i ON_i}{100} \tag{6.4}$$

- ON_b: octane number of blend.
- V_i: percentage of volume of i component (fuel).
- ON_i: octane number of i component (fuel).

$$CN_b = \sum_{i=1}^{j} \frac{V_i CN_i}{100} \tag{6.5}$$

- CN_b: cetane number of blend.
- V_i: percentage of volum of *i* component (fuel).
- CN_i: cetane number of *i* component (fuel).

If the fuels are supplied separately, with one injected via port fuel injection (PFI) and the other via DI, the concentration of each fuel will vary across different regions of the cylinder. This results in non-homogeneous octane or cetane values throughout the combustion chamber. However, considering a widely studied case in advanced combustion modes, mainly RCCI and PCCI, where the fuel supplied in DI is injected well in advance into the cylinder, there may be enough time to promote mixing between the two fuels, allowing Equations (6.4) and (6.5) to be applied to estimate octane and cetane numbers.

6.2.4 Boiling Point and Self-Ignition Temperature

Boiling point and self-ignition temperature are relevant fuel properties that impact combustion. When evaporation and autoignition occur at high temperatures, achieving a homogeneous air-fuel mixture becomes more challenging, especially when the fuel is supplied via DI early in the compression stroke or through PFI, as is common in advanced combustion modes. Table 6.4 presents typical values from the literature.

Alcohol fuels generally have lower boiling points compared to petroleum-derived fuels and biodiesel. These lower boiling points facilitate evaporation inside

TABLE 6.4
Typical Boiling Point and Self-Ignition Temperature of Fuels for Advanced Combustion Modes

Fuel	Boiling Point (°C)	Ref.	Self-Ignition Temperature (°C)	Ref.
Methanol	64.7	Ghadikolaei et al. (2018)	463–470	Gwalwanshi et al. (2022)
Ethanol	78.4	Gawale and Naga Srinivasulu (2020)	423–434	Gwalwanshi et al. (2022)
Butanol	117.5	Ghadikolaei et al. (2018)	385–397	Gwalwanshi et al. (2022)
Gasoline	30–125	Gawale and Naga Srinivasulu (2020)	247–280	Gawale and Naga Srinivasulu (2020)
Diesel	188–360	Gawale and Naga Srinivasulu (2020)	235	Telli et al. (2018b)
Biodiesel	339.8	Reddy et al. (2014)	523	Elumalai and Ravi (2023)
PODE	180	Wang et al. (2022)	–	–
Dimethyl ether (DME)	−23	Park et al. (2016)	508	Park et al. (2016)
Ammonia	−33.34	Elumalai and Ravi (2023)	650	Uddeen et al. (2024)
CNG	–	–	540	Imioek et al. (2010)

the cylinder, even at the beginning in the compression stroke, promoting better conditions for air-fuel mixture formation. In contrast, petroleum-based fuels exhibit a wide range of boiling points due to their composition of molecules with varying molar masses. For example, gasoline consists of hydrocarbons (HC) with six to ten carbon atoms, where the lighter compounds evaporate first. In the case of fuels like alcohols, shown in Table 6.4, the molecular composition is more uniform, resulting in a single boiling point or a narrower temperature range. However, alcohols have significantly higher autoignition temperatures than petroleum-derived fuels and biodiesel, which complicates the combustion process. Coupled with their higher HoV and high octane number, these properties collectively make it more difficult to ignite the air-fuel mixture.

6.2.5 Kinematic Viscosity and Density

In general, the density (ρ) of a fuel is specified for a temperature of 20°C. For a mixture of two or more liquid fuels, we can calculate the density as

$$\rho_b = \frac{\Sigma V_i \rho_i}{\Sigma V_i} \tag{6.6}$$

- ρ_b: density of blend (kg/m^3).
- V_i: percentage of volum of i component (fuel).
- ρ_i: density of i component (fuel) (kg/m^3)

The kinematic viscosity for a mixture of two or more fuels can be calculated by

$$v_b = \frac{\Sigma V_i \rho_i v_i}{\Sigma V_i \rho_i} \tag{6.7}$$

- ρ_b: kinematic viscosity of blend (mm^2/s).
- V_i: percentage of volum of i component (fuel).
- v_i: kinematic viscosity of i component (fuel) (mm^2/s)

For an engine operating in advanced combustion mode with distinct fuel injection systems, it makes no sense to determine a resulting density or kinematic viscosity value. In such cases, an analysis of the effects of these parameters on engine operation must be conducted individually. Table 6.5 shows typical values of kinematic viscosity and density of various fuels used in advanced combustion modes.

6.3 LITERATURE REVIEW

Some publications that employed two different fuels in advanced combustion modes are presented in Table 6.6.

Regarding the use two different fuels in RCCI combustion, Kakati et al. (2021), explored an RCCI engine operating with mahua biodiesel and methanol under different split injection strategies to address emissions, performance, and operational

TABLE 6.5
Typical Kinematic Viscosity and Density of Fuels for Advanced Combustion Modes

Fuel	Kinematic Viscosity (mm^2/s) at 40°C	Ref.	Density (kg/m^3) at 20°C	Ref.
Methanol	0.58	Ghadikolaei et al. (2018)	791.3	Ghadikolaei et al. (2018)
Ethanol	1.13	Ghadikolaei et al. (2018)	789.4	Ghadikolaei et al. (2018)
Butanol	2.63–3.70	Gwalwanshi et al. (2022)	808–809	Gwalwanshi et al. (2022)
Gasoline	0.61	Turkcan et al. (2013)	720–780	Pan et al. (2020)
Diesel	4.22	Zuo et al. (2022)	840	Rosa et al. (2021)
Biodiesel	4.60	Zuo et al. (2022)	871	Ghadikolaei et al. (2018)
PODE	1.05	J. Li et al. (2022)	1046	J. Li et al. (2022)
Dimethyl ether	0.20	Sun et al. (2022)	660	Park et al. (2016)
Ammonia	–	–	0.732	Elumalai and Ravi (2023)
CNG	–	–	0.87	Garlet et al. (2024)

stability challenges. Split injection strategies involve any strategy involving multiple injections of a fuel during the same cycle. With modern fuel injection systems, it is possible to divide the fuel injection into up to three or four different injections. The results indicated that the methanol/biodiesel RCCI operation under split injection strategy improved the exergy efficiency of around 86% and significantly reduced NOx and soot emissions by around 86% and 96%, respectively, compared to single injection baseline operation with methanol/biodiesel. The authors use the properties of the fuels to explain some of the results. For example, they observed that the CA50 combustion phase occurs at earlier before top dead center (BTDC) positions when there is lower methanol participation, while higher methanol participation rates lead the CA50 phase position beyond TDC. These position changes are attributed to the low reactivity of methanol, which leads to ignition delay. The authors also found that when using biodiesel and methanol, the ignition delay is shorter than when using diesel and methanol, and they attribute this to the higher cetane number of biodiesels. However, the property tables shown in the paper do not demonstrate the higher cetane number.

Wenming and Meng (2019), also investigated an RCCI engine fueled with biodiesel and methanol. The study shows that increasing methanol fraction in the combustion improves engine performance at full load conditions but deteriorates combustion at low load conditions due to misfire at high methanol percentages. In addition, increasing the methanol fraction enhances combustion efficiency and reduces soot formation but leads to a slight increase in NOx emissions due to higher temperatures. Finally, the authors explain that at high loads, the negative effects of the high HoV of methanol are reduced because of the improved premixed combustion.

TABLE 6.6
Papers Describing Dual-Fuel Use in Advanced Combustion Modes

		RCCI	HCCI	PCCI
Methanol	Gasoline	–	Kale and Krishnasamy (2023) Turkcan et al. (2013)	–
	Biodiesel	Kakati et al. (2021) Wenming and Meng (2019)	–	Gowtham et al. (2021)
	Diesel	Y. Li et al. (2016) Duraisamy et al. (2020)	Panda and Ramesh (2022) Randive and Thombare (2016)	–
	CNG	–	–	–
	DME	–	Yao et al. (2007)	Sun et al. (2022)
	PODE	Wang et al. (2022, 2023)	–	J. Li et al. (2022)
Ethanol	Gasoline	–	Turkcan et al. (2013) Coskun et al. (2018) He et al. (2015)	–
	Biodiesel	Biswas et al. (2022) Z. Zheng et al. (2018)	Gawale and Naga Srinivasulu (2020)	–
	Diesel	Pan et al. (2021) Rosa et al. (2021)	Mosbah et al. (2023)	Elzahaby et al. (2018) Natarajan et al. (2017)
	CNG	Biswas et al. (2022, 2023)	–	Biswas et al. (2022, 2023)
	DME	Park et al. (2016)	–	–
	PODE	–	–	–
Butanol	Gasoline	–	He et al. (2015)	–
	Biodiesel	Zhao et al. (2021) Z. Zheng et al. (2018)	–	Soloiu et al. (2018)
	Diesel	Pan et al. (2017, 2020)	M. Zheng et al. (2015)	Karwade et al. (2020)
	CNG	–	–	–
	DME	–	–	–
	PODE	–	–	–

Duraisamy et al. (2020), experimentally investigated the performance, combustion, and emission characteristics of methanol/polyoxymethylene dimethyl ether (PODE) dual-fuel RCCI combustion and compared it with methanol/diesel dual-fuel RCCI combustion. The experiments were conducted at a 3-cylinder, turbocharged, common rail diesel injection (CRDI) diesel engine by varying methanol premixed mass fraction at a constant CA50 of about 10 CA ATDC (after the top dead center). It was found that increasing the m EMF prolonged ignition delay and decreased in-cylinder pressure for both methanol/diesel and methanol/PODE RCCI combustion. At 80% premixed mass fraction, both methanol/diesel and methanol/PODE achieved a maximum brake thermal efficiency of about 31%, which is 3.5% higher than conventional diesel combustion. Both dual-fuel RCCI combustion methods showed significantly reduced NO_X and soot emissions. Methanol/PODE dual-fuel RCCI combustion resulted in more NO emissions, but lower HC, CO, and soot emissions compared to methanol/diesel RCCI combustion. The authors mention that the advancement of the CA50 position, observed with the increase in ethanol participation, may be due to the presence of more premixed methanol, which reduces the cylinder temperature by absorbing the heat of compression due to its greater HoV.

Wang et al. (2022), evaluated the effects of load extension ratio (LER) on in-cylinder combustion, emission characteristics, and fuel economy of a PODE/methanol RCCI engine with an optimized injection strategy. The authors showed that with an optimized fuel injection strategy, it is possible to reduce peak in-cylinder pressure and pressure rise rate, extending high load operation. In another work from Wang et al. (2023), they numerically studied the in-cylinder combustion of PODE/methanol RCCI engine. They explored the effects of exhaust gas recirculation (EGR) ratio and CR on engine performance, combustion, and emission characteristics. It was found that increasing the EGR led to a decrease in peak in-cylinder pressure, longer combustion duration, and reduced NOx emissions by 47.3%. Additionally, the study shows that a reduction in CR postpones the start of combustion, decreases peak heat release rate, and improves combustion stability. The optimal CR found was 16.5 for stable operation of the PODE/methanol dual-fuel RCCI engine under high load conditions.

Zheng et al. (2018), investigated the combustion and emissions characteristics of a RCCI engine using n-butanol, DMF, and ethanol combined with biodiesel. The biofuels were port fuel injected at ratios of 20%, 50%, and 80% of energy fraction. Biodiesel/n-butanol operation achieved the highest indicated thermal efficiency of 47.5%. Biodiesel/ethanol demonstrates potential in reducing NOx and soot emissions simultaneously, with soot levels below 0.3 FSN and NOx below 1.5 g/kWh at a specific start of injection (SOI) of −8°CA ATDC. The authors also found that employing 80% biofuels by port injection results in lower soot emissions and higher CO and HC emissions compared to neat biodiesel.

Rosa et al. (2021), studied the feasibility of using ethanol as a partial fuel in a diesel CI engine, analyzing technical performance, combustion characteristics, emissions, and economic viability. The ethanol was injected in six different proportions and achieved the maximum mass fraction of 0.634. The use of ethanol increased ignition delay and reduced combustion duration. CO emissions decreased by up to 43% with ethanol use and resulted in about 2.4 g/kWh, while NOx emissions reduced by up to 20% compared to the diesel fuel test. Pan et al. (2021) analyzed the effects of diesel injection timing and ethanol-premixed ratio on the cyclic variations of RCCI combustion. The results showed that with the advancement of diesel injection timing, the coefficient of variation (COV) of parameters initially decreases and then increases, affecting key operating parameters like maximum in-cylinder pressure and heat release rate. It is clear from the results that optimizing diesel injection timing and ethanol-premixed ratio can achieve smaller cyclic variation of ethanol/diesel RCCI engines.

Biswas et al. (2022), explored the impact of split injection strategies on emissions, performance, and the stability of a RCCI engine fueled with biodiesel-CNG/ethanol. It was reported that the split injection strategies with biodiesel-CNG/ethanol can decrease NOx, soot, and unburned hydrocarbons (UHC) emissions while improving fuel conversion efficiency. In another work from Biswas et al. (2023), it was performed a multi-objective optimization to find a parametric solution for combustion, performance, stability, and emissions of a RCCI engine powered by mahua biodiesel mixed with CNG/ethanol. The optimization results in significant improvements in HC, CO, PM emissions, and brake thermal efficiency for different fuel mixtures and injection strategies.

Park et al. (2016), investigated the effect of the ethanol fraction and DME injection timings on the combustion and emission characteristics of an RCCI engine. The DME-ethanol dual-fuel combustion strategy reduced NOx and soot emissions with lower HC and CO levels compared to diesel or biodiesel dual-fuel combustion. This dual-fuel combustion approach showed a significant reduction in indicated specific of emissions of nitrogen oxides (ISNOX) without a deterioration in soot emissions. Additionally, HC and CO emissions slightly increased with an increase in ethanol fraction but remained lower than those observed in biodiesel-ethanol and diesel-ethanol dual-fuel combustion modes. Zhao et al. (2021), compared the engine combustion and emission characteristics of blend fuel technique, RCCI mode, and intelligent charge compression ignition (ICCI) mode using butanol and biodiesel in a diesel engine at different engine loads, energy ratios, and EGR rates. At low engine load, blend fuel exhibited a higher heat release rate and in-cylinder pressure, leading to better combustion characteristics compared to RCCI and ICCI modes. It maintained relatively higher indicated thermal efficiency. As engine load increased to medium and high loads, ICCI and RCCI modes showed advantages in maintaining higher thermal efficiency and more stable combustion.

Pan et al. (2017), explored the combustion and emission characteristics of 2-butanol/diesel RCCI mode on a light-duty diesel engine under various engine loads. The 2-butanol/diesel combustion decreased NOx and PM emissions, along with an increase in HC and CO emissions. Pan et al. (2020), experimentally studied the combustion and emission characteristics of iso-butanol/diesel RCCI combustion under low loads and compared it with gasoline/diesel RCCI. The RCCI engine fueled with iso-butanol/diesel showed longer ignition delay, combustion duration, and higher in-cylinder temperature and indicated thermal efficiency compared to gasoline/diesel RCCI. It was also reported that increasing the premixed ratio decreased the CO, NOx, and PM emissions, while HC emissions increased for both RCCI modes, with iso-butanol/diesel RCCI achieving lower emissions than gasoline/diesel RCCI.

As reported in Table 6.6, several works also evaluated the use of different fuel combinations in the HCCI combustion mode. Kale and Krishnasamy (2023), investigated HCCI combustion using methanol-gasoline blends. The increase in methanol in the fuel blend improved the indicated thermal efficiency at specific loads; 30% of improvement at 60% of load. HCCI combustion resulted in lower CO, UHC, NOx, and soot emissions compared to RCCI at all loads. Additionally, the research emphasized that increasing methanol content in fuel for HCCI combustion resulted in more efficient combustion than using EGR, as methanol is an oxygenated fuel that enhances combustion efficiency. Turkcan et al. (2013), investigated the effects of different alcohol-gasoline fuel blends using two-stage direct injection (TSDI) to control the HCCI combustion phasing. The authors explored five different fuel blends with pure gasoline; 10% and 20% ethanol in gasoline; 10% and 20% methanol in gasoline in energy fraction. The study showed that by adjusting the second fuel injection timing, better combustion characteristics, lower NOx, UHC, and CO emissions, higher indicated mean effective pressure (IMEP), and indicated efficiency values were achieved for alcohol-gasoline fuel blends compared to gasoline at a low equivalence ratio.

Panda and Ramesh (2022), studied the HCCI operation with methanol being port fuel injected and diesel being directly injected. The study varied the methanol energy share and the first and second pulse in the DI. It was reported that the methanol-diesel HCCI-DI mode of combustion has the potential for higher efficiency, lower NOx, and reduced soot emissions compared to the conventional dual fuel (CDF) mode. At 75% and 50% of load, the thermal efficiencies were higher by about 2.9% and 1.9%, respectively, in the HCCI-DI mode compared to the CDF mode. In addition, soot emissions were significantly reduced by 94%, 87.6%, and 80.3% in the HCCI-DI mode, while NOx levels were reduced by 77.5%, 65.2%, and 29.9% at different IMEPs.

Yao et al. (2007), explored the use of DME and methanol in HCCI combustion. The stable HCCI operation was achieved over a broad speed and load range for DME/methanol combustion, with ultra-low NOx emissions and high indicated thermal efficiency (~28%–46%). Both DME and methanol strongly influenced HCCI combustion, with DME advancing ignition timing and shortening combustion duration, while methanol delayed ignition timing and decreased burning rate.

Coskun et al. (2018), investigated the effects of different injection parameters on a DI HCCI combustion fueled with ethanol-gasoline blends (E10 and E20 blends) using astochastic reactor model (SRM). The results indicated that the variation of the second injection timing and injection ratios significantly impacts peak pressure in HCCI combustion, while the first injection variation does not have a significant effect on controlling HCCI combustion. He et al. (2015) evaluated the combustion characteristics of n-butanol, ethanol, and their blends with gasoline in a single-cylinder PFI gasoline engine under stoichiometric conditions highly diluted by exhaust gases. It was reported that the autoignition timing for alcohol-gasoline blends is dependent on alcohol types, concentration, engine speed, and intake valve opening/exhaust valve closing timing. N-butanol-gasoline blends ignite earlier than ethanol-gasoline blends at 1500 rpm, while the timings are close at 2000 rpm. Combustion stability improves with advanced exhaust valve closing timing. However, advanced autoignition timing for alcohol-gasoline blends negatively impacts thermal efficiency.

Gawale and Naga Srinivasulu (2020), studied different mass flow rates of ethanol on a dual-fuel mode HCCI engine under various load conditions, focusing on reducing NO_X and soot emissions. The study supplied the ethanol through the carburetor and injected a diesel-biodiesel blend to initiate combustion, with varying mass flow rates achieved by different fuel jets to optimize engine performance. It was found that increasing the mass flow rate of ethanol led to an increase in ignition delay, reduction in combustion duration, in-cylinder pressure, and temperature, resulting in lower NO_X emissions and smoke opacity for ethanol-diesel and ethanol-biodiesel dual-fuel mode HCCI engines compared to neat diesel engines. On the other hand, HC and CO emissions were higher for dual-fuel mode HCCI engines compared to net diesel engines.

Finally, several works also evaluated the combination of different fuels in PCCI combustion modes. Sun et al. (2022), experimentally investigated the influence of EGR on the combustion, performance, and emissions of a turbocharged common-rail DME engine with methanol blends. The methanol was mixed with DME in proportions of

0%, 5%, 10%, and 15% by mass. Methanol addition delays the combustion phasing, shortened combustion duration, and increases pressure rise rate, with M5 showing the lowest fuel consumption. In the PCCI mode, NOx emissions are lower, CO and HC emissions are higher, and particle emissions are slightly lower compared to the diesel conventional combustion. Li et al. (2022), evaluated the pilot injection strategy on the combustion and emission characteristics of PODE/methanol blends in a common-rail engine. Increasing the methanol ratio in PODE/methanol blends achieved longer ignition duration, increased premixed combustion ratio, and reduced NOx emissions. Adjusting pilot injection timing increased thermal efficiency and decreased HC and CO emissions but increased NOx and soot emissions. The maximum thermal efficiency for the blends was 46.6% while maintaining low emission levels.

Natarajan et al. (2017), explored the PCCI combustion with early DI of neat diesel and bio-ethanol-diesel blend (E20) fuels at an equivalence ratio of 0.6. The advanced injection timing was set at 18° BTDC compared to the normal 23° BTDC. NO_X emissions were significantly reduced by 77% for Diesel PCCI and 80% for E20 PCCI operation compared to conventional diesel-fueled operation. In addition, soot emissions were reduced by 24.5% in diesel-fueled PCCI operation and 46.5% in bio-ethanol diesel (E20) PCCI operation. The results also indicated that CO emissions were lower in E20 PCCI operation due to the higher oxygen concentration in bio-ethanol.

Soloiu et al. (2018), studied the PCCI combustion using cottonseed biodiesel and n-butanol. The study varied the fuel injection strategies, comparing PFI of n-butanol with DI of biodiesel and binary mixtures (biodiesel/n-butanol) to analyze combustion characteristics and emissions. Additionally, the study explored the impact of EGR rates on emissions and combustion efficiency. The ignition delay was reduced by 17% with the use of 50% n-butanol PFI and 50% biodiesel DI compared to ULSD2, while the binary mixture increased it by 12%. Soot and NO_X emissions were significantly reduced by 84% and 17%, respectively, using the PCCI strategy, which was attributed to lower peak in-cylinder temperatures and increased oxygenation of the mixture. Karwade et al. (2020), investigated the combustion characteristics and performance of n-butanol-diesel blends (B10, B20, B30, B40) in a PCCI engine. The B40 blend showed higher brake thermal efficiency at high loads, with a slight increase in brake-specific fuel consumption. It also demonstrated lower levels of CO and soot emissions, but higher NO and HC emissions compared to other blends.

6.4 CONCLUSIONS

In recent years, numerous studies have highlighted the potential of advanced combustion modes in CI engines to enhance efficiency and reduce emissions. These studies often base their findings on the specific properties of the fuels used. From the review of the studies discussed earlier in this chapter, several conclusions can be drawn based on the authors' analyses:

- Cetane and octane numbers play an important role in influencing the combustion phases, particularly the position of CA50, which is ideally optimized to maximize the cycle's work. Fuels with higher cetane numbers, indicating easier ignition, tend to advance this phase, while fuels with higher octane numbers delay it. Notably, the CA50 position can be controlled by precisely timing the injection of HRF.

- The high HoV of certain fuels, particularly alcohols, leads to a cooling effect on the intake charge, extending the ignition delay. Multiple authors have experimentally demonstrated the relationship between HoV and intake temperature by observing a decrease in cylinder pressure during the compression stroke as the ethanol proportion increases in ethanol-diesel RCCI mode. Moreover, a cooler intake charge results in lower combustion temperatures, reducing NOx formation but potentially increasing emissions of CO and UHC due to incomplete combustion.
- The lower calorific value of alcohols requires more fuel to be consumed compared to conventional fuels. However, their higher cetane numbers allow for operation at higher CRs, which, from a thermodynamic perspective, improves the engine's thermal efficiency.
- In terms of boiling points, all the fuels examined in this chapter have boiling points below the temperatures reached inside the cylinder near the end of compression. However, with respect to autoignition temperature, alcohols are at a disadvantage compared to petroleum-derived fuels and biodiesel.

Beyond these conclusions, there are numerous other insights regarding the influence of fuel properties on engine performance in advanced combustion modes. A wide range of fuel combinations, including methanol, ethanol, and butanol, have already been studied and listed in Table 6.6. Additionally, other factors related to the air-fuel mixture significantly affect combustion parameters, emissions, and performance—not only in advanced combustion modes but across all combustion types (CI, spark ignition (SI), etc.). These factors include the actual air-fuel ratio compared to the stoichiometric ratio and the homogeneity of the fuel mixture. Achieving an adequate (homogeneous) mixture depends on various technical aspects, such as injection pressure, as well as fuel properties like viscosity and boiling point.

REFERENCES

Ashok, B., Jeevanantham, A. K., Nanthagopal, K., Saravanan, B., Senthil Kumar, M., Johny, A., Mohan, A., Kaisan, M. U., & Abubakar, S. (2019). An experimental analysis on the effect of n-pentanol- Calophyllum Inophyllum Biodiesel binary blends in CI engine characteristcis. *Energy*, *173*, 290–305. https://doi.org/10.1016/j.energy.2019.02.092

Ashour, M. K., & Elwardany, A. E. (2020). Addition of two kerosene-based fuels to diesel–biodiesel fuel: Effect on combustion, performance and emissions characteristics of CI engine. *Fuel*, *269*. https://doi.org/10.1016/j.fuel.2020.117473

Biswas, S., Kakati, D., Chakraborti, P., & Banerjee, R. (2022). Performance-emission-stability mapping of CI engine in RCCI-PCCI modes under varying ethanol and CNG induced reactivity profiles: A comparative study through experimental and optimization perspectives. *Energy*, *254*. https://doi.org/10.1016/j.energy.2022.124231

Biswas, S., Sengupta, A., Kakati, D., Chakraborti, P., & Banerjee, R. (2023). Parametric optimization of the CNG/ethanol-induced RCCI profiles in biodiesel combustion through a robust design space foray. *Fuel*, *332*. https://doi.org/10.1016/j.fuel.2022.126203

Coskun, G., Demir, U., Soyhan, H. S., Turkcan, A., Ozsezen, A. N., & Canakci, M. (2018). An experimental and modeling study to investigate effects of different injection parameters on a direct injection HCCI combustion fueled with ethanol–gasoline fuel blends. *Fuel, 215*, 879–891. https://doi.org/10.1016/j.fuel.2017.11.126

Druzgalski, C. L., Lapointe, S., Whitesides, R., & McNenly, M. J. (2019). Predicting octane number from microscale flame dynamics. *Combustion and Flame, 208*, 5–14. https://doi.org/10.1016/j.combustflame.2019.06.019

Duraisamy, G., Rangasamy, M., & Govindan, N. (2020). A comparative study on methanol/diesel and methanol/PODE dual fuel RCCI combustion in an automotive diesel engine. *Renewable Energy, 145*, 542–556. https://doi.org/10.1016/j.renene.2019.06.044

Elumalai, R., & Ravi, K. (2023). A hybrid calibration scheme for developing hydrogen enrichment ratio control map using RSM and ANN technique to enhance the characteristics of an ammonia biodiesel RCCI combustion engine. *Case Studies in Thermal Engineering, 49*. https://doi.org/10.1016/j.csite.2023.103257

Elzahaby, A. M., Elkelawy, M., Bastawissi, H. A. E., El Malla, S. M., & Naceb, A. M. M. (2018). Kinetic modeling and experimental study on the combustion, performance and emission characteristics of a PCCI engine fueled with ethanol-diesel blends. *Egyptian Journal of Petroleum, 27*(4), 927–937. https://doi.org/10.1016/j.ejpe.2018.02.003

Garlet, R. A., Fagundez, J. L. S., Hausen, R. B., Roso, V. R., Lanzanova, T. D. M., Salau, N. P. G., & Martins, M. E. S. (2024). Prospects of performance, emissions and cost of methane-based fuels in a spark-ignition engine compared to conventional Brazilian fuels. *Fuel, 357*. https://doi.org/10.1016/j.fuel.2023.129788

Gawale, G. R., & Naga Srinivasulu, G. (2020). Experimental investigation of ethanol/diesel and ethanol/biodiesel on dual fuel mode HCCI engine for different engine load conditions. *Fuel, 263*. https://doi.org/10.1016/j.fuel.2019.116725

Ge, J. C., Wu, G., & Choi, N. J. (2022). Comparative study of pilot–main injection timings and diesel/ethanol binary blends on combustion, emission and microstructure of particles emitted from diesel engines. *Fuel, 313*. https://doi.org/10.1016/j.fuel.2021.122658

Ghadikolaei, M. A., Cheung, C. S., & Yung, K. F. (2018). Study of combustion, performance and emissions of diesel engine fueled with diesel/biodiesel/alcohol blends having the same oxygen concentration. *Energy, 157*, 258–269. https://doi.org/10.1016/j.energy.2018.05.164

Goldsworthy, L. (2013). Fumigation of a heavy duty common rail marine diesel engine with ethanol-water mixtures. *Experimental Thermal and Fluid Science, 47*, 48–59. https://doi.org/10.1016/j.expthermflusci.2012.12.018

Gowtham, R., Akash, V., Ganesh Kumar, P., & Bhaskar, K. (2021). An experimental investigation on performance and emission characteristics of PCCI engine using BiodieselEthanol blends in dual fuel mode. *IOP Conference Series: Materials Science and Engineering, 1059*(1). https://doi.org/10.1088/1757-899X/1059/1/012036

Gwalwanshi, M., Kumar, R., & Kumar Chauhan, M. (2022). A review on butanol properties, production and its application in internal combustion engines. *Materials Today: Proceedings, 62*(P12), 6573–6577. https://doi.org/10.1016/j.matpr.2022.04.573

He, B. Q., Liu, M. B., & Zhao, H. (2015). Comparison of combustion characteristics of n-butanol/ethanol-gasoline blends in a HCCI engine. *Energy Conversion and Management, 95*, 101–109. https://doi.org/10.1016/j.enconman.2015.02.019

Imioek, M., Pitak, A., & Wierzbicki, S. (2010). The effect of fuel injection parameters on the combustion process in a self-ignition engine. *Journal of KONES Powertrain and Transport, 17*(3).

Jing, L., Zhao, J., Wang, H., Li, W., Du, Y., Zhu, Q., & Zayed, M. E. (2022). Numerical analysis of the effect of swirl angle and fuel equivalence ratio on the methanol combustion characteristics in a swirl burner. *Process Safety and Environmental Protection, 158*, 320–330. https://doi.org/10.1016/j.psep.2021.11.019

Kakati, D., Biswas, S., & Banerjee, R. (2021). Parametric sensitivity analysis of split injection coupled varying methanol induced reactivity strategies on the exergy efficiency enhancement and emission reductions objectives in a biodiesel fuelled CI engine. *Energy*, *225*. https://doi.org/10.1016/j.energy.2021.120204

Kale, A. V., & Krishnasamy, A. (2023). Experimental optimization of homogeneous charge compression ignition through fuel modifications and a relative comparison with reactivity controlled compression ignition. *Energy Conversion and Management*, *276*. https://doi.org/10.1016/j.enconman.2022.116439

Kandasamy, S. K., Selvaraj, A. S., & Rajagopal, T. K. R. (2019). Experimental investigations of ethanol blended biodiesel fuel on automotive diesel engine performance, emission and durability characteristics. *Renewable Energy*, *141*, 411–419. https://doi.org/10.1016/j.renene.2019.04.039

Karwade, A., Thombre, S., & Bhiogade, G. (2020). Investigations on premixed charge compression ignition type combustion using butanol-diesel blends. *Journal of Thermal Science and Technology*, *15*(3), 1–11. https://doi.org/10.1299/jtst.2020jtst0026

Kukana, R., & Jakhar, O. P. (2022). Effect of ternary blends diesel/n-propanol/composite biodiesel on diesel engine operating parameters. *Energy*, *260*. https://doi.org/10.1016/j.energy.2022.124970

Li, J., Liu, J., Ji, Q., Sun, P., Wei, M., Liu, S., & Fang, J. (2022). Effects of pilot injection strategy on in-cylinder combustion and emission characteristics of PODE/methanol blends. *Fuel Processing Technology*, *228*. https://doi.org/10.1016/j.fuproc.2022.107168

Li, Y., Jia, M., Chang, Y., Xie, M., & Reitz, R. D. (2016). Towards a comprehensive understanding of the influence of fuel properties on the combustion characteristics of a RCCI (reactivity controlled compression ignition) engine. *Energy*, *99*, 69–82. https://doi.org/10.1016/j.energy.2016.01.056

Li, Y. L., Yu, Y. C., Wang, Z. W., & Wang, J. F. (2015). Physical and chemical properties of isobutanol-gasoline blends. *Environmental Progress and Sustainable Energy*, *34*(3), 908–914. https://doi.org/10.1002/ep.12061

Morsy, M. H. (2015). Assessment of a direct injection diesel engine fumigated with ethanol/water mixtures. *Energy Conversion and Management*, *94*, 406–414. https://doi.org/10.1016/j.enconman.2015.01.086

Mosbah, M., Boutera, Z., Rezgui, Y., Guemini, M., & Tighezza, A. (2023). Effect of ethanol enrichment and engine parameters on the performance of an HCCI engine fuelled with biodiesel/ethanol mixtures. *Combustion Theory and Modelling*, *27*(7), 925–948. https://doi.org/10.1080/13647830.2023.2248954

Natarajan, S., Shankar, S. A., & Sundareswaran, A. U. M. (2017). Early injected PCCI engine fuelled with bio ethanol and diesel blends – An experimental investigation. *Energy Procedia*, *105*, 358–366. https://doi.org/10.1016/j.egypro.2017.03.326

Pan, S., Cai, K., Cai, M., Du, C., Li, X., Han, W., Wang, X., Liu, D., Wei, J., Fang, J., & Bao, X. (2021). Experimental study on the cyclic variations of ethanol/diesel reactivity controlled compression ignition (RCCI) combustion in a heavy-duty diesel engine. *Energy*, *237*. https://doi.org/10.1016/j.energy.2021.121614

Pan, S., Li, X., Han, W., & Huang, Y. (2017). An experimental investigation on multi-cylinder RCCI engine fueled with 2-butanol/diesel. *Energy Conversion and Management*, *154*, 92–101. https://doi.org/10.1016/j.enconman.2017.10.047

Pan, S., Liu, X., Cai, K., Li, X., Han, W., & Li, B. (2020). Experimental study on combustion and emission characteristics of iso-butanol/diesel and gasoline/diesel RCCI in a heavy-duty engine under low loads. *Fuel*, *261*. https://doi.org/10.1016/j.fuel.2019.116434

Panda, K., & Ramesh, A. (2022). HCCI combustion of methanol along with diesel through novel injection strategies and its potential over conventional dual fuel combustion. *Fuel*, *324*. https://doi.org/10.1016/j.fuel.2022.124766

Park, S. H., Shin, D., & Park, J. (2016). Effect of ethanol fraction on the combustion and emission characteristics of a dimethyl ether-ethanol dual-fuel reactivity controlled compression ignition engine. *Applied Energy, 182*, 243–252. https://doi.org/10.1016/j.apenergy.2016.07.101

Paul, A., Bose, P. K., Panua, R. S., & Banerjee, R. (2013). An experimental investigation of performance-emission trade off of a CI engine fueled by diesel-compressed natural gas (CNG) combination and diesel-ethanol blends with CNG enrichment. *Energy, 55*, 787–802. https://doi.org/10.1016/j.energy.2013.04.002

Rajesh Kumar, B., & Saravanan, S. (2016). Use of higher alcohol biofuels in diesel engines: A review. *Renewable and Sustainable Energy Reviews, 60*, 84–115. Elsevier Ltd. https://doi.org/10.1016/j.rser.2016.01.085

Randive, S., & Thombare, D. (2016). Modelling and simulation of methanol and diesel fuelled HCCI engine for improved performance and emission characteristics. *International Journal of Science and Research (IJSR), 5*(6), 112–115. https://doi.org/10.21275/v5i6.NOV163994

Reddy, V. M., Biswas, P., Garg, P., & Kumar, S. (2014). Combustion characteristics of biodiesel fuel in high recirculation conditions. *Fuel Processing Technology, 118*, 310–317. https://doi.org/10.1016/j.fuproc.2013.10.004

Rosa, J. S., Altafini, C. R., Wander, P. R., Telli, G. D., & Rocha, L. A. O. (2019). Wet ethanol fumigation on a compression ignition engine: Effects of air intake throttled. *Journal of the Brazilian Society of Mechanical Sciences and Engineering, 41*(11). https://doi.org/10.1007/s40430-019-2023-1

Rosa, J. S., Lorenzini, G., Altafini, C. R., Wander, P. R., Telli, G. D., & Rocha, L. A. O. (2019). Performance effects and economic viability of high-hydrated ethanol fumigation and diesel direct injection in a small compression ignition engine. *Mathematical Modelling of Engineering Problems, 6*(1), 1–9. https://doi.org/10.18280/mmep.060101

Rosa, J. S., Martins, M. E. S., Telli, G. D., Altafini, C. R., Wander, P. R., & Rocha, L. A. O. (2020). Exploring the effects of diesel start of injection and water-in-ethanol concentration on a reactivity controlled compression ignition engine. *Fuel, 281*. https://doi.org/10.1016/j.fuel.2020.118751

Rosa, J. S., Telli, G. D., Altafini, C. R., Wander, P. R., & Oliveira Rocha, L. A. (2021). Dual fuel ethanol port injection in a compression ignition diesel engine: Technical analysis, environmental behavior, and economic viability. *Journal of Cleaner Production, 308*. https://doi.org/10.1016/j.jclepro.2021.127396

Soloiu, V., Moncada, J. D., Gaubert, R., Muiños, M., Harp, S., Ilie, M., Zdanowicz, A., & Molina, G. (2018). LTC (low-temperature combustion) analysis of PCCI (premixed charge compression ignition) with n-butanol and cotton seed biodiesel versus combustion and emissions characteristics of their binary mixtures. *Renewable Energy, 123*, 323–333. https://doi.org/10.1016/j.renene.2018.02.061

Sun, C., Liu, Y., Qiao, X., Ju, D., Han, W., Fang, X., & Zhou, F. (2022). Experimental study of effects of exhaust gas recirculation and combustion mode on combustion, emission, and performance of DME-methanol-fueled turbocharged common-rail engine. *International Journal of Energy Research, 46*(3), 2385–2402. https://doi.org/10.1002/er.7315

Telli, G. D., Altafini, C. R., Costa, C. A., Rosa, J. S., Rocha, L. A. O., & Lorenzini, G. (2021). Experimental study of a dual-fuel generator set operating on diesel fuel direct injected and hydrous ethanol fumigation at different loads. *International Journal of Design and Nature and Ecodynamics, 15*(6), 777–784. https://doi.org/10.18280/ijdne.150602

Telli, G. D., Altafini, C. R., Rosa, J. S., & Costa, C. A. (2018a). Experimental analysis of a small engine operating on diesel–natural gas and soybean vegetable oil–natural gas. *Journal of the Brazilian Society of Mechanical Sciences and Engineering, 40*(11). https://doi.org/10.1007/s40430-018-1469-x

Telli, G. D., Altafini, C. R., Rosa, J. S., & Costa, C. A. (2018b). Experimental investigation of a compression ignition engine operating on B7 direct injected and hydrous ethanol fumigation. *Energy, 165*, 106–117. https://doi.org/10.1016/j.energy.2018.09.171

Telli, G. D., Zulian, G. Y., Lanzanova, T. D. M., Martins, M. E. S., & Rocha, L. A. O. (2023). Investigation of the use of wet ethanol in an HCCI engine using water injection and direct exhaust heat recovery. *Energy Conversion and Management: X, 18*(January), 100377. https://doi.org/10.1016/j.ecmx.2023.100377

Turkcan, A., Ozsezen, A. N., & Canakci, M. (2013). Effects of second injection timing on combustion characteristics of a two stage direct injection gasoline-alcohol HCCI engine. *Fuel, 111*, 30–39. https://doi.org/10.1016/j.fuel.2013.04.029

Uddeen, K., Tang, Q., Shi, H., & Turner, J. (2024). Performance and emission analysis of ammonia-ethanol and ammonia-methane dual-fuel combustion in a spark-ignition engine: An optical study. *Fuel, 358*. https://doi.org/10.1016/j.fuel.2023.130296

Wang, L., Liang, W., Ma, H., Ji, Q., Sun, P., & Liu, J. (2023). Simulation study on effects of EGR ratio and compression ratio on combustion and emission characteristics of PODE/methanol RCCI engine. *Fuel, 334*. https://doi.org/10.1016/j.fuel.2022.126593

Wang, L., Liu, J., Ji, Q., Sun, P., Li, J., Wei, M., & Liu, S. (2022). Experimental study on the high load extension of PODE/methanol RCCI combustion mode with optimized injection strategy. *Fuel, 314*. https://doi.org/10.1016/j.fuel.2021.122726

Wenming, Y., & Meng, Y. (2019). Phi-T map analysis on RCCI engine fueled by methanol and biodiesel. *Energy, 187*. https://doi.org/10.1016/j.energy.2019.115958

Yao, M., Zheng, Z., Chen, Z., & Zhang, B. (2007). Experimental study on homogeneous charge compression ignition operation by burning dimethyl ether and methanol. *International Journal of Green Energy, 4*(3), 283–300. https://doi.org/10.1080/15435070701193134

Zhao, W., Zhang, Y., Huang, G., He, Z., Qian, Y., & Lu, X. (2021). Experimental investigation on combustion and emission characteristics of butanol/biodiesel under blend fuel mode, dual fuel RCCI and ICCI modes. *Fuel, 305*. https://doi.org/10.1016/j.fuel.2021.121590

Zheng, M., Han, X., Asad, U., & Wang, J. (2015). Investigation of butanol-fuelled HCCI combustion on a high efficiency diesel engine. *Energy Conversion and Management, 98*, 215–224. https://doi.org/10.1016/j.enconman.2015.03.098

Zheng, Z., Xia, M., Liu, H., Wang, X., & Yao, M. (2018). Experimental study on combustion and emissions of dual fuel RCCI mode fueled with biodiesel/n-butanol, biodiesel/2,5-dimethylfuran and biodiesel/ethanol. *Energy, 148*, 824–838. https://doi.org/10.1016/j.energy.2018.02.015

Zuo, L., Wang, J., Mei, D., Dai, S., & Adu-Mensah, D. (2022). Experimental investigation on combustion and (regulated and unregulated) emissions performance of a common-rail diesel engine using partially hydrogenated biodiesel-ethanol-diesel ternary blend. *Renewable Energy, 185*, 1272–1283. https://doi.org/10.1016/j.renene.2021.12.085

7 Alcohol Fuel Injection Strategies in Compression Ignition Engines

Mandeep Singh, Akash Deep, Vijayanta Ahuja, Vinutha Ashokan, and Sarbjot Singh Sandhu

7.1 INTRODUCTION

Compression ignition (CI) engines, also known as diesel engines, are widely used in the transportation industry and agriculture due to their high performance and high power consumption. However, these engines have faced problems in complying with stringent regulations, especially regarding small particles and nitrogen oxides [1–4]. In addition, the unpredictability and variability of fossil fuel prices create serious problems for importing countries. Due to environmental concerns, the engine emits more pollutants such as nitrogen oxides (NO_x), smoke (the most polluting), carbon dioxide (CO_2), and particulate matter (PM) when burning gasoline.

Many countries have been utilizing their resources to conduct extensive research on alternative energy sources derived from biological materials. The United States is the largest producer of ethanol, primarily derived from corn. The Renewable Fuel Standard (RFS) mandates the blending of ethanol with gasoline, commonly at 10% (E10) and increasingly at 15% (E15) levels. Flex-fuel vehicles (FFVs) that can run on up to 85% ethanol (E85) are widespread, promoting higher ethanol usage. Brazil is a pioneer in ethanol use, producing it mainly from sugarcane. The country has a long history of ethanol utilization, with gasoline typically blended with 27.5% ethanol (E27).

Among these countries, India stands out as one of the most populous and highly reliant on petroleum, with a five-fold increase in diesel consumption compared to gasoline [5]. According to 'Statista 2024', the number of registered motor vehicles in India in fiscal year 2020 was approximately 326 million, a 9% increase from the previous decade. Sadly, India ranks third in the world for air pollution, following closely behind China and the United States, with 14 out of the 15 most contaminated cities located within its borders (source: UNHCR, World Bank). Despite this alarming fact, India's oil consumption has continued to rise, with an 8.1% increase in 2015 compared to 2014. While India only produced 0.2% of the world's biofuel in 2014, this number rose significantly to 13.1% in 2015. To promote the use of renewable energy, the Government of India (GOI) has proposed a 20% blending

DOI: 10.1201/9781003482277-7

of biofuel with fossil fuels by 2017. However, more serious measures need to be taken in order to truly support the development of alternative energy sources. Currently, the country has a total ethanol production capacity of 13,800 million liters, with 8750 million liters being derived from molasses and 5050 million liters from grains. The use of alcohol as fuel in internal combustion engines (ICE) can be done through various methods, each with its own unique advantages. In an effort to reduce the import bill, India has turned to using ethanol as both a blending agent in gasoline (ranging from E5 to E20) and as a standalone fuel in the form of E85 (which is nominally 85% ethanol and 15% gasoline), E100, and ED95. The GOI has also implemented the Ethanol Blended with Gasoline (EBP) Programme nationwide, where oil marketing companies (OMCs) sell gasoline blended with ethanol. As part of this program, the government has set a target of achieving 20% blending of ethanol with gasoline by 2025. In recent years, biodiesel has emerged as a popular alternative fuel for CI engines, undergoing extensive research and development in various countries [6–8].

Additionally, in the realm of CI engines, there has been an introduction of the low-temperature combustion (LTC) concept, which aims to achieve exceedingly minimal levels of engine-out NO_x and PM emissions while maintaining high engine efficiencies [9]. A promising solution for this is the utilization of alcoholic fuels, which belong to the new generation of biofuels and offer the potential for lower-temperature combustion as well as the benefits of renewability and carbon neutrality [10]. These biofuels possess favorable physicochemical properties, a diverse range of raw materials, and a well-developed production process, making them effective in reducing soot formation and PM emissions while preserving brake thermal efficiency (BTE) due to their oxygen-containing nature. As a result, there has been a growing interest in alternative fuels, including alcoholic fuels, to mitigate emissions and enhance engine performance. In CI engines, alcoholic fuels can either be used as a direct substitute for diesel fuel or in combination with diesel. Both higher alcohols (such as butanol, pentanol, and hexanol) and lower alcohols (such as methanol, ethanol, and propanol) have been investigated in CI engines. However, since these alcohols possess different properties compared to conventional diesel fuel, special considerations must be taken into account for optimal combustion.

The utilization of alcohol as a fuel in diesel engines relies heavily on the implementation of fuel injection. The injection strategy is a critical factor to consider when using alcohol as a fuel in diesel engines, as it has different combustion properties compared to traditional diesel fuel. To optimize its performance, specific injection techniques must be employed. By carefully designing the fuel injection system to accommodate alcohol, combustion efficiency can be improved, and emissions can be reduced. This chapter thoroughly outlines the various strategies for alcohol fuel injection in CI engines. It also covers the advancements in reactivity-controlled CI (RCCI) and premixed charge CI (PCCI) engines when using alcohols in a dual-fuel mode. Furthermore, in this chapter the enhancements made to higher-alcohol engines through the addition of cetane number improvers, adjustments to exhaust gas recirculation (EGR) rates, and changes to injection strategies are discussed.

7.2 APPLICATION OF DIFFERENT TYPES OF FUEL INJECTION STRATEGIES FOR ALCOHOL FUELS IN CI ENGINES

Diverse approaches exist for implementing fuel injection when utilizing alcohol as a fuel in diesel engines, each presenting its own set of benefits and challenges. These methods may involve modifying the timing, duration, quantity, and pressure of injection, all of which are controlled by the electronic control unit (ECU), which activates or deactivates the injectors as needed. Detailed studies of the aforementioned methodologies and their ramifications on engine performance are imperative for the advancement of efficient and reliable diesel engines fueled by alcohol. Some of the fuel injection strategies for alcohol fuel in diesel engines include direct injection (DI), port fuel injection (PFI), and dual-fuel injection (DFI). To optimize engine performance parameters, combustion stability, and emissions, a single injection can be divided into multiple stages through a multi-injection strategy. This involves breaking down the full injection into a pilot injection and a post-injection, where a small amount of fuel is injected before and after the main injection, respectively. However, the purposes of these injection strategies differ. The pilot injection elevates the combustion chamber temperature, consequently reducing the ignition delay, whereas post-injection aids in oxidizing the by-products stemming from incomplete combustion. Following are the studies that examine these injection strategies in further detail.

7.2.1 Direct Injection (DI)

DI is the primary method for utilizing alcohol as fuel in diesel engines. It involves injecting alcohol directly into the combustion chamber at high pressure, leading to improved mixing with air and controlled combustion.

Li et al. [11] conducted a study on the combustion and emission properties of E13, NB20, IB20, and P25 (ethanol-diesel, n-butanol-diesel, iso-butanol-diesel, and n-pentanol-diesel, respectively) with the same oxygen content. The experiment was carried out on a naturally aspirated, water-cooled, four-cylinder diesel engine. The fuel pump and injector settings remained uniform across all fuels, with the injection timing consistently maintained at 8°CA before top dead center (BTDC). The results revealed that blending multiple alcohols led to a higher peak heat release rate (HRR), prolonged ignition delays, and reduced combustion duration. Notably, higher-alcohol blends (NB20, IB20, and P25) exhibited more pronounced alterations, including augmented premixed combustion fractions, decreased peak particle size distribution, and diminished total particle number emissions, in contrast to lower alcohol blends.

The primary focus of the study was on the combustion characteristics of diesel-methanol and diesel-ethanol blends. These blends were examined at volume percentages of 10, 15, 20, 25, 30, 35, and 40% in both cases.

A team of researchers led by F.J. Martos et al. [12] developed a semi-empirical model (based on experimental data) to forecast in-cylinder growth of soot primary particles in engines powered by alcohol-diesel fuel blends. The model incorporates data from engine parameters, such as in-cylinder pressure, at a macroscopic level. Additionally, the study presents an empirical correlation that estimates the average size of soot primary particles, considering properties of the alcohol-diesel fuel blend and the fuel/air ratio. The mean size of primary soot particles is determined through high-resolution transmission electron microscope (HT-TEM) micrographs of soot particles collected through thermophoresis.

As a representative example, Figure 7.1 illustrates the results obtained from the combustion process of a butanol-diesel fuel blend under two engine operating conditions. The graph illustrates the mean in-cylinder pressure at any given moment (black), the rate of heat release (green), the oxygen fraction from air within the cylinder (red), and the mass fraction of hydroxyl radical (blue). The same trends are observed for the other fuel blends, but for clarity, they are not included in Figure 7.1. The primary distinguishing characteristic among the fuels lies in the mass fraction of hydroxyl radicals. A comparison between Figures 7.1(a) and 7.3(b) demonstrates how a decrease in engine load leads to an increase in the final oxygen mass fraction in the cylinder.

The data presented in Figure 7.2 clearly demonstrates that there is a considerable increase in the spread of primary particle sizes as engine load rises when operating on alcohol fuel mixtures. Conversely, the dispersion of primary particle sizes remains relatively constant when the engine is running on pure diesel. This can be attributed to the significant changes in local conditions within the combustion chamber caused by the higher mass of alcohol fuel, resulting in the production of both small and large soot particles in similar proportions.

DI of alcohols has some limitations also despite the benefits it offers which still remain a challenge and need to be reviewed before adoption of the technology. This has issues like infrastructure limitations, availability of alcohol fuels, economic factors, and regulatory compliance. Adding to this, current research is needed to enhance the durability, reliability, efficiency, and compatibility of Data Acquisition Unit (DAU) technology with distinct engine layouts and operating factors and conditions. Efforts from academia, industry, and policymakers collaboratively can

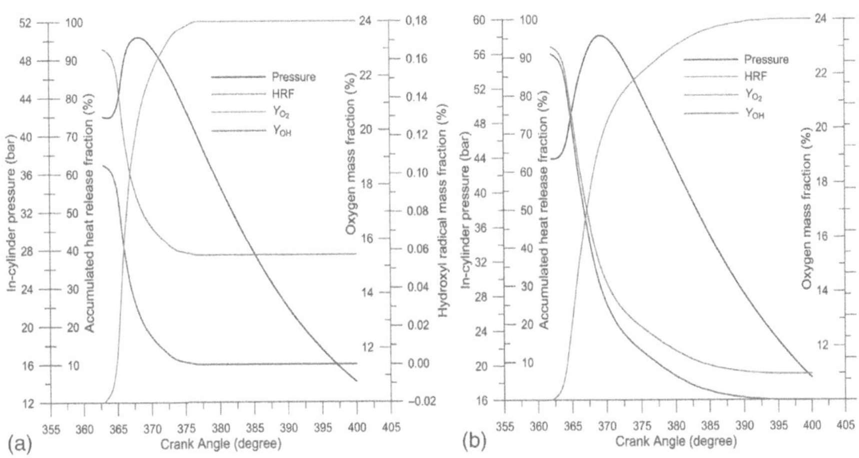

FIGURE 7.1 Time evolution of pressure, heat release fraction (HRF), Y_{O_2} and Y_{OH} vs. crank angle degree. (a) B20L, (b) B20M. (Source: F.J. Martos et al. [12].)

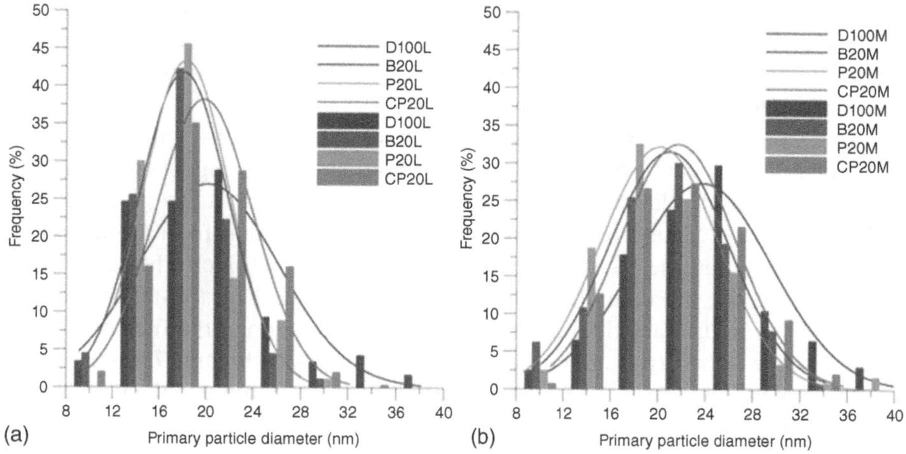

FIGURE 7.2 Histograms illustrating the mean diameters of primary particles measured experimentally are presented for (a) low load modes, L, and (b) medium load modes, M. (Source: F.J. Martos et al. [12].)

overcome these limitations and provide a path for DAI (direct alcohol injection) technology for sustainability and transportation.

7.2.2 Port Fuel Injection (PFI)

In PFI, alcohol is injected into the intake port before the air enters the combustion chamber. This approach facilitates enhanced air-fuel mixing and combustion regulation, though it might not attain the identical efficiency level as in DI. A.F. López et al. [12] examined the multipoint port-injection system's capability to replace 10% of diesel fuel on an energy basis with hydrous ethanol (H-Et10) and n-butanol (n-Bu10) at an injection pressure of 300 kPa. Alcohol fumigation promoted the cooling effect due to its higher latent heat; hence, the in-cylinder temperature decreased because of which ignition delay increased and both alcohols induced high combustion instabilities in comparison to diesel fuel. The obvious result of this was increase in brake specific fuel consumption (BSFC) and lowered BTE because of lower heating value (LHV) which has been observed only at low loads. However, both alcohols performed better at high loads as BTE remained unaffected due to alcohol molecular oxygen content. Similarly, CO, PM, and THC emissions were less during higher loads as compared to lower load conditions. On the other hand, at higher loads, NO_x emissions were high.

In Ankur Kalwar et al. [13], the study concentrates on investigating the utilization of DFI technology, where alcohols are injected into the port, while gasoline is directly injected into the engine combustion chamber.

7.2.3 Dual-Fuel Injection (DFI)

DFI involves using both alcohol and diesel fuel simultaneously, allowing for flexibility in fuel choice and potentially reducing cold-start issues. However, this strategy requires complex engine modifications and control systems to optimize performance.

Michael Saccullo et al. [14] implemented a dual-fuel direct injection system (DIS). This system utilized a small diesel pilot injection, delivered through a separate direct injector, to ignite the main alcohol fuel (methanol or ethanol). The main injector, located at the center, remained unchanged, while a light-duty passenger car diesel injector was added at a slight angle and offset from the cylinder center. The nozzle tip of this secondary injector (also known as the side injector) extended into the chamber between the two inlet valves. To prevent direct collision between the sprays, the spray orientations of the two injectors were adjusted. This system also utilized separate common rail and fuel pump to supply fuel to the main and side injectors. The pilot injection parameters included a dwell time of 3CAD, injection pressure of 500 bar, and a duration of 0.4 ms. The main fuel injection pressure was maintained at 1250 bar to ensure similar conditions in terms of turbulence and fuel-air mixing. To achieve a similar combustion pattern to that of methanol, the ethanol injection pressure was reduced at medium load (indicated by the dash-dotted AROHR curve). Engine testing was carried out at three different conditions: (i) low speed and load (871 rpm, 86 Nm); (ii) medium speed and load (1262 rpm, 172 Nm); (iii) high speed 75% load condition (1508 rpm, 285 Nm). Figure 7.3 shows the combustion process at medium speed and load and is divided into five phases:

Phase 1: It represents the combustion by using diesel as a fuel in pilot injection. In this phase minor rise in the HRR was observed, which remains consistent across all fuels.

Phase 2: It initiates with the premixed combustion stage, succeeded by the propagation of the free-flame jet from the nozzle toward the piston bowl. This phase concludes as the combustion speed of the fuel notably decreases near the peak of the heat release curve, occurring when the flame jet collides with the walls of the piston bowl, inducing a deceleration in the combustion process. All three fuels exhibit a brief premixed combustion stage, where the initially vaporized air-fuel mixture

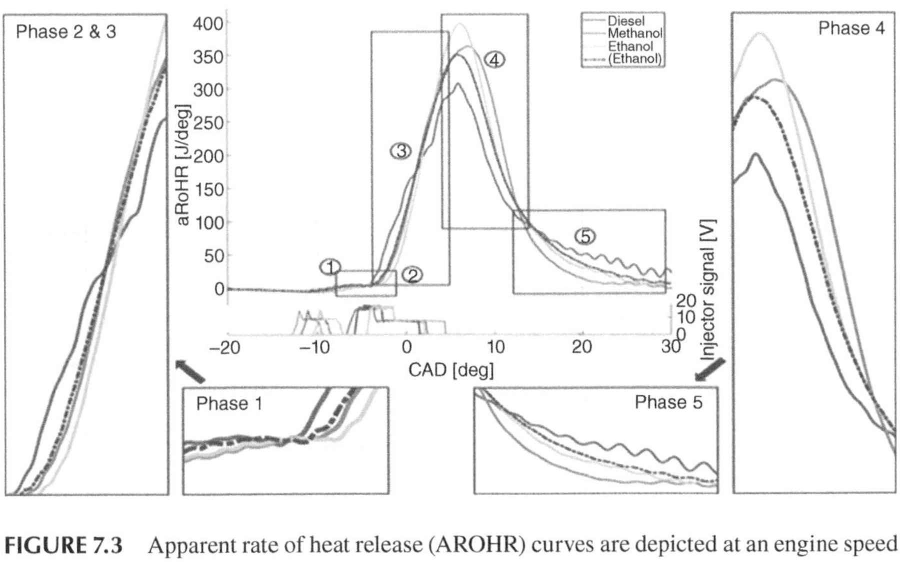

FIGURE 7.3 Apparent rate of heat release (AROHR) curves are depicted at an engine speed of 1262 rpm and a load of 172 Nm. (Source: Michael Saccullo et al. [14].)

ignites. Following the premixed stage, the slope of the alcohol curves is notably steeper compared to that of diesel, indicating a faster combustion rate.

Phase 3: This phase, referred to as the flame-wall interaction stage, occurs when the flame front encounters the walls of the piston bowl and is compelled to deviate from its initial spray direction, eventually traversing inside the piston bowl. This transition is discernible by a decrease in the slope of the apparent rate of heat release (AROHR) curve, attributed to a deceleration in the combustion process due to a reduction in flame surface area and turbulent kinetic energy. Nevertheless, the rate of heat release continues to escalate. Identifying the precise moment at which the free flame impacts the piston bowl walls in the heat release curves of the alcohol fuels proves challenging.

Phase 4: This phase is marked by the convergence of flames from distinct sprays, leading to the formation of flame interaction zones around the peaks of each fuel. Earlier research suggests that combustion decelerates upon the collision of two flame fronts due to a reduction in oxygen content within the flames, particularly following the establishment of the radial mixing zone as discussed by Eismark et al. This phenomenon is evident in both apparent rates of heat release (AROHR) curves at approximately 5° after top dead center (TDC) (ATDC).

Phase 5: Methanol has the shortest combustion tail, with ethanol and diesel following closely behind. The ethanol curves all have similar tails, suggesting a uniform end of combustion behavior.

In Figure 7.4 the energy balances for diesel, methanol, and ethanol fuels under medium load conditions (1262 rpm and 172 Nm) with and without EGR are illustrated. When EGR is excluded, ethanol exhibits a smaller proportion of heat losses, indicated by energy loss in Figure 7.4, at 22.5% compared to methanol at 24.2%. Although ethanol shows a slightly higher energy exiting the system via exhaust gases, its efficiency is higher.

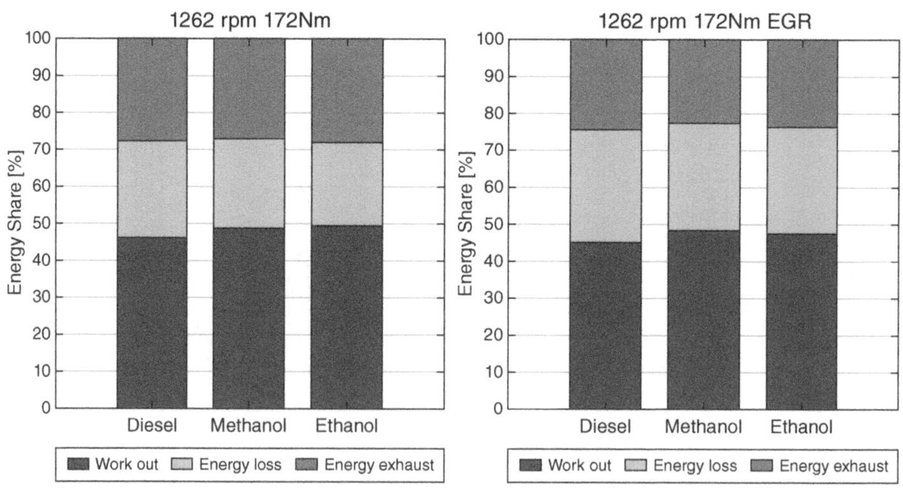

FIGURE 7.4 Energy balances of methanol, ethanol, and standard diesel fuel at 1262 rpm 172 Nm with and without EGR. (Source: Michael Saccullo et al. [14].)

V.M. Domínguez et al. [15] investigated overall engine performance, emissions, combustion patterns, and energy substitution in a dual-fuel mode, specifically focusing on the impact of two short-chain alcohols, methanol and ethanol. These alcohols served as pilot fuels in conjunction with either diesel or hydrotreated vegetable oil (HVO) as the main fuel. Initially, experiments were conducted solely using diesel or HVO fuel to achieve the desired indicated mean effective pressure (IMEP). Subsequently, the amount of diesel/HVO fuel was limited to the required quantity for the test. Methanol or ethanol was then supplied to the intake manifold as needed until the desired IMEP was restored. Pilot alcohol injection was capped at a maximum of 5 bar, while main injection was performed at a pressure of 2500 bar.

The use of EGR rates helped offset the anticipated NO_x emission penalty resulting from the combustion shift, especially under dual-fuel conditions with higher EGR rates. Combustion was advanced to capitalize on the expected increase in thermal efficiency from centered combustion, while the burning of premixed alcohol was enhanced due to higher temperatures near TDC. Regardless of the combination of low-reactivity fuels (LRFs) (methanol and ethanol) with high-reactivity fuels (HRFs) (diesel and HVO), careful consideration of fuel selection is necessary depending on the desired outcome, particularly at low load. Methanol is preferred over ethanol for reducing particle and NO_x emissions due to its higher oxygen content and greater cooling effect. However, achieving particle-free combustion requires a high alcohol substitution ratio when using HVO as the HRF, as it lacks soot precursors. The impact of the fuel was most significant in determining the maximum alcohol substitution ratio achievable at low load. The order of effectiveness was as follows: methanol-diesel < ethanol-diesel < methanol-HVO < ethanol-HVO, highlighting the importance of chemical reactivity of the LRF and HRF. HVO emerges as the preferred option due to its very high cetane number.

$$\text{Objective Function, OF} = \frac{ITE - ITEmin}{ITEmax - ITEmin} - \frac{NOx - NOx\,min}{NOx\,max - NOx\,min}$$

$$\text{Substitution ratio } SR = \frac{\dot{m}_{alc.} \times LHV_{alc.}}{\dot{m}_{alc.} \times LHV_{alc.} + \dot{m}_{\frac{D}{HVO}} \times LHV_{\frac{D}{HVO}}} \times 100$$

where ITE = indicated thermal efficiency;
$LHV_{alc.}$ = lower heating value; and
$\dot{m}_{alc.}$ = mass flow rate of the alcoholic fuel.

Detailed results have been discussed below:

It is clear from the data presented in Figures 7.5 and 7.6 that in order to achieve the optimal balance between engine efficiency and NO_x emissions, different levels of EGR and combustion centering α50 (50% of apparent heat release at a specific crank angle [CA]) were necessary for conventional diesel combustion using either fossil diesel fuel or renewable HVO, and for dual-fuel combustion. In particular, dual-fuel combustion required a combustion phasing closer to TDC to take advantage of favorable thermodynamic conditions for more complete and efficient burning of the

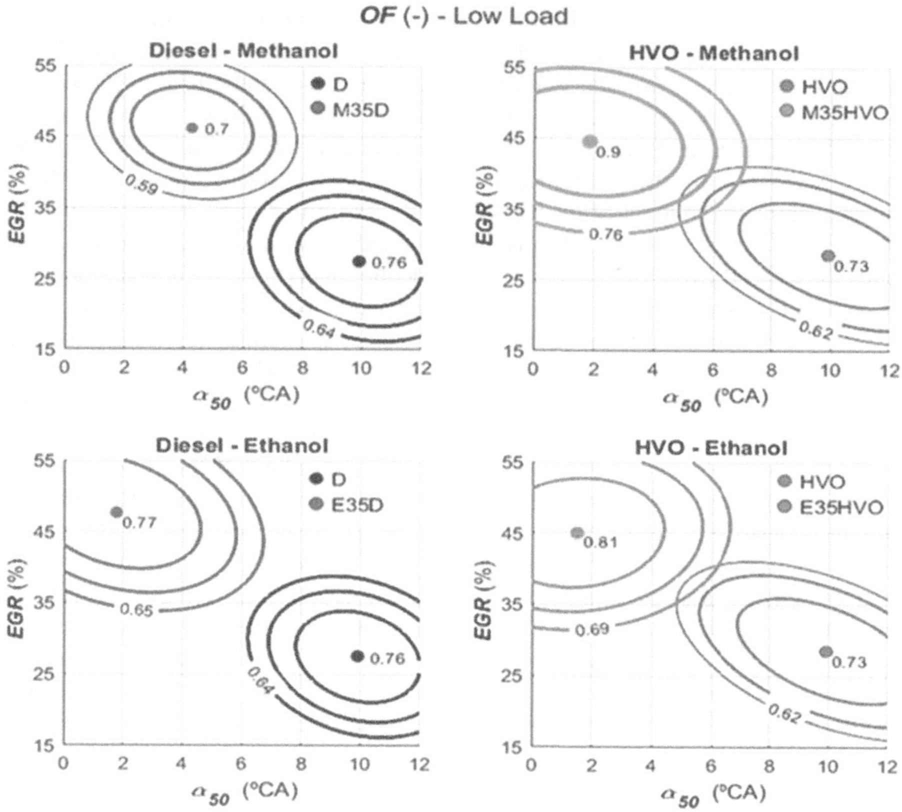

FIGURE 7.5 Contour plots of the response surface, low load. (Source: V.M. Domínguez et al. [15].)

premixed alcohol. The expected increase in NO_x emissions due to this combustion shift was mitigated by a higher EGR rate. This was especially beneficial for dual-fuel combustion, as previous studies [16] have shown that high EGR rates can improve engine thermal efficiency when using alternative fuels.

Under dual-fuel conditions, methanol necessitated increased EGR rates and centered combustion processes, regardless of the load [Figure 7.7 (a)–(h))]. This was particularly evident at low load, likely due to the lower equivalence ratio and pressure rise rate (PRR), allowing for more flexibility in engine parameter adjustments. When alcohols were utilized in dual-fuel mode, exceptionally high EGR rates were attainable at both low and high loads, resulting in notably higher equivalence ratios compared to conventional modes. Equivalence ratios ranged from 0.62 to 0.75 at low load and approximately 0.6 at high load. Optimal outcomes were observed at low load with EGR rates nearing 50%, which proved unattainable at high engine loads, resulting in equivalence ratios of around 0.75. This also proved advantageous in elevating intake temperatures and mitigating the cooling effects associated with alcohols possessing high vaporization enthalpy. Moreover, the prolonged ignition

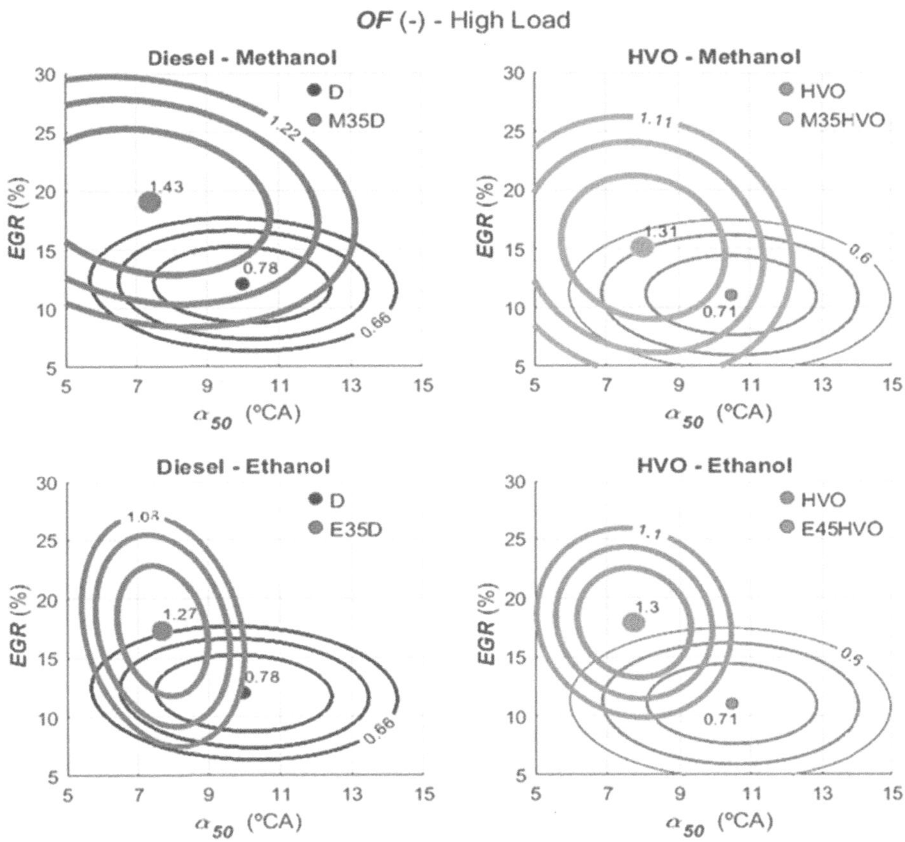

FIGURE 7.6 Contour plots of the response surface, high load. (Source: V.M. Domínguez et al. [15].)

delays at low load during dual-fuel combustion predominantly resulted in premixed combustion, markedly differing from compression-detonation combustion (CDC) and contributing to observed variances at high engine loads.

At elevated loads, it is noteworthy that methanol exhibited less susceptibility to changes in EGR rate and combustion centering compared to ethanol. This is evident in Figure 7.8 (a)–(h), which shows contour lines extending over a wider range in the EGR-α50 map. This can be attributed to the varying oxygen content and cooling capabilities of the alcohols, which hold greater significance at higher loads due to the increased temperature and alcohol-to-charge (air + EGR) ratio. The EGR rate plays a major role in determining the availability of oxygen for combustion, and the higher oxygen content of methanol may make the EGR rate less influential compared to ethanol. On the other hand, delaying the combustion timing is a well-known method for reducing NO_x emissions by lowering the combustion temperature. In this case, the greater cooling effect of methanol (due to a higher mass flow rate for the same energy substitution ratio and higher vaporization enthalpy)

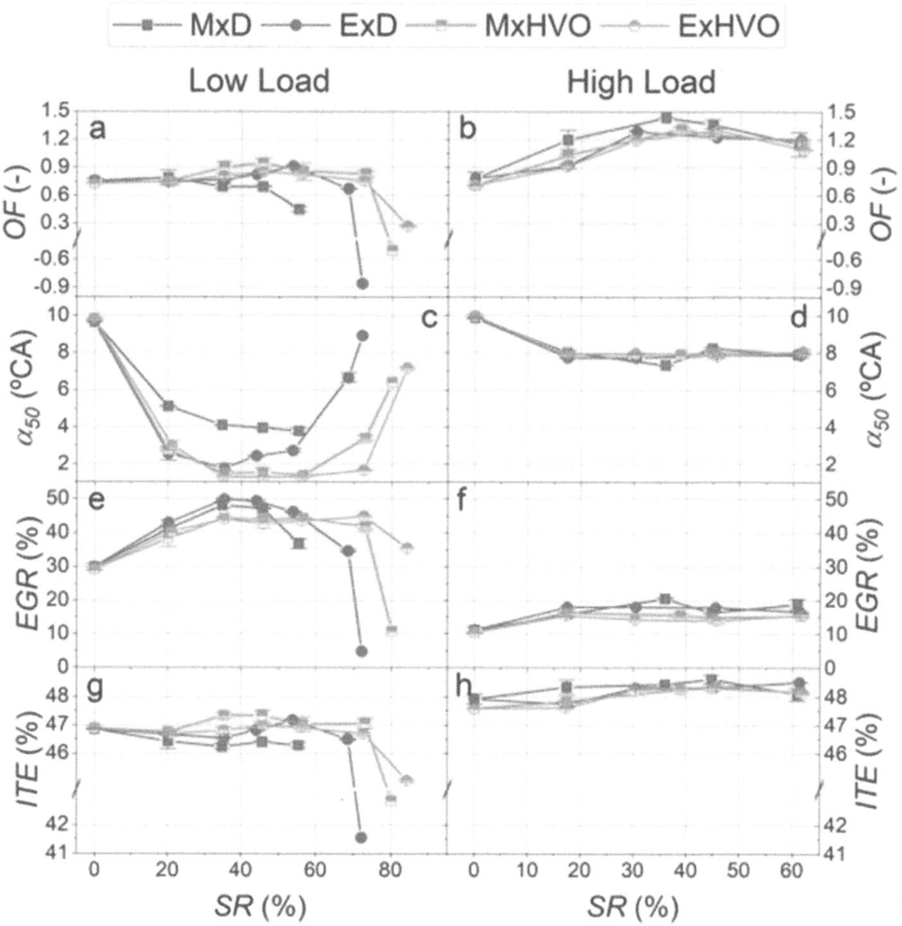

FIGURE 7.7 (a)–(h) Effect of the substitution ratio on operating variables, OF and ITE. (Source: V.M. Domínguez et al. [15].)

provides sufficient cooling, making the combustion centering less significant in terms of NO_x formation.

7.2.4 Other Advanced Strategies

The use of RCCI engines has the potential to reduce both PM and nitrogen oxide (NO_x) emissions while also improving engine performance, although the degree of efficiency improvement is dependent on the injection ratio of higher alcohols. To further enhance the performance and emissions of higher-alcohol RCCI engines, measures such as increasing injection advance angle, implementing EGR, utilizing multiple injections, and adding hydrogen can be implemented [9]. In a study by Sabu et al. [17], the effects of a split injection strategy on a single-cylinder CI engine using a blend of B20 Karanja biodiesel as the HRF and n-pentanol as the LRF were investigated. Three different

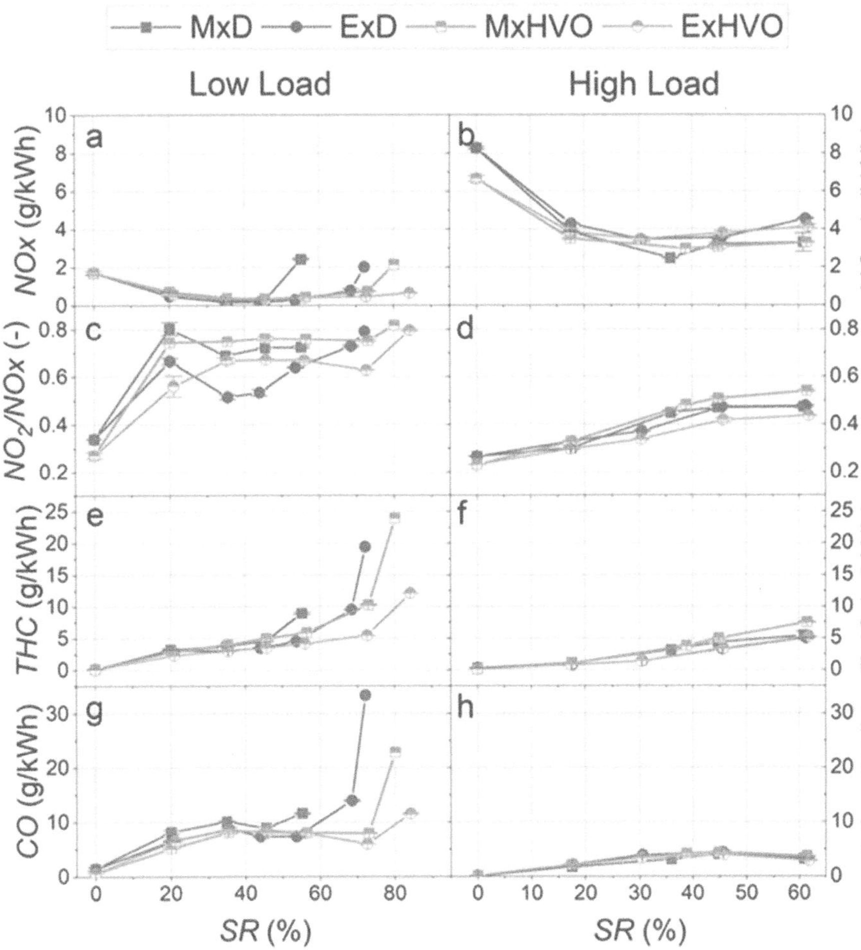

FIGURE 7.8 (a)–(h) Effect of the substitution ratio on gaseous emissions. (Source: V.M. Domínguez et al. [15].)

starts of injection (SOI) sweeps (SOI 1 sweep from 41° to 49°, SOI 2 sweep from 21° to 29°, and SOI 3 sweep from 11° to 19° before top dead center [BTDC]) were studied by varying one of the injections while keeping the other two constant. The results showed a decrease in smoke emissions and an increase in NO_x emissions with advances in SOI 2 and SOI 3 angles. Conversely, in the case of SOI 1, an advanced injection angle resulted in an increase in smoke emissions and a decrease in NO_x due to lower pressure and temperature during the early compression stroke. Another study by Şen et al. [18] demonstrated a prolonged ignition delay with an advanced injection timing from 20° BTDC to 24° BTDC and an increased ratio of n-pentanol from 10% to 30% by volume. Across all operating conditions, the smoke emissions of n-pentanol-diesel mixtures (10%–30% n-pentanol) were lower than those of diesel fuel, and the smoke further decreased with an advance in injection timing.

J. Li, et al. [19] examined the performance and emissions of a diesel (D100)-fueled CI engine, as well as blends of n-octanol/diesel (D50O50) and n-butanol/diesel (D50B50), using both single and split injection strategies. Three different start of injection (SOI) timings (–5, 0, and 3 CA ATDC) were selected for the single injection, while for split injection, the first SOI was set at –20°CA ATDC and the second SOI at 0°CA ATDC. The experimental results showed that D50O50 and D50B50 could cause severe knocking combustion under certain conditions when using the single injection strategy. However, the split injection strategy was found to reduce peak pressure and peak HRR, thus mitigating the knocking tendency for D50O50 and D50B50. Additionally, the researchers observed a distinct two-stage combustion for D100 and D50O50 due to the NTC behavior of diesel and n-octanol. Overall, due to the lower reactivity of n-octanol and n-butanol, the ignition delay periods for D50O50 and D50B50 were significantly prolonged with single injection, but this difference was not significant with the split injection strategy. Furthermore, D50O50 showed the highest thermal efficiency with proper injection pressure and combustion phasing. The use of split injection at an injection pressure of 60 MPa also resulted in simultaneous reductions in NO_x and black carbon emissions for D50O50. However, the addition of n-octanol and n-butanol increased HC emissions at all tested conditions due to their lower fuel reactivity.

Another combustion concept, i.e. homogeneous charge CI (HCCI), is also being explored by many researchers with expected advantage of achieving low NO_x, smoke, and PM emissions simultaneously. In this concept, a very lean and nearly homogeneous air-fuel mixture is directly injected into combustion chamber to reduce peak combustion temperature. In this way, several small homogeneous mixture zones are formed which leads to simultaneous combustion throughout, and hence overall combustion rate is high. Due to this advantage, HCCI engine widens the possibility of using high compression ratios, and PRR is below the maximum limits. Undoubtedly, this concept reduces NO_x emissions, but on the other hand, high HC and CO levels are because of low combustion temperatures and low post-oxidation of these emission components. In one of the studies of K. Mathivanan et al. [14], the impact of injecting diesel fuel in five separate pulses has been studied in a CI engine running in HCCI mode. The findings have been compared with HCCI operation utilizing a single pulse (SP) of diesel fuel injected at 100° CA BTDC, as well as with the standard CI mode at similar outputs. The results indicate that multiple pulse (MP) injection outperforms SP injection, resulting in reduced emissions and increased thermal efficiency. This can be attributed to improved combustion timing and higher rates of heat release.

Despite some drawbacks in terms of emissions and thermal efficiency at medium load conditions, the post-injection strategy demonstrated effectiveness in reducing the peak PRR and could prove advantageous for applications demanding higher loads, where this parameter is a greater concern [20].

Y. Qian et al. [21] conducted an experiment to study the combustion and emission characteristics of dual-fuel sequential combustion (DFSC) mode. This mode utilizes PFI of n-heptane combined with in-cylinder directly injected ethanol, n-butanol, and n-amyl alcohol in a single-cylinder engine with a fixed injection timing. The results indicated that the heat release can be divided into three stages: low-temperature reaction, high-temperature reaction of n-heptane, and the directly injected fuel combustion stage. The maximum in-cylinder pressure (P_{max}), maximum in-cylinder mass-averaged temperature (T_{max}), and maximum PRR were greatly influenced by the amount of port-injected n-heptane. When there was a high overall LHV per cycle, the CO emissions decreased as the premixed ratio increased, but when the overall LHVs per cycle were at medium or low levels, the CO emissions increased with the premixed ratio. However, the NO_x and soot emissions remained at low levels for all experimental conditions. Notably, the use of ethanol, with its higher latent heat, lower cetane values, and shorter carbon chains, resulted in lower NO_x and soot emissions compared to n-butanol and n-amyl alcohol. At low loads and with optimized premixed ratio, directly injecting n-butanol resulted in an indicated thermal efficiency higher than 46% while maintaining low emissions.

Y. Bai et al. [22] explored the impact of injection timing and n-butanol energy ratio (BER) on combustion and emissions in a modified single-cylinder engine equipped with dual common rail direct injection (CRDI) systems. The n-butanol fuel was injected through an additional gasoline direct injection (GDI) injector, while the original diesel injector was used for diesel fuel injection. The total fuel energy was kept constant at 510 J/cyc during low load and 680 J/cyc during medium load. The researchers examined how the n-BER and fuel injection timing affected engine combustion and emission characteristics under both low and medium load conditions. As the n-BER increased, the combustion phase was delayed, the indicated thermal efficiency (ITE) decreased, and the Coefficient of Variation (COV) of IMEP increased at both loads. For low load, an n-butanol injection timing of approximately 125° CA BTDC was recommended in order to achieve higher ITE by adjusting the stratification of n-butanol to an optimal level. When the diesel injection timing was advanced, the combustion phase was initially advanced and then delayed, resulting in an initial increase and then decrease in ITE. Under medium load, a two-stage high-temperature heat release event was observed. To achieve higher ITE and lower COV of IMEP, early n-butanol injection was preferred. Additionally, for improved ITE and reduced nitrogen oxide and soot emissions, early diesel injection was recommended for medium load. Overall, the use of n-butanol/diesel in Dual-fuel direct injection (DFDI) mode resulted in stable and highly efficient combustion, with a 2.8% increase in ITE and an 82% and 79% decrease in NO_x and soot emissions, respectively, compared to pure diesel operation.

According to numerical studies, it has been confirmed that the mixing of fuel has a significant impact on the combustion state switching of methanol/diesel direct dual-fuel stratification (DDFS) engines [23]. The premature injection of methanol using a single-stage injection strategy can cause the ringing intensity (RI) to exceed the engine's limit. Additionally, if the start of diesel injection (SOID) occurs earlier than the start of methanol injection (SOIM), the shorter interval between methanol and diesel injection (MDI) can negatively affect the premixed combustion of diesel due to interference from the methanol spray. Therefore, it is recommended to use an injection strategy where SOID is earlier than SOIM in combination with an appropriately extended MDI to achieve better overall engine performance. Further optimization of the methanol/diesel DDFS engine using a two-stage injection strategy has shown that it can improve fuel economy while maintaining acceptable levels of RI. By increasing the methanol pre-injection ratio (MR1) and optimizing the interval between the two methanol injections (MMI), the engine's economy can be further enhanced. The optimized two-stage injection strategy resulted in a 3.95% reduction in equivalent indicated specific fuel consumption (EISFC) compared to the single-stage strategy. However, it should be noted that the combustion completeness is reduced under this strategy, resulting in higher emissions of hydrocarbons (HC), carbon monoxide (CO), and soot. On the other hand, this strategy leads to lower NO_X emissions due to more effective premixed combustion. Utilizing a three-dimensional simulation model in conjunction with chemical reaction kinetics, the impact of dual-fuel injector parameters, specifically nozzle number and diesel/methanol spray spatial angle (β), is investigated on a diesel/methanol dual-fuel DI engine [24]. The results demonstrate that increased nozzle quantity and decreased nozzle diameter contribute positively to fuel efficiency while also reducing emissions of soot, THC, CO, and CH_2O. However, these changes also result in higher RI and NO_x emissions. Further analysis reveals that the most optimal configuration is achieved with eight nozzles, taking into account fuel economy, RI, and emissions. Additionally, it is discovered that the in-cylinder airflow has a significant impact, with an optimal spray spatial angle ($\beta = 10°$) yielding the best balance of fuel efficiency and emissions.

7.3 CONCLUSIONS

By promoting more complete combustion of the fuel-air mixture, multiple fuel injections can improve the combustion process. The fuel is distributed more evenly within the combustion chamber by splitting up the fuel injection into several pulses, hence leading to better air mixing and efficient combustion. If we break up all these fuel quantities into smaller injections, then peak PRR will be reduced, thereby resulting in smoother combustion and quieter engine operation particularly at cold starts plus low-speed operations. Cleaner exhaust gases are achieved because multiple injections help in optimizing the combustion process as well as reducing formation of soot and NO_x. Several injections depending on engine operating conditions may lead to improvements in efficiency of fuel use; there can be a number of injections employed by an injector. Some examples include port-injection systems for gaseous fuels, direct-injection systems for both gasoline and diesel engines which are mostly used in modern vehicles like MPVs or SUVs. When it comes to certain terms such as precisely controlling the timing and quantity of fuel injection for optimal power generation, there is a need to have various types of gaugeable injectors that allow for improved atomization along with better miscibility during burning stage, thereby consuming relatively less amounts under specified conditions. Multiple injections are aimed at contributing toward improving engine performance through enhancing control over cylinder pressure as well as combustion phasing. Turbocharged diesel engines especially show increased torque, power output, and throttle response due to changes made on them like addition or removal of certain parts inside their cylinders. In general, multi-fuel injection engines can provide many benefits such as better combustion, fewer emissions, better fuel economy, and engine quality. However, the best injection depends on many factors such as engine design, performance, and emissions requirements.

REFERENCES

1. Kumar P, Singh M, Sandhu SS. Wavelet analysis for cyclic combustion dynamics of a multi-cylinder CRDI diesel engine fuelled with a blending of argemone biodiesel–diesel oil. Chaos An Interdiscip J Nonlinear Sci 2022;32:043107. doi:10.1063/5.0080910.
2. Sandhu A, Chander SS. Experimental investigations on the influence of fuel injection timing and pressure on single cylinder C.I. engine fueled with 20% blend of castor biodiesel in diesel. Fuel 2017;210:15–22. doi:10.1016/j.fuel.2017.08.023.
3. Singh M, Sandhu SS. Performance and emission characteristics of an indirect injection (IDI) multi-cylinder compression ignition (CI) engine using diesel/*Argemone maxicana* biodiesel blends. RSC Adv 2015;5:91069–81. doi:10.1039/C5RA10506D.
4. Singh M, Gandhi SK, Mahla SK, Sandhu SS. Experimental investigations on performance and emission characteristics of variable speed multi-cylinder compression ignition engine using Diesel/Argemone biodiesel blends. Energy Explor Exploit 2018; 36:535–55. doi:10.1177/0144598717738573.
5. Patel RL, Sankhavara CD. Biodiesel production from Karanja oil and its use in diesel engine : A review. Renew Sustain Energy Rev 2020;0–1. doi:10.1016/j.rser.2016.12.075.
6. Singh M, Sandhu SS. Performance, emission and combustion characteristics of multi-cylinder CRDI engine fueled with argemone biodiesel/diesel blends. Fuel 2020;265:117024. doi:10.1016/j.fuel.2020.117024.
7. Singh M, Sandhu SS. Effect of boost pressure on combustion, performance and emission characteristics of a multicylinder CRDI engine fuelled with argemone biodiesel/diesel blends. Fuel 2021;300:121001. doi:10.1016/j.fuel.2021.121001.
8. Akash D, Sandhu SS, Chander S. Experimental Investigations on Castor Biodiesel as an Alternative Fuel for Single Cylinder Compression Ignition Engine 2016;00. doi:10.1002/ep.12518.
9. Hua Y. Research progress of higher alcohols as alternative fuels for compression ignition engines. Fuel 2024;357:129749. doi:10.1016/j.fuel.2023.129749.
10. Han J, Wang S, Somers B. Effects of different injection strategies and EGR on partially premixed combustion. SAE Tech Pap 2018;2018-Septe:1–12. doi:10.4271/2018-01-1798.
11. Li X, Guan C, Yang K, Cheung CS, Huang Z. Impact of lower and higher alcohol additions to diesel on the combustion and emissions of a direct-injection diesel engine. Environ Sci Pollut Res 2019;26:21001–12. doi:10.1007/s11356-019-05275-y.
12. López AF, Cadrazco M, Agudelo AF, Corredor LA, Vélez JA, Agudelo JR. Impact of n-butanol and hydrous ethanol fumigation on the performance and pollutant emissions of an automotive diesel engine. Fuel 2015;153:483–491. https://doi.org/10.1016/j.fuel.2015.03.022
13. Kalwar A, Singh AP, Agarwal AK. Utilization of primary alcohols in dual-fuel injection mode in a gasoline direct injection engine. Fuel 2020;276:118068. doi:10.1016/j.fuel.2020.118068.
14. Mathivanan K, Mallikarjuna JM, Ramesh A. Influence of multiple fuel injection strategies on performance and combustion characteristics of a diesel fuelled HCCI engine - An experimental investigation. Experimental Thermal and Fluid Science 2016; 77: 337–346. https://doi.org/10.1016/j.expthermflusci.2016.05.010
15. Domínguez VM, Hernández JJ, Ramos Á, Giménez B, Rodríguez-Fernández J. Exploring the effect of methanol and ethanol on the overall performance and substitution window of a dual-fuel compression-ignition engine fueled with HVO. Fuel 2024;359. doi:10.1016/j.fuel.2023.130529.
16. Dahham RY, Wei H, Pan J. Improving thermal efficiency of internal combustion engines: Recent progress and remaining challenges. Energies 2022;15. doi:10.3390/en15176222.

17. Sabu VR, Thomas JJ, Nagarajan G. Experimental investigation on the effects of multiple injections and EGR on: N-pentanol-biodiesel fuelled RCCI engine. RSC Adv 2020;10:29498–509. doi:10.1039/d0ra03723k.
18. Şen M, Emiroğlu AO, Keskin A. Impact of pentanol addition and injection timing on the characteristics of a single-cylinder diesel engine. Energy and Fuels 2019;33: 9224–31. doi:10.1021/acs.energyfuels.9b01759.
19. Li J, Yu W, Yang W. Evaluating performance and emissions of a CI engine fueled with n-octanol/diesel and n-butanol/diesel blends under different injection strategies. Fuel 2021;284:119085. doi:10.1016/j.fuel.2020.119085.
20. Jeftić M, Zheng M. Postinjection strategy for the reduction of the peak pressure rise rate of neat n-butanol combustion. J Eng Gas Turbines Power 2016;138:1–9. doi:10.1115/1.4032765.
21. Qian Y, Zhou Q, Wang X, Zhu L, Lu X. Enabling dual fuel sequential combustion using port fuel injection of high reactivity fuel combined with direct injection of low reactivity fuels. Appl Therm Eng 2016;103:399–410. doi:10.1016/j.applthermaleng.2016.04.122.
22. Bai Y, Wang Y, Hao L. Experimental study on the effects of injection timing and n-butanol energy ratio on combustion and emissions of n-butanol/diesel DFDI engine. Fuel 2022;324. doi:10.1016/j.fuel.2022.124654.
23. Sun W, Jiang M, Guo L, Zhang H, Jia Z, Qin Z, et al. Numerical study of injection strategies for marine methanol/diesel direct dual fuel stratification engine. J Clean Prod 2023;421:138505. doi:10.1016/j.jclepro.2023.138505.
24. Li Z, Wang Y, Yin Z, Gao Z, Wang Y, Zhen X. An exploratory numerical study of a diesel/methanol dual-fuel injector: Effects of nozzle number, nozzle diameter and spray spacial angle on a diesel/methanol dual-fuel direct injection engine. Fuel 2022;318:123700. doi:10.1016/j.fuel.2022.123700.

8 Impact of Biobutanol Addition into Waste Seed Oils on Performance, Combustion and Emissions of Compression Ignition Engines

Prabakaran B.

8.1 INTRODUCTION

Recent trends indicate that the availability of resources for fossil fuels such as diesel is emptying out and may not be sufficient to meet the demand from sectors such as industrial power packs, sectors of transportation [1] and power backs used for backups. Also, wasting resources such as peel-off from fruits and vegetables used for the supply of domestic and commercial food products are, as of now, wasted without proper use, which in turn increases the disposal cost of the collected garbage. This also increases the contamination of land. There is the possibility of using the fruit peel-offs to convert bioalcohols and the seeds from the waste fruits seeds as oils, which can be used as fuel for power packs instead of fossil fuel resources. Previous research has reported the utilization of biodiesel from vegetable oils and bioalcohols from biomass as fuels through blending with diesel or biodiesel. In this line [2–5], another study utilized 50–50 proportions of butanol and diesel by feeding various proportions of oxygen for the combustion. The emissions from the test such as nitrogen oxides and monoxides of carbon are within the limits of base fuel diesel. The proportions used in this study were the following: equal volume of diesel and butanol up to 50% and ternary blends containing diesel, biodiesel and butanol in which butanol was 25%. In a controlled combustion chamber, the atomization of the blends was tested. In this blend of biodiesel and isobutanol were tested. The higher the volume of isobutane, the higher the value of ignition delay. The emissions such as nitrous oxide and monoxide of carbon were reported as higher than diesel. The average diameter of the particles from the exhaust was reported as low. This study was conducted by using 10% of butanol and 10% of biodiesel as fuel in diesel fuel engine. It was also reported that the emissions from the power source are found [6, 7] to be low in low powers. This research work includes the analysis of ignition properties of the blends in a

DOI: 10.1201/9781003482277-8

chamber and the dissipation of the heat by using the isomers of butanol. The results depicted that isobutanol was best with respect to the performance when fueled under low temperatures and tert-butanol showed the best in higher temperature operation. Up to 30% of isobutanol was tested in this study. Overall 20% butanol was blended with biodiesel and used as fuel for power source; low gas temperature from exhaust was reported up to utilization of 20% of butanol in the blend. Low oxides of nitrogen and increase in hydrocarbons were reported in this study. Rice bran oil was used [8–11] by converting into biodiesel, up to a volume of 20%of biodiesel was used along with diesel and the report depicted that the fuel used was found higher, lesser efficiency of the engine, with higher emissions comparing the base reading of diesel fuel. In the similar manner, karanja oil [12–14] was used along with butanol as an additive for fueling compression ignition engine. The report showed higher efficiency, low CO, HC with high NO_X and EGT (exhaust gas temperature) by the increase in butanol (up to 30%) content in the blend. The delay in the ignition of the injected fuel was [15–17] reported longer and shorter combustion duration of the engine. Increase in the volume of butanol increased the hydrocarbons, carbon monoxide and lowers the soot emissions. Biodiesel along with 20% of butanol was tested in transient state of the engine. Low soot emissions – up to 3% low with respect to diesel when operating under ambient conditions, with the increase of injection pressure the soot emissions were lowered up to 45.5%. The increase in the pressure of the injected fuel increased the oxidation in the chamber. Another study [9, 10, 18] was conducted by using diesel-biodiesel along with butanol up to 10% in CI engine. The results indicated that out of the blends used, blend containing 10% of butanol performed the best that produced better performance and higher CO_2 emissions which is an indication of complete combustion. This test reported low efficiency, low CO and HC high nitrous oxide and CO_2 from the engine during operation.

As an extract, the previous research outputs stated that blends of diesel, biodiesel and butanol can be used as fuel for engine. This research work includes the biobutanol (BB) from waste fruits and vegetables cut wastes, oils extracted from waste seeds for fueling the engine. The stages of this study are the following:

1. Blend preparation of waste seed oil of Amla, seed oil from Jamun, BB from the waste fruits and vegetables with oil of castor from waste seeds followed by the testing of required properties.
2. Choosing the suitable blends and testing of performance of the chosen blends for performance.
3. Analyzing the results by the comparison of base readings of diesel operated engine as reference.

8.2 MATERIALS AND METHODS

Commercial diesel rated as low sulfur is received with test reports for basic properties and for the reference use. From the authorized manufacturer, the BB and seed oils, who is extracting these from the wastes are being received with the test reports of the required properties. As per the extract from the previous literature, approximately 100 million metric tons of Amla [19] seeds, 10 million metric tons of Jamun seeds [20–22] and 170 million tons of seeds of castor [23] are from the fruits in India per annum. Out of these 40% are wasted in the garbage without properly used. This research study aims to utilize these wastes into energy to decrease the import of crude from the other supplying countries.

8.2.1 Blend Preparation and Choosing Suitable Blends

BB along with waste seed oil of Amla (Amla seed oil (ASO)), waste seed oil of Jamun (Jamun seed oil (JSO)) were proportionally (in increments of 5%) prepared. By using 20% BB and diesel, the base blend was prepared, and waste seed oil from castor is being blended in increments of 2% for various proportions. Stirring was performed by using a magnetic stirrer for a 3-minute cycle. This stirring was repeated for five to six cycles for each blend proportion. The prepared blends are kept undisturbed in controlled storage, which is maintained under 25°C, which is generally specified as ambient for most experiments. Hourly monitoring of the status of the prepared blends is performed and recorded to ensure the blends are in a single liquid phase without being separated. This is continued for a period of 21 days, and the blends are taken for further study. The required properties are measured by testing the blends following the previous step. For each blend, five samples are taken and tested for their properties. The average of these values is taken for comparison to the reference readings. For the testing of required properties, a redwood viscometer with digital recording, digitally supported closed cup apparatus for the flash point of the blends, ignition delay method for the calculated cetane index of the blends and digital bomb calorimeter for the energy content of the blends. These equipment are periodically calibrated and maintained.

8.2.2 Extraction of Waste Seed Oils and Biobutanol

Initially, the cut vegetable wastes are collected and segregated, followed by pre-treatment, and proceeded for the fermentation process. The cut vegetables include cumulative collection from food outlets and home wastes. Clostridia are used to extract BB from fruit peel-offs and waste fruits. The yield is found to be approximately 50%–55%, 10%–12% bioethanol and 26%–29% acetone as combined output. Out of the available waste peel-offs for BB, potato peel-off give out a maximum of 26 g/L, and apple a maximum of 2.5%. Similarly, Jamun waste seeds segregated from food industry outlets and domestic kitchens are dried in a temperature-controlled furnace of 60–70°C. The ether from petroleum is added to the powdered seeds. Using a Soxlet apparatus by maintaining the temperature at 540–660°C for up to a maximum of 40 minutes, the oil from the waste seeds of Jamun is extracted. Waste Amla seeds from various places are properly segregated, dried and powdered. The agent used for this process is also ether from petroleum; this process includes treating by acid, alkali and neutralization. This extraction consumed time of up to 4.2 hours. The fatty acid content was approximated up to 26%. The mean output of this process is maximum of 17% with respect to the powders from the waste seeds of Jamun. Ricinus communis.L–castor seeds from waste are being collected from oil producers and using n-hexane, procedural trans-esterification castor oil (CO) is being extracted. The maximum output is approximately

35% of biodiesel and 2 mg of NaOH is used per gram approximately. Major uses of this research work are the following:

1. Using the wasted seeds of various fruits, preventing from more garbage.
2. Decrease in disposal cost of garbage, saving land from deterioration.
3. Using power sources, in turn to decrease in imports of crude.
4. Improving the quality of combustion by blending renewable fuel of butanol, by reducing the viscous nature of waste oils.

Also,

1. These fuel possess higher vaporization value comparing regular fuels, increasing the volume of fuel used.
2. A higher flash point of the fuels comparatively necessitates maintaining the minimum temperature of intake air higher than 25°C, which is indicated as ambient (Figures 8.1 and 8.2).

The schematic layout includes the air preheater for maintaining the intake temperature of 26°C which is 1 degree higher than the ambient, to ensure the air inducted is free from moisture to the engine. The outlet status of the air source

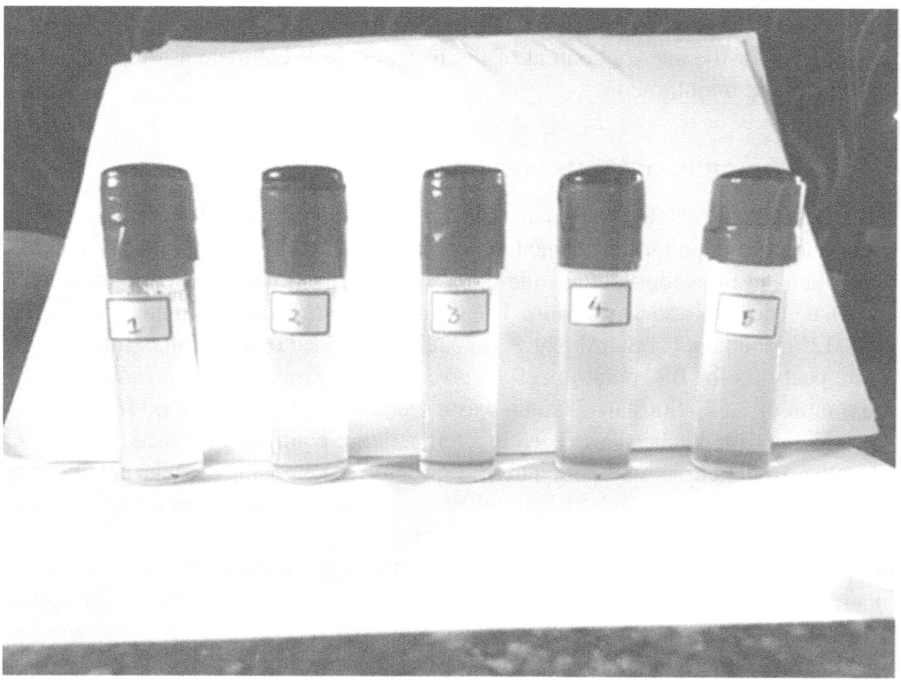

FIGURE 8.1 Representative blends kept for solubility test.

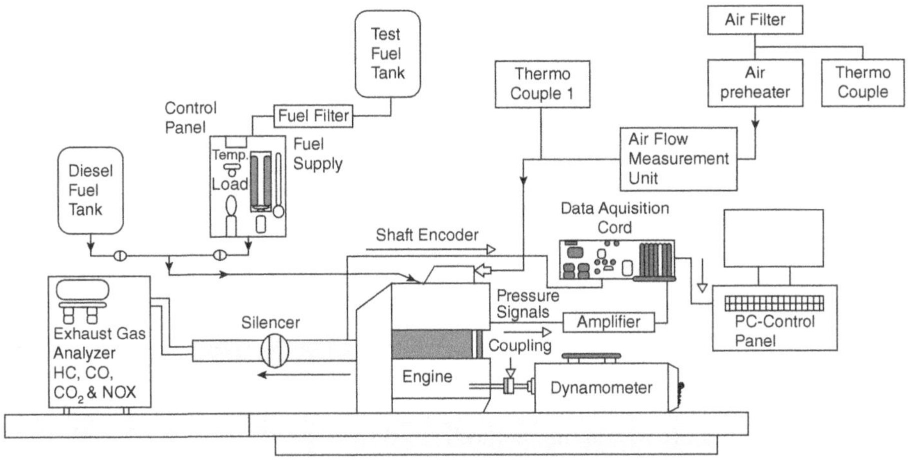

FIGURE 8.2 Layout of the test set up with equipment.

TABLE 8.1
Base Properties of Fuels

Property/Fuel	Diesel (D)	Biobutanol (BB100)	Amla Seed Oil (ASO)	Jamun Seed Oil (JSO)	Castor Oil (CO)	20BB80D
Flash point °C	62	38	151	75	192	58
Density kg/m³	826	814	888	871	945	825
Kinematic viscosity cst.	4.05	3.65	4.65	4.73	14	3.85
Lower calorific value kJ/kg	42901	35746	40105	40705	39165	41655
Cetane number	51	27	64	59	45	47
Heat of vaporization kJ/kg	271	580				
Boiling point °C	185–225	116				

TABLE 8.2
Properties of Chosen Representative Fuels

Property/Fuel	45ASO55BB	55ASO45BB	65JSO35BB	20BB80D-10CO
Flash point °C	82	88	63	70
Density kg/m³	842	848	847	837
Kinematic viscosity cst.	4.05	4.08	4.3	4.7
Lower calorific value kJ/kg	38015	38345	38950	41565
Cetane number	42	45	44	47

is periodically monitored by digitally recording thermocouples for temperature. Proper filtration is performed for fuel feed to the system and metered by digital flow rate recorder connected in line. Data logger performing the collection of pressure data from sensor of chamber of test engine is of AVL Di-Com and this also receives signals from crank about the position. The test engine has a rated power of 4400 W and 1500 revolutions per minute of speed which is properly cooled to keep the engine is permissible temperature. This engine is controlled by dynamometer and power input set up. The exhaust of the engine is being monitored, measured and recorded using AVL digital gas analyzer which is periodically calibrated. This includes the smoke meter for recording the smoke volume from the engine. For every test, the steady-state time of 20 minutes is provided, followed by five sets of values and the average arriving out of five sets is being used for further analysis. An uncertainty calculation for individual parameters and the total uncertainty is obtained using square averaging method.

8.3 RESULTS AND DISCUSSION

Sequentially, three results segments are presented, viz., property testing and comparison, parameters of performance, and parameters of emissions from the engine. Choosing the suitable blend is obtained from various properties, and the results are compared with reference values of diesel.

8.3.1 Properties Testing Results and Comparison

As presented in Tables 8.1 and 8.2, the blends containing 45% waste of ASO with 55% of BB [45ASO55BB], 65% of waste seed oil from Jamun with 35% of BB [65JSO35BB] and [20BB80D-10CO] 10% waste seed oil of castor with 20% BB and 80% diesel (as base) possess nearer values of properties considering diesel values. The chosen blends have approximately 32%–45AS055BB, 4.6%–65JSO35BB and 17.1%–20BB80D-10CO higher flash points with respect to diesel values. This ensures the safer handling of the fuel blend during operation. Similarly, the blends' calorific value is less than 5.9%, 3.7% and 0.05%, respectively. Also these blends have low calculated cetane index as 7.3%, 6.4% and 3.9%, respectively.

8.3.2 Comparison Analysis of Performance

This research study presents brake thermal efficiency (BTE), specific fuel consumption (SFC), in cylinder maximum pressure, EGT as comparison for performance. Totally six power operating conditions are presented as no load, 25%, 50%, 75% and 100% with respect to the rated power of the power source. Chosen fuel blends depict significantly lesser efficiency in no load, 25 and up to 60% brake powers (BP). Above 60% of BP, there is proportionally (Figure 8.3) increase in the trend of the efficiency up to the maximum power of the engine. The lesser efficiency in low BPs is due to the low rate of combustion, resulting from low oxidation which is in turn the presence of low temperature. This low temperature is created by the presence of higher amount of BB. This dominance is overcome by the increase in power, in turn higher

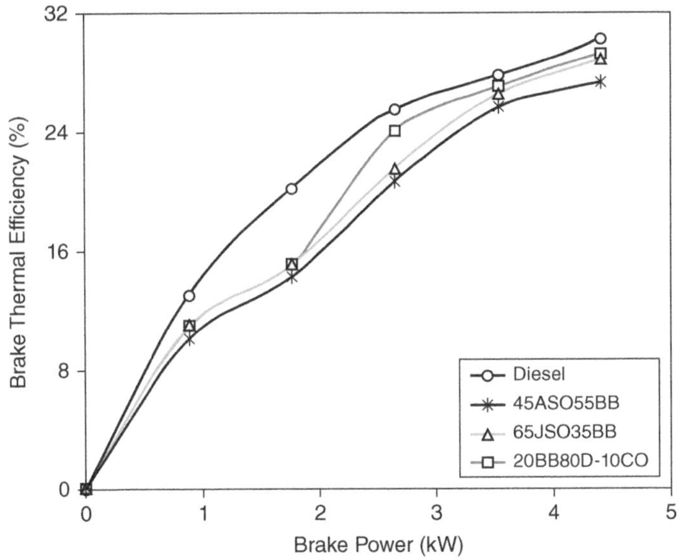

FIGURE 8.3 Variation in the engine BTE with respect to brake power on the engine.

surrounding temperature in the chamber around the fuel injected that improves the rate of combustion. This improvement in combustion and higher thermal release results in higher BTE. The three blends produce BTE lesser by 9.5%–45ASO55BB, 6.5%–65JSO35BB and 2.8%–20BB80-10CO against reference of diesel fuel. Second parameter of performance in this study is the consumption of the fuel blends used for operating the engine. The low energy content, higher vaporization of fuel blend and comparatively higher flash point cumulatively increasing the fuel used operating under 55% of BPs with respect to the maximum power. The higher consumption percentages are 21%, 7.4% and 3.9%, respectively, for (Figure 8.4) 45ASO55BB, 65JSO35BB and 20BB80D-10CO. The previous researchers also [24–27] reported that the blends containing alcohol as one of the fuels in the blend resulted in lesser BTE in low BP. Maximum pressure of any diesel fueled engine is a good indication, as the pressure and temperature are proportional to each other; this is the base for the higher complete combustion. The chosen three blends produce improved maximum pressure operated under higher BPs (higher than 70% of rated power); this is the operating range for most diesel-powered power sources. The significant difference (Figure 8.5) in low BPs (in lower than 70% of rated power) can be attributed to the low energy of the fuel blends, longer time consumed for atomization, higher dense and slightly higher viscous comparing base values of diesel. The temperature of the gases from the exhaust of the power sources is one of the parameters indicating the performance, as this is an indication of the level of combustion occurring in the chamber during the conversion of hydrocarbons and oxygen into water and carbon dioxide. A higher temperature is also necessary to initiate the operation after treatment is set in the exhaust of any power source.

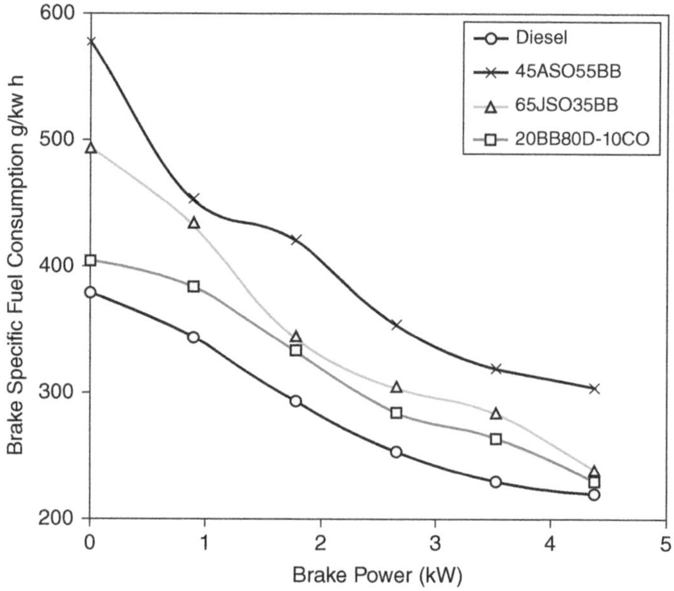

FIGURE 8.4 Variation of fuel consumed with respect to brake powers of the engine.

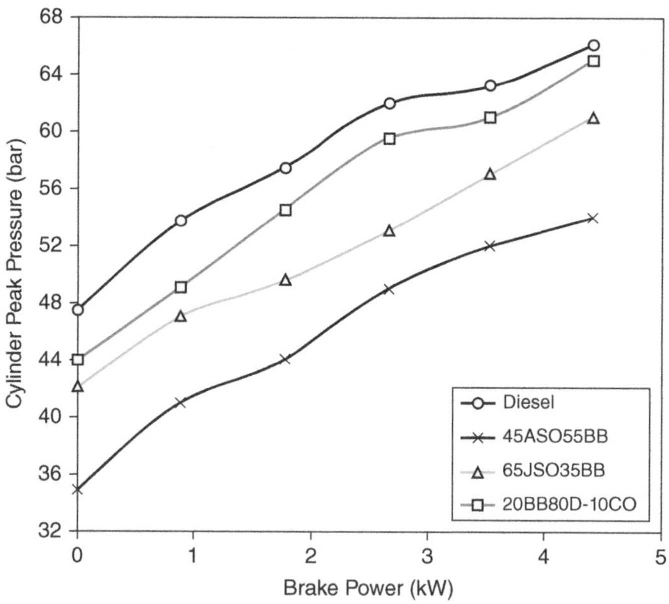

FIGURE 8.5 Variation of cylinder peak pressure versus brake power of the engine.

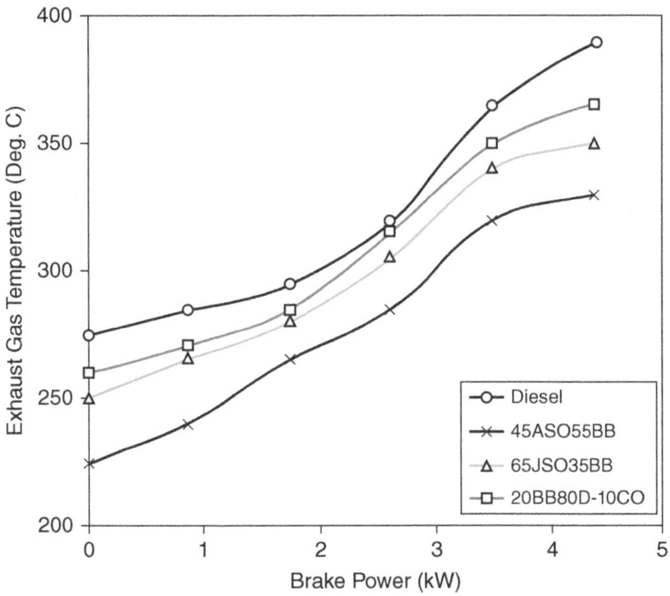

FIGURE 8.6 Variation of the gas temperature of exhaust from the engine by fuels.

Figure 8.6 indicates the change in the EGT proportion to the BP of the engine. The EGT of the chosen blends are found to be less by 11%, 10% and 7.5% for the blends 45ASO55BB, 65JSO35BB and 20BB80D-10CO, respectively, when fueling in the power source against the reference values of base diesel. Above 50% of BPs, the chosen three blends produce a higher temperature of the exhaust, more than 270°C, which is higher than the required minimum triggering temperature for the after-treatment system of the diesel-powered power pack. Also, previous research studies reported that power sources operating from alcohol-blended fuels produced similar trends. The net heat release rate is the base indication of the thermal energy release in the combustion. In this research study (Figure 8.7) at all operating BPs, the heat release rate of the power source is being captured and presented for analysis (one that is being captured under maximum BPs). Out of the three chosen fuel blends, the start of combustion for 20BB80-10CO is closest to that of diesel, followed by 65JSO35BB and 45ASO55BB. This is found to be proportional to the values of the calculated cetane index. The higher the value of the calculated cetane index, the nearer the start of combustion of the fuel comparing the value of diesel fuel. The decreases in percentages of the net heat release rate of the test engine are very marginal when operated at maximum BP. This ensures the suitability of the chosen blends for replacement of fossil fuel diesel.

8.3.3 Comparison Analysis–Emission Parameters

Emissions from the test power source is captured, measured and analyzed for NO_X, SMOKE HC and CO in various operating powers by considering the reference fuels as diesel. It can be seen (Figure 8.8) from the results that the blend

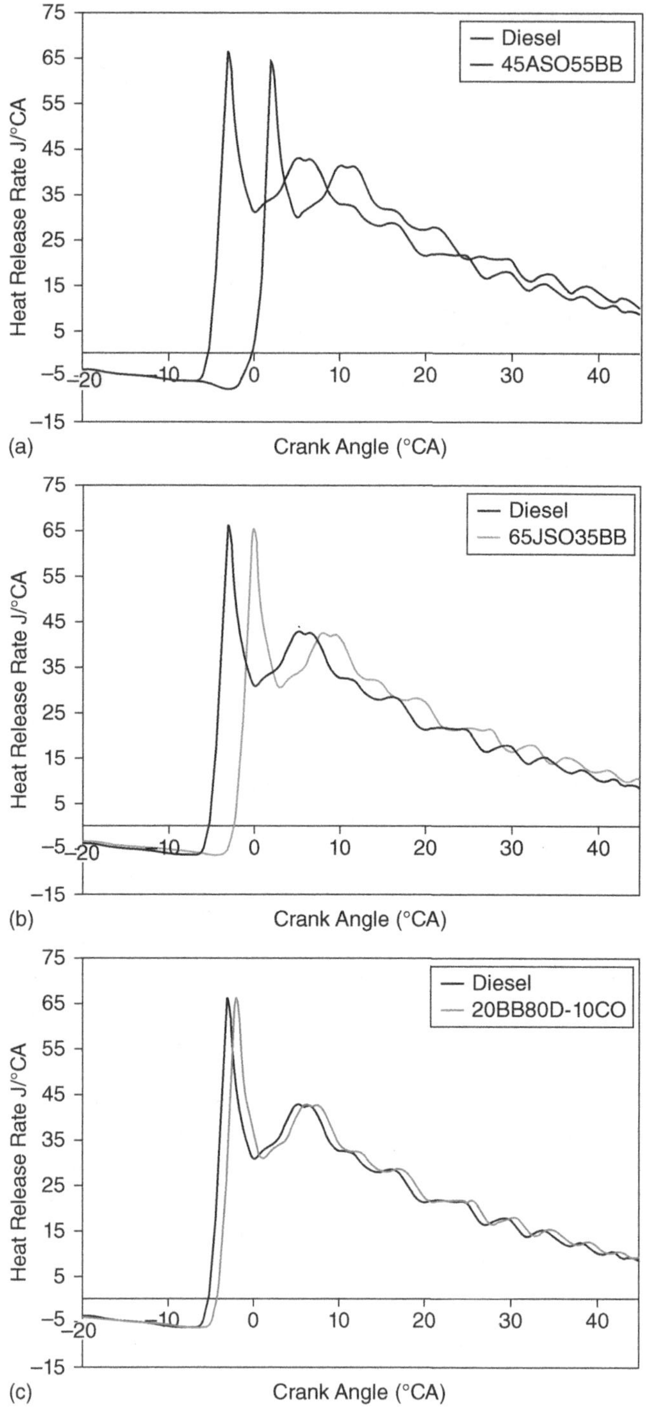

FIGURE 8.7 Heat release rate versus Crank Angle for different fuel blends: (a) Diesel and 45AS055BB, (b) Diesel and 65JSO35BB, and (c) Diesel and 20BB80D–10CO.

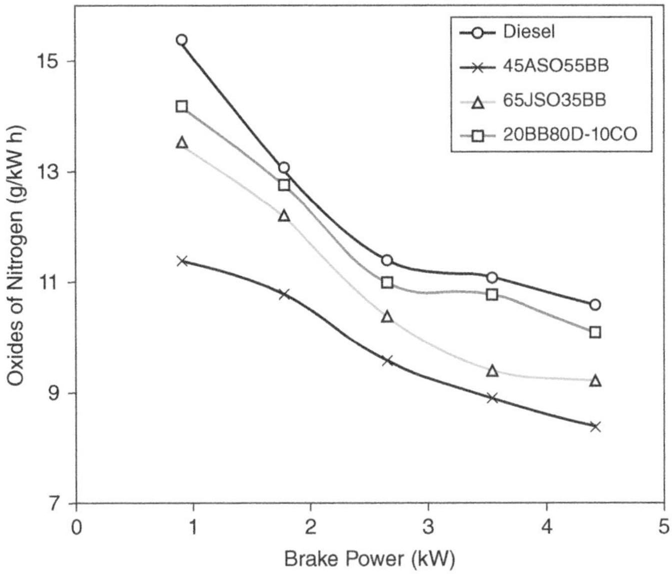

FIGURE 8.8 Variation in NO_X emissions of the test engine under various brake powers.

20BB80D-10CO emits highest NO_X and 45ASO55BB lowest among the chosen fuels. Mainly less NO_X is due to the low temperature which is from low energy content and higher vaporization of this blend. However, this blend produces 1.2% only low NO_X under maximum operating power. This is due to the higher rate of combustion under high BP that the influence of BB is overcome by the higher temperature of chamber. 6.5% low NO_X and 8.4% low NO_X by 45ASO55BB with respect to the reference fuel diesel. The previous study reports indicated that using biodiesel, diesel along with BB emitted similar trends of nitrous oxide while fueled in CI engine. However, nitrous oxides higher than limits are not advisable. Smoke from any power pack is one of the scales of incomplete combustion occurring during the operation. Higher smoke indicates a higher rate of incompleteness of combustion. From the present results in Figure 8.9, higher BPs higher than 40%, the emissions of smoke are trending closer to diesel base values comparing those in low BPs. At maximum BP of power pack, 6.1%, 12.3% and 29.6% of higher smokes by 20BB80D-10CO, 65JSO35BB and 45ASO55BB, respectively, by considering reference diesel engine fuel values (Figure 8.10). Previous reports indicated that blends of diesel and biodiesel containing butanol had similar trends when fueled in compression ignition engine. One more emission due to incomplete combustion is hydrocarbons, which is the base reaction-hydrocarbons conversion to complete combustion products and again into work on the piston. Highest HC is from 45ASO55BB at lowest BP and lowest HC is from 20BB80-10CO operating the power pack in highest BP. The increase of HC from the blend is seen as 8.6% with reference values of diesel-powered power source. High HC by 45ASO55BB can be due to higher flash point, low energy content and high viscous, specifically at low BP. With an increase in

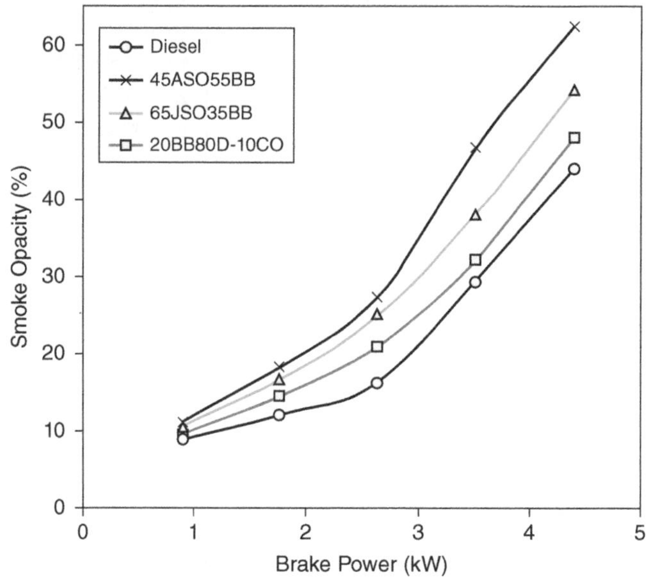

FIGURE 8.9 Smoke emissions under various brake powers of the test engine fueled with fuel blends.

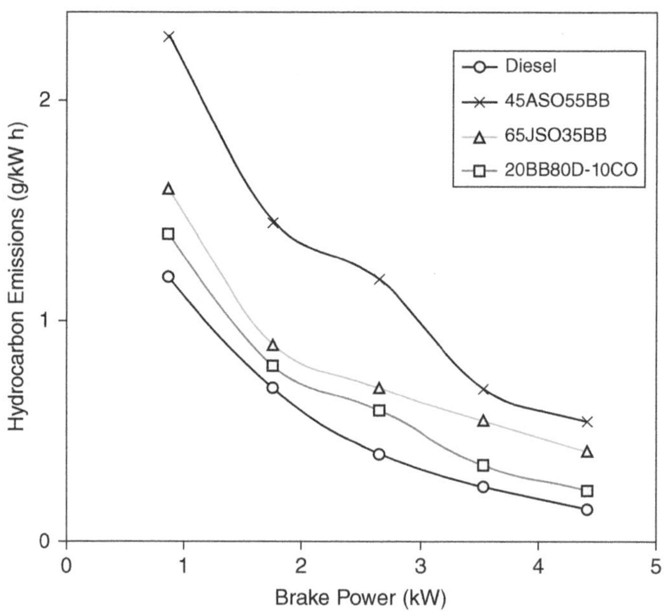

FIGURE 8.10 Variation of hydrocarbon emissions from test engine under various brake powers.

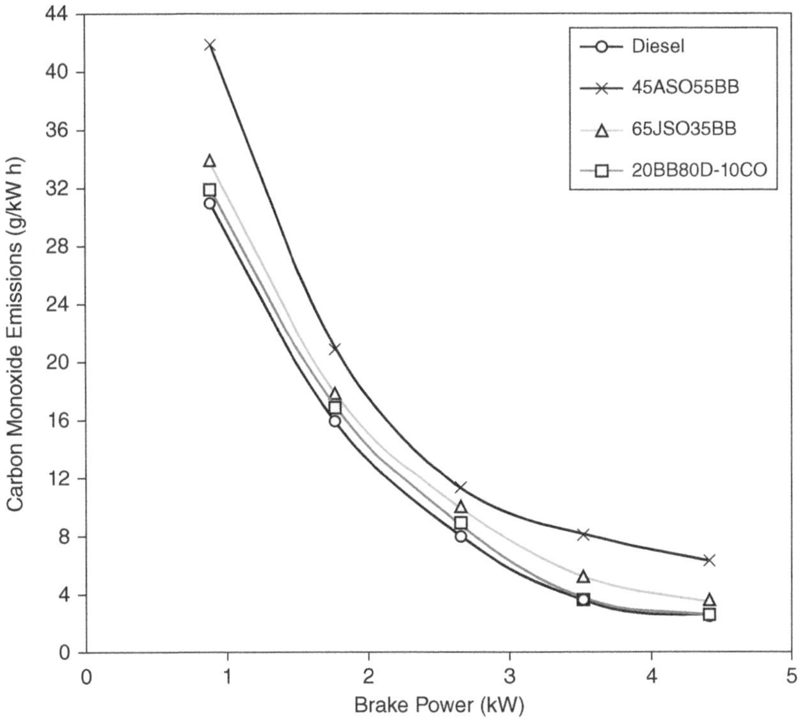

FIGURE 8.11 CO emissions from the fuel blends in various brake powers.

BP, the impact is decreasing and significant reduction is seen at maximum BP. Above 65% of BP the three chosen blends are showing good decreasing trend of emission of HC. Similar trends are being reported by previous researchers when fueled with butanol blended-fueled engine. Carbon monoxide emissions are also an incomplete combustion product. These emissions are unstable and search for oxygen to get converted into carbon dioxide by oxidation. Out of three blends two blends–65JSO35BB and 20BB80D-10CO are producing comparatively closer CO in all BP values and from Figure 8.11 the closer curve trends are seen. Closer CO is due to better combustion and comparatively better properties of the two fuel blends comparing 45ASO55BB. Also, less CO emissions by 45ASO55BB during the operation of the power pack in high BP is that the increase of temperature increases the rate of reaction of the fuel, conversion to thermal and then to mechanical work exhaling low CO. The research output of the previous studies also reported similar trends by fueling butanol as one of the blends along with diesel and biodiesel.

Above 75% of the BP, the blend 20BB80D-10CO was found to be releasing similar CO emissions in all operated BP. In general the power packs powered by diesel are being operated near to 75% or higher BPs for the operation.

8.4 CONCLUSION

Research study of suing waste seed oils and waste vegetables and fruits are being utilized by converting them into useful fuels, fueling in power pack and the results are being presented. The following points are the essence of the study:

- The chosen representative three blends are found possessing competent properties comparing those of other proportions.
- Decrease of the percentages in power pack efficiency and increase in fuel used for operation are 9.1% and 20.1% – waste ASO blend; 6.3% and 8.1% – waste Jamun seed oil blend; and 2.2% and 9.8% – waste CO blend, respectively, at maximum BPs.
- Peak pressures of the power pack are less by 10.1%, 2.1%, and 0.4% using waste ASO blend, waste Jamun seed oil blend and waste castor seed oil blend, respectively.
- There is slightly low NOx in using these fuel blends and a maximum decrease of 9.4% by waste ASO blend.
- Significantly high smoke of 28.7%, 13.2%, and 6.2% higher by waste ASO blend and other fuels, respectively.
- Finally, in this study 58% approximately is being saved by using the waste products, in turn decreasing the garbage disposal cost and saving the land being polluted. Also, this work has not included the variation in parameters and can be progressed in further study for the improvement of the performance of the power pack.

ACKNOWLEDGEMENT

The author is sincerely thanking the management and staff of Hindustan Institute of Technology and Science, Chennai, Tamil Nadu, India for the support in providing the facility to perform this experiment. Also, the sponsor for using the lab for testing the properties of the fuel blends.

ABBREVIATIONS

Amla seed oil	**ASO**
Biobutanol	**BB**
Carbon monoxide	**CO**
Castor oil	**CO**
Diesel	**D**
Jamun seed oil	**JSO**
Hydrocarbons	**HC**
kW	**kilowatt**

REFERENCES

[1] Sayyed, S., Kulkarni, K., Das, R. K., Alam, T., and Eldin, S. M. "Direct injection diesel engine characteristics fuelled with diesel, biodiesel and 1-butanol blends." *Case Studies in Thermal Engineering* 53 (2024): 103826.

[2] Hassan, T., Rahman, M. M., Adib, A. R., Meraz, R. M., Rahman, M. A. and Tushar, M. S. "Effect of Ni and Al nanoadditives on the performance and emission characteristics of a diesel engine fueled with diesel-castor oil biodiesel-n-butanol blends." *Case Studies in Chemical and Environmental Engineering* 8 (2023): 100531.
[3] Han, J., and Somers, L. M. T. "Comparative investigation of ignition behavior of butanol isomers using constant volume combustion chamber under diesel-engine like conditions." *Fuel* 304 (2021): 121347.
[4] Tipanluisa, L., Thakkar, K., Fonseca, N., & López, J. M. "Investigation of diesel/n-butanol blends as drop-in fuel for heavy-duty diesel engines: Combustion, performance, and emissions." *Energy Conversion and Management* 255 (2022): 115334.
[5] Darwish, M., Hidegh, G., Csemány, D., & Józsa, V. "Distributed combustion of diesel–butanol fuel blends in a mixture temperature-controlled burner." *Fuel* 307 (2022): 121840.
[6] Yilmaz, Nadir, and Stephen M. Davis. "Diesel blends with high concentrations of biodiesel and n-butanol: Effects on regulated pollutants and polycyclic aromatic hydrocarbons." *Process Safety and Environmental Protection* 166 (2022): 430–439.
[7] Xiao, Helin, Fengyun Guo, Ru Wang, Xiaolong Yang, Shengjun Li, and Jie Ruan. "Combustion performance and emission characteristics of diesel engine fueled with iso-butanol/biodiesel blends." *Fuel* 268 (2020): 117387.
[8] Yilmaz, Nadir, Francisco M. Vigil, Kyle Benalil, Stephen M. Davis, and Antonio Calva. "Effect of biodiesel–butanol fuel blends on emissions and performance characteristics of a diesel engine." *Fuel* 135 (2014): 46–50.
[9] Goga, Geetesh, Bhupendra Singh Chauhan, Sunil Kumar Mahla, and Haeng Muk Cho. "Performance and emission characteristics of diesel engine fueled with rice bran biodiesel and n-butanol." *Energy Reports* 5 (2019): 78–83.
[10] Killol, A., Reddy, N., Paruvada, S., and Murugan, S. "Experimental studies of a diesel engine run on biodiesel n-butanol blends." *Renewable Energy* 135 (2019): 687–700.
[11] Xiao, Helin, Fengyun Guo, Shengjun Li, Ru Wang, and Xiaolong Yang. "Combustion performance and emission characteristics of a diesel engine burning biodiesel blended with n-butanol." *Fuel* 258 (2019): 115887.
[12] Xu, Min, Peng Jiang, Wenjun Zhong, Feibin Yan, Xu Liu, and Qian Wang. "Experimental investigation combined with steady-state and transient-state tests on soot characteristics of hydrogenated catalytic biodiesel/n-butanol blends." *Energy* 282 (2023): 128929.
[13] Razak, Nurul Hanim, Haslenda Hashim, Nor Alafiza Yunus, and Jiří Jaromír Klemeš. "Integrated linear programming and analytical hierarchy process method for diesel/biodiesel/butanol in reducing diesel emissions." *Journal of Cleaner Production* 337 (2022): 130297.
[14] Singh, Rickwinder, Sandeep Singh, and Mukesh Kumar. "Impact of n-butanol as an additive with eucalyptus biodiesel-diesel blends on the performance and emission parameters of the diesel engine." *Fuel* 277 (2020): 118178.
[15] Liang, Marissa S. "Assessing emission and power tradeoffs of biodiesel and n-butanol in diesel blends for fuel sustainability." *Fuel* 283 (2021): 118861.
[16] Zhao, Wenbin, Zilong Li, Guan Huang, Yaoyuan Zhang, Yong Qian, and Xingcai Lu. "Experimental investigation of direct injection dual fuel of n-butanol and biodiesel on intelligent charge compression ignition (ICCI) combustion mode." *Applied Energy* 266 (2020): 114884.
[17] Zhao, Wenbin, Yaoyuan Zhang, Guan Huang, Zhuoyao He, Yong Qian, and Xingcai Lu. "Experimental study of butanol/biodiesel dual-fuel combustion in intelligent charge compression ignition (ICCI) mode: A systematic analysis at low load." *Fuel* 287 (2021): 119523.
[18] Sarıkoç, Selçuk, İlker Örs, and Sebahattin Ünalan. "An experimental study on energy-exergy analysis and sustainability index in a diesel engine with direct injection diesel-biodiesel-butanol fuel blends." *Fuel* 268 (2020): 117321.

[19] Mathur, Akleshwar. "A comparative physicochemical analysis of crude and refined jamun (*Syzygium cumini*) seed oil." *Iconic Research and Engineering Journals* 1 (2017): 1–5.

[20] Mo, Jun, Chenglong Tang, Junge Li, Li Guan, and Zuohua Huang. "Experimental investigation on the effect of n-butanol blending on spray characteristics of soybean biodiesel in a common-rail fuel injection system." *Fuel* 182 (2016): 391–401.

[21] Liu, Haifeng, Xin Wang, Zunqing Zheng, Jingbo Gu, Hu Wang, and Mingfa Yao. "Experimental and simulation investigation of the combustion characteristics and emissions using n-butanol/biodiesel dual-fuel injection on a diesel engine." *Energy* 74 (2014): 741–752.

[22] Singh, Anshu, Arindam Kuila, Sunita Adak, Moumita Bishai, and Rintu Banerjee. "Utilization of vegetable wastes for bioenergy generation." *Agricultural Research* 1 (2012): 213–222.

[23] Singh, Shubham, Er K. Lakshmi Bala, Shanker Suwan Singh, and Er Ankita Kumari. "Effect of processing conditions on solvent extraction of Amla (Emblica officinalis) seed oil." PhD diss., Doctoral dissertation, Department of Food Process Engineering Vaugh Institute of Agricultural Engineering & Technology Sam Higginbottom University of Agriculture, Technology and Sciences Prayagraj (Allahabad) (UP), India–211007, 2019.

[24] Thakkar, Kartikkumar, Surendra Singh Kachhwaha, Pravin Kodgire, and Seshasai Srinivasan. "Combustion investigation of ternary blend mixture of biodiesel/n-butanol/diesel: CI engine performance and emission control." *Renewable and Sustainable Energy Reviews* 137 (2021): 110468.

[25] Jeevahan, J., Sriramanjaneyulu, G., Durairaj, R. B., and Mageshwaran, G. "Experimental investigation of the suitability of 1-butanol blended with biodiesel as an alternative biofuel in diesel engines." *Biocatalysis and Agricultural Biotechnology* 15 (2018): 72–77.

[26] El-Seesy, Ahmed I., Hamdy Hassan, and Hidenori Kosaka. "Improving the performance of a diesel engine operated with jojoba biodiesel-diesel-n-butanol ternary blends." *Energy Procedia* 156 (2019): 33–37.

[27] Nanthagopal, K., Ashok, B., Saravanan, B., Patel, D., Sudarshan, B., and Ramasamy, R.A. "An assessment on the effects of 1-pentanol and 1-butanol as additives with *Calophyllum inophyllum* biodiesel." *Energy Conversion and Management* 158 (2018): 70–80.

9 Enhancing LTC Engine Performance and Reducing Emissions with Alcohol Fuels

Giovani Dambros Telli, Josimar Souza Rosa, and Luiz Alberto Oliveira Rocha

9.1 INTRODUCTION

The reduction of greenhouse gas emissions (GHGs) is one of the great priorities of our century because of its contribution to global warming and climate change. Fossils contribute significantly to GHG emissions as a result of the higher amount of carbon in their chemical composition. The transportation sector is one of the greatest fossil fuels since it is powered by internal combustion engines, making it a major contributor to the carbon dioxide emissions emitted into the atmosphere, approximately 23% of the total carbon dioxide emitted in 2022 (IEA, 2023). These issues, coupled with stringent emission legislation, are driving various studies on internal combustion engines to transform the transportation sector into a more environmentally friendly system, using low-carbon technologies and novel combustion concepts.

Diesel engines are developed with expensive technologies to enhance the premixed charge, such as high injection pressures, dual-fuel injection, multiple injection, and a high swirl ratio, promoting lower in-cylinder temperatures and control of combustion emissions (Harari, 2018). Modern diesel engines also need an expensive aftertreatment system to reduce pollutant emissions. Recently, advanced combustion technologies have been proposed to overcome these limitations and improve the relation between emissions and efficiency in conventional combustion modes.

Among advanced combustion concepts, the low-temperature combustion (LTC) concept is considered one of them. Generally, the LTC mode operates unthrottled, globally lean with a highly diluted charge (high levels of exhaust gas recirculation [EGR] or lean mixtures) and near-homogeneous, maintaining the maximum in-cylinder temperature below ~1800K. As a result, it can achieve high engine efficiency with very low NO_X, particulate matter (PM), and soot emissions (Gainey et al., 2021; Krishnamoorthi et al., 2019; Saxena & Bedoya, 2013; Srivastava et al., 2018). Homogeneous charge compression ignition (HCCI) (Kale & Krishnasamy, 2024; Morales Rojas et al., 2024; Spanò et al., 2024), premixed charge compression ignition (PCCI) (Wu et al., 2024; Xu et al., 2024), and reactivity-controlled compression ignition (RCCI) (Dempsey et al., 2012; Hossein Fakhari et al., 2024; Reitz &

DOI: 10.1201/9781003482277-9

Duraisamy, 2015; Rosa et al., 2020; Yadav et al., 2024) engines are defined as LTC strategies and have been investigated in several studies. Since LTC strategies do not have a direct mechanism to control combustion, such as spark ignition (SI) or conventional diesel combustion (CDC) engines, they are usually coupled with some combustion control techniques, especially to control the combustion phasing and ensure that the start of combustion occurs near the top dead center of the engine; some of them are injection timing, EGR, and variable valve actuation (VVA) (Bharath et al., 2015; Panda & Ramesh, 2024; Punasiya & Sarangi, 2024). Another possibility that many studies have suggested is fuel modification, combination, or blending, which can be an easier way to control LTC (Gharehghani, 2019). In general, LTC engines can run on flexible fuels. However, they prefer higher volubility and lower reactive fuels or fuel blends than diesel or biodiesel (Taghavi et al., 2019).

Alcohol fuels are commonly adopted as an interesting option for SI and compression ignition (CI) engines. They can be obtained from many fossil or renewable sources, such as natural gas, coal, and biomass (sugarcane and corn as examples of ethanol sources) (Telli et al., 2018b, 2020). Alcohols show the advantage of being oxygenated fuels with a hydroxyl group (OH), which improves the amount of oxygen throughout the combustion process, resulting in cleaner combustion with lower soot emissions levels as a result of a later combustion phase and lower combustion temperature. Furthermore, the lower in-cylinder temperature is also a consequence of the cooling effect due to its high latent heat of vaporization, causing a decrease in the in-cylinder temperature and more delayed combustion (Chen et al., 2018; Nour et al., 2019). Alcohols such as ethanol can be used directly in conventional SI engines in single-fuel mode or even mixed with gasoline due to their high octane number (Gong et al., 2019, 2020). However, the lower cetane number, lower viscosity, and high resistance of alcohols present certain complications when used directly in diesel engines without ignition assistance (Çelebi & Aydın, 2019; Rajesh Kumar & Saravanan, 2016; Rosa et al., 2021). Thus, alcohol fuels are usually used in PCCI or RCCI combustion modes, with diesel fuel being directly injected or applied to HCCI engines directly. Table 9.1 presents the fuel properties of diesel and alcohol fuels.

In this chapter, the performance and emissions characteristics of LTC strategies (HCCI, PCCI, and RCCI) fueled by alcohols were discussed in detail and compared. This chapter focused on three main alcohol types, which are ethanol, methanol, and butanol, used in LTC combustion modes. The influence of each of the alcohols on performance and emission characteristics was investigated separately for each LTC combustion mode.

9.2 LOW-TEMPERATURE COMBUSTION TECHNIQUES

LTC has gained attention from researchers, and new concepts have been proposed in the last few decades to meet the current and future emission requirements. LTC concepts demonstrated excellent potential to reduce NO_X and soot emissions simultaneously (Agarwal et al., 2022; No, 2016). As mentioned above, all LTC modes achieve relatively low combustion temperatures, which means that the NO_X activation energy is not reached and reduces the production of NO_X by the thermal mechanism. Furthermore, soot emissions are reduced due to longer ignition time and longer

TABLE 9.1
Fuel Properties (Munuswamy et al., 2022; Rajesh Kumar & Saravanan, 2016)

Properties	Diesel	Methanol	Ethanol	Butanol
Chemical formula	C_XH_Y	CH_3OH	C_2H_5OH	$C_4H_{10}OH$
Molecular weight, kg/kmol	190–211.7	32.04	46.7	74.12
Carbon, mass %	86.13	37.5	52.2	64.91
Hydrogen, mass %	13.87	12.48	34.7	13.49
Oyxgen, mass %	0	49.93	34.7	21.6
Density, kg/m^3 at 15°C	835	791.3	789.4	803.7
Boiling temperature, °C	180–360	78	78.25	117.25
Research octane no.	20–30	108.7	108.6–110	98
Cetane number	52	2	11	17
Lower heating value, MJ/kg	42.5	20.1	26.9	33.2
Latent heat, kJ/kg	270–375	1162.6	918.4	581.4
Viscosity, mm^2/s at 40°C	2.72	0.58	1.2–1.5	2.22
Flashpoint, °C	>55	11	11–20	–
Autoignition temperature, °C	254–300	423	425	385

mixing time, decreasing the rich areas, and avoiding soot formation. Additionally, engine efficiency is improved as thermal losses due to heat transfer are reduced, and since they operate unthrottled, pumping losses are also reduced (Kakoee et al., 2018). Figure 9.1 presents the main characteristics of the three low-temperature strategies.

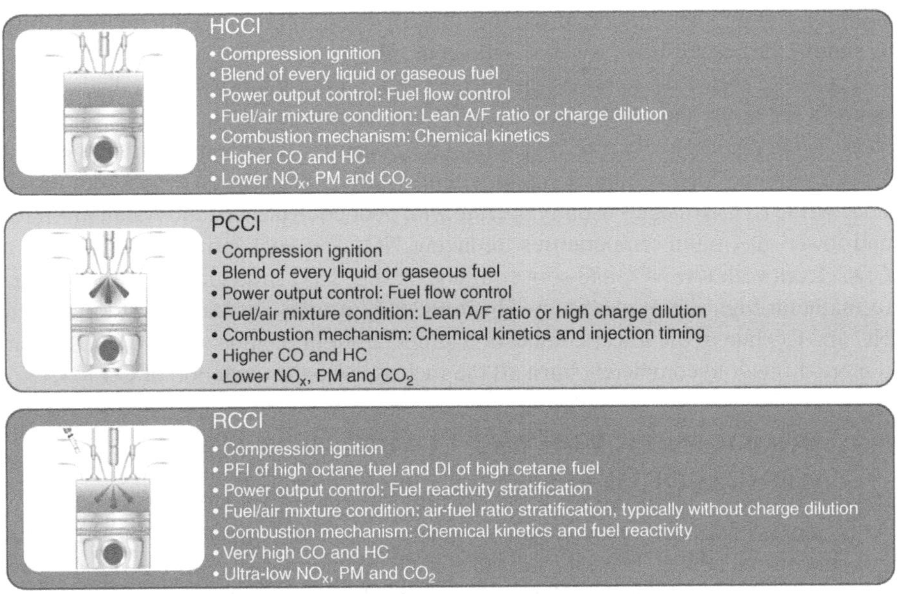

FIGURE 9.1 Main characteristics of the three low-temperature strategies. (Adapted from Krishnasamy et al. (2021), Paykani et al. (2016), and Rameez and Mohamed Ibrahim (2024).)

HCCI combustion, also known as Controlled Auto-Ignition (CAI) combustion, was the first LTC strategy. Typically, HCCI uses highly diluted charge or high residual gas fractions to achieve a diluted near-homogeneous fuel-air mixture (Telli et al., 2021). In this LTC mode, the fuel injection can be injected as port fuel injection (PFI), late or early direct injection (DI), and it helps the charge homogeneity (Saxena & Bedoya, 2013). During the compression stroke, temperature stratification controls the ignition process. When the hot spots reach autoignition conditions, they release heat and lead to the ignition of the entire cylinder charge at multiple locations – sequential autoignition (Gainey & Lawler, 2021). It results in very fast LTC with high thermal efficiency and low levels of NO_X and PM emissions (Agarwal et al., 2017). However, since HCCI combustion is short and there is no direct mechanism to control combustion (completely governed by chemical kinetics), it leads to a limited engine operational load map and a lack of combustion control. In the literature, many techniques have been developed to overcome these challenges, such as the use of residual gas trapping, EGR, or water injection (Akkoli et al., 2022; Duan et al., 2021; Shere & Subramanian, 2024; Telli, Rocha, et al., 2022; Telli, Zulian, et al., 2022; Zulian et al., 2022). Furthermore, other LTC concepts have been proposed to address these problems, such as PCCI and RCCI. The PCCI introduces a reactivity gradient within the cylinder for the combustion process. The PCCI can be achieved with a single fuel and its control the burning rate using the stratification of the equivalence ratio. In this case, the fuel should have a high sensitivity to equivalence ratio and should exhibit cool flame reactions (Sjöberg & Dec, 2006). On the other hand, RCCI controls the combustion process using two different fuels. One low-reactivity fuel (LRF) is port fuel injected into the engine intake manifold and premixed with air, while the second fuel is a high-reactivity fuel (HRF), that is (DI during the compression stroke. The combustion starts in the local with the higher amount of the HRF, following by the autoignition of the other areas. Thus, the injection timing and duration of the fuel directly injected provide a direct mechanism to control the combustion process (Gohn et al., 2020).

Figure 9.2 shows the relationship between NO_X and soot formation for distinct local equivalence ratios and combustion temperatures for different LTC concepts and CDC. What is interesting to observe is that the CDC operates in regions with high equivalence ratios and high combustion temperatures, which promote NO_X and soot formation. While LTC strategies usually operate with poor equivalence ratios (lean mixtures) and lower maximum temperatures, reducing NO_X and soot formation compared to CDC. Even with low NO_X and soot emissions levels, LTC strategies have the potential to maintain high diesel efficiency. However, they usually increase the formation of HC and CO due to the lower combustion temperature in relation to the CDC, making it more difficult to completely burn all the fuel and also the oxidation of CO to CO_2.

9.3 HOMOGENEOUS CHARGE COMPRESSION IGNITION WITH ALCOHOL FUELS

As mentioned above, HCCI combustion does not have any direct mechanism to control HCCI combustion. Thus, it limits the HCCI operational map, where at high loads it is restricted by high levels of heat release rates because of the fast combustion and at low loads it is limited by misfire. In general, studies in the literature adopted some

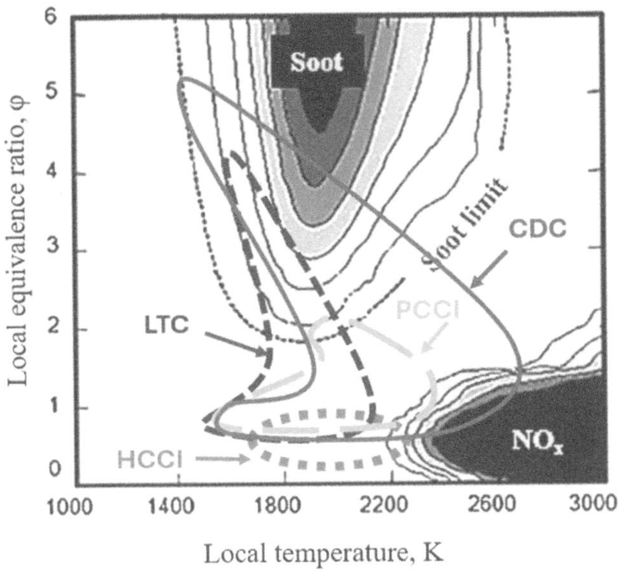

FIGURE 9.2 Equivalence ratio versus local temperature. (Adapted from Kamimoto and Bae (1988), Krishnasamy et al. (2021), and Telli et al. (2018a).)

strategies to control HCCI combustion and explore the operational map. The operational map of the HCCI engine is usually described in Figure 9.3. The blue area represents the operating conditions (indicated mean effective pressure [IMEP] and CA50) under which the HCCI engine can operate with good combustion stability. The ringing intensity serves as an indicator to limit the high-load operation. RI < 6 MW/m^2 while COVimep is usually used as a low load criterion (COVimep < 5%). In particular, the operational map in Figure 9.3 was achieved on an ethanol HCCI engine using the water injection strategy to control the combustion phasing (Telli et al., 2022).

Gharehghani (2019) studied an HCCI engine fueled with ethanol, metanol, and natural gas, varying the air-fuel ratio and also the intake temperature. The author identified the operational map for each fuel in relation to lambda and charge intake temperature. The highest indicated mean effective pressure (IMEP) found was 6.68 bar for the natural gas case at an intake temperature between 410 and 430K and a lambda of around 1.7. For the ethanol and methanol cases, the maximum IMEP was 5.4 and 5.9 bar for lower intake temperature values and lambda between 1.5 and 2.0. It was reported that the combustion tended to start earlier as the mixture became richer and the intake temperature and pressure increased. The author explains that as the intake pressure and temperature increase, the in-cylinder temperature rises, which accelerates the charge autoignition. The highest gross indicated thermal efficiency (ITE) was around 51.0% for the ethanol case at an intake temperature of 450K and lambda < 3.3. For a higher intake temperature, the combustion was advanced, increasing the negative work and reducing the thermal efficiency. Similar trends were noticed for methanol, which achieved a thermal efficiency close to 51.0%. The

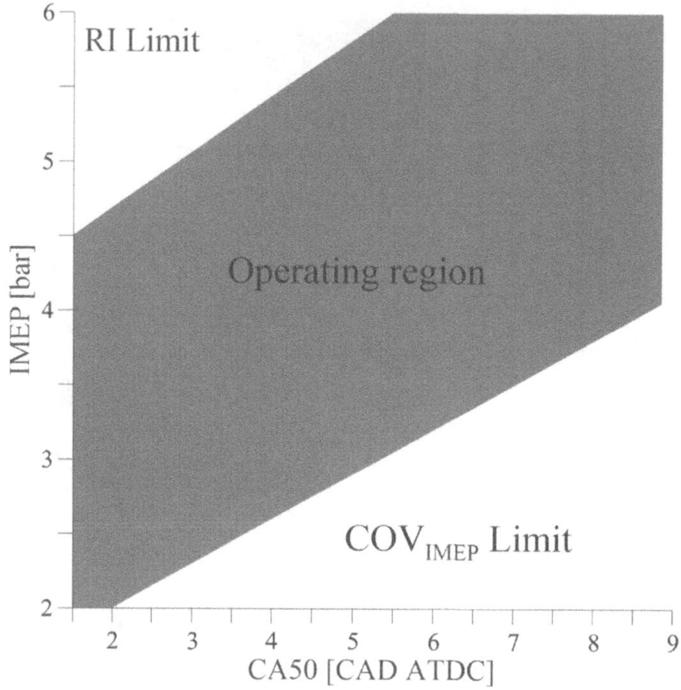

FIGURE 9.3 HCCI operational map. (Reprinted from Telli, Zulian, et al. (2022) with permission from Elsevier.)

octane number is lower for methanol than ethanol; thus, methanol needs lower intake temperatures for complete combustion. In another paper by Gharehghani et al. (2021), the same trends in results were reported. Figure 9.4 shows the gross thermal efficiency of ethanol (a) and methanol (b) in relation to lambda and intake temperature (Tivc). Data were extracted from (Gharehghani, 2019).

Regarding emissions, Gharehghani (2019) reported that for the ethanol case, NO_X emissions were between 0.06 and 0.26 g/kWh and for methanol, the values ranged from 0.05 to 0.45 g/kWh. CO and HC decreased with the increase in intake temperature due to the higher combustion temperature and increased with increasing lambda due to lower combustion temperatures, resulting in values for ethanol between 0.4 and 4.9 g/kWh and 0.8 and 5.8 g/kWh for HC and CO, respectively. For the methanol case, CO emissions ranged from 0.7 to 4.0 g/kWh and HC emissions from 0.35 to 4.2 g/kWh. On the other hand, the CO emissions for the natural gas ranged from 1.5 to 6.5 g/kWh, HC emissions from 0.8 to 4.2 5 g/kWh and NO_X emissions from 0.02 to 0.12 g/kWh.

Maurya and Agarwal (2014) studied an HCCI engine operating with gasoline, ethanol, and methanol. The charge autoignition conditions were achieved using preheated intake air. The authors explored and plotted the HCCI engine operational map for several intake temperature values and lambda values. The gasoline autoignition conditions demand a higher intake temperature, while ethanol and methanol demand lower intake

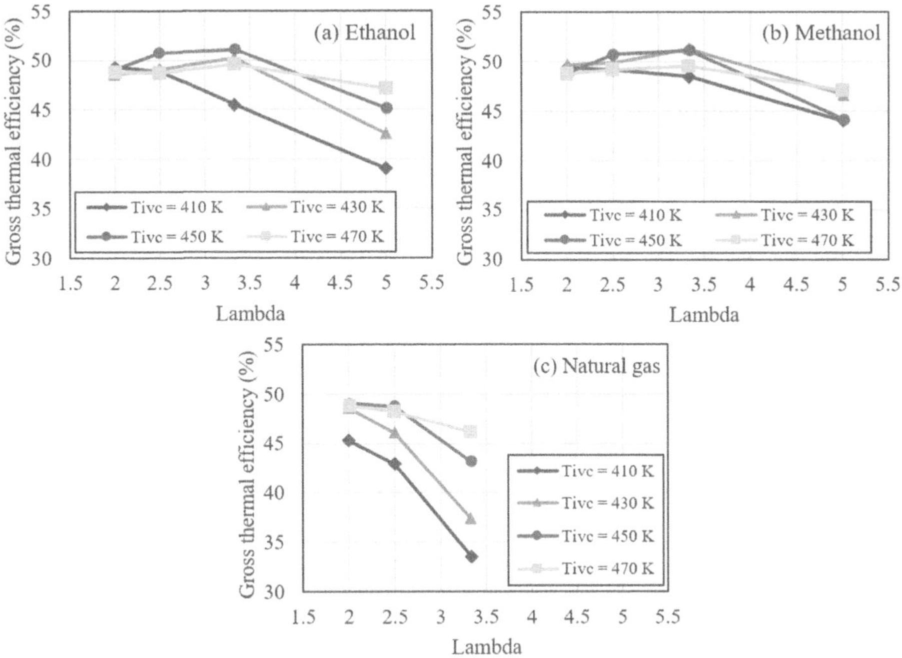

FIGURE 9.4 Gross thermal efficiency of HCCI combustion with ethanol (a) and methanol (b) and natural gas (c) in relation to lambda and intake temperature (Tivc). (Data extracted from Gharehghani (2019).)

temperatures. Ethanol and methanol reached higher IMEP values for the combustion phasing, which corresponded to the highest thermal efficiencies found. The highest thermal efficiencies found followed the order of about 48.8% for ethanol, about 48% for gasoline, and about 45% for methanol. With respect to emissions, NO_X emissions were higher for points close to the high load limit (richer air-fuel mixtures), where the combustion temperature is higher than at low loads and lean mixtures. The maximum NO_X emissions found were 0.26 g/kWh for the gasoline case, while the NO_X emissions for ethanol varied between 0.024 and 0.24 g/kWh. CO emissions ranged from 13.3 to 54.4 g/kWh for the methanol case, from 7.4 to 52.9 g/kWh for the ethanol case, and the lowest values were for the gasoline case – 10.0–35.3 g/kWh. The same trend was found for HC emissions, which increase with increasing lambda values and intake air temperature, especially at low loads. This can be explained because of the low heat release rate and low combustion temperatures, which lead to a lower fuel oxidation rate and incomplete combustion, resulting in higher HC emissions. The highest HC emissions were 18.4 g/kWh for gasoline and ranged from 7.1 to 18.2 g/kWh for ethanol and from 7.0 to 11.0 g/kWh for methanol. In the work by Bendu and Sivalingam (2016) and Bendu et al. (2016), it was reported that there was a decrease in HC, CO and an increase in smoke opacity emissions with an increase in intake temperature. In addition, CO, HC, and smoke opacity emissions reduced, whereas NO_X emissions increased with the increase in engine load.

Mack et al. (2016) investigated an HCCI engine operating with ethanol, gasoline, n-butanol, and isobutanol. The authors explored how the intake temperature influences the combustion timing (CA50). The higher the intake temperature, the lower the CA50, i.e., when the CA50 increases, the IMEP tends to increase with an equivalence ratio between 0.40 and 0.45 for all the fuels tested. The authors also evaluated the increase in intake pressure from 1.0 bar to 1.8 bar, that led to an increase in gross IMEP. The highest gross IMEP found was 7.2 bar for n-butanol at 1.8 bar intake pressure and CA50 = 8 CAD ATDC. Ethanol showed the lowest IMEP values because of its lower calorific value in relation to the other fuels. An increase in NO_X emissions with lower CA50 values, higher equivalence ratios, and intake pressures. The authors quote that the NO_X emission results can be attributed to the peak of in-cylinder temperature since the formation of NO_X depends on temperature. According to the authors, CO and HC emissions increased with delayed combustion phasing and decreased with increased intake pressure. Figure 9.5 presents the CO, NO_X, and HC emissions for the combustion of HCCI using gasoline, ethanol, isobutanol, and n-butanol with an intake pressure of 1.0 bar and phi = 0.40. The data was extracted from Mack et al. (2016). These results are in agreement with those reported by Viggiano and Magi (2012), who explained that CO and HC emissions decreased with the increase in intake pressure due to an improvement in fuel oxidation with pressure. However, the total amount of NO_X increases with increasing intake pressure, although it remains at very low levels compared to conventional combustion strategies. Viggiano and Magi (2012) also reported an increase in fuel consumption with the increase in the intake pressure.

In the work of He et al. (2015), the effect of distinct alcohol fuels on HCCI combustion under stoichiometric conditions was studied. The authors evaluated the n-butanol and ethanol. They also evaluated their blends with gasoline (30% vol. of n-butanol/ethanol in gasoline). The authors investigated HCCI combustion by changing the valve timing to alter the residual gas fraction inside the cylinder during HCCI combustion in order to control the autoignition timing. The results indicated that alcohol-gasoline blends with residual gases higher than 38% by mass tend to autoignite easier than gasoline. Furthermore, the n-butanol-gasoline mixtures showed an advanced autoignition timing compared to the ethanol-gasoline mixtures at 1500 rpm.

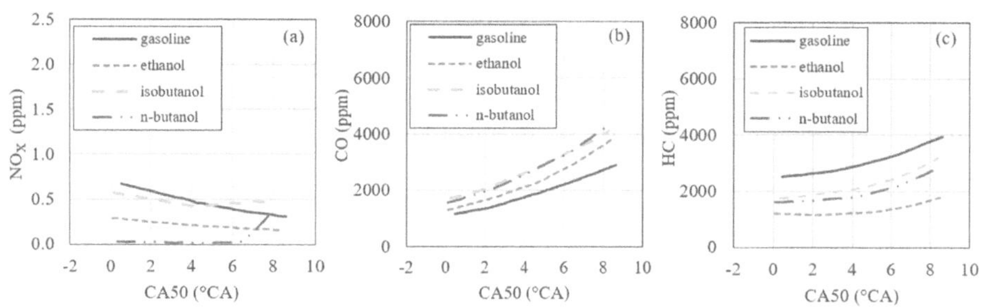

FIGURE 9.5 NO_x (a), CO (b), and HC (c) emissions as a function of CA50 for gasoline, ethanol, n-butanol, and isobutanol at intake pressure = 1.0 bar and phi = 0.40. (Data extracted from Mack et al. (2016).)

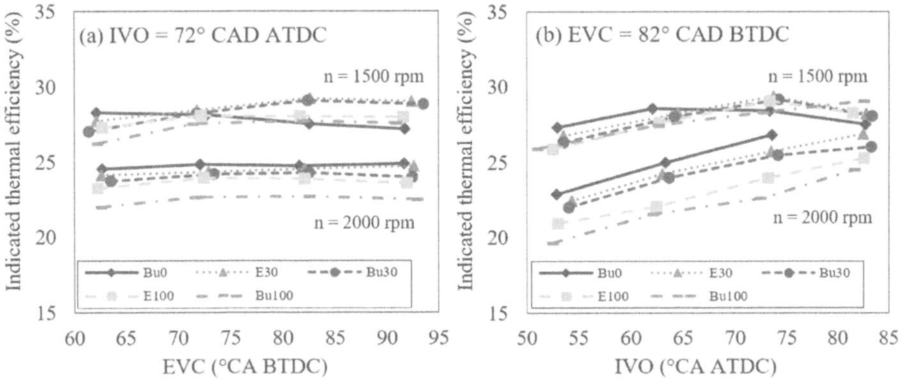

FIGURE 9.6 Influence of (a) EVC and (b) IVO timing on the thermal efficiency of etanol, butanol, gasoline, and its blends – Lambda = 1.0. (Data extracted from He et al. (2015).)

The autoignition timings for both mixtures were close for the same operating conditions at 2000 rpm. The authors also evaluated the combustion stability, which was enhanced with advanced exhaust valve closing (EVC) timing at fixed intake valve opening (IVO) timing. Figure 9.6 illustrates the ITE for EVC (a) and IVO (b), which depend on fuel types, engine speeds, and autoignition timing. It is possible to note that the thermal efficiency slightly increases with the increase in EVC timing, except for the Bu0 at 1500 rpm. In general, according to He et al. (2015), the thermal efficiency tends to improve in the order of the different fuels Bu0 > E30 > Bu30 > E100 > Bu100, except for Bu0 at 1500 rpm. The authors justified it to the autoignition timing, which gradually advances and changes the negative compression work and losses due to the heat transfer.

In the work of Bissoli et al. (2018), they explored the effect of EGR on the performance and emissions parameters of an engine running with ethanol and PRF100. The results showed that increasing the percentage of EGR while maintaining a fixed lambda value reduced the IMEP values. The maximum value obtained was around 2.5 bar for a lambda close to 2.5. At 0% of the EGR and λ around 3.5, the ITE was around 37% and decreased to around 30% with 60% EGR. It was justified due to the deterioration of combustion efficiency when EGR was employed. In addition, NO_X emissions were reduced with higher EGR values, ranging from 0.1 to 0.01 g/kWh for ethanol. However, CO emissions increased with a higher percentage of EGR, ranging from 10 to 300 g/kWh. The authors explain that for a fixed lambda value, a higher EGR rate reduces the combustion temperature (due to dilution), resulting in lower NO_X emissions and higher CO emissions.

As previously presented, the use of ethanol in HCCI combustion was thoroughly investigated. The reason is that because ethanol has good availability, it is easy to store and handle. Nevertheless, the production process of anhydrous ethanol requires a significant amount of energy due to the distillation and dehydration steps (Gainey et al., 2020). Specifically, the energy required for ethanol-in-water mixes above 80% is exponential (Lanzanova et al., 2016; Martinez-Frias et al., 2007). Then, it affects the ethanol energy balance and compromises the final price of the market. In this regard, the utilization of ethanol with more than 5% vol. of water (called wet ethanol) is investigated in the literature. Direct use of wet ethanol could decrease the energy consumed during the distillation process from 23% to 3% and prevent the dehydration process, representing an energy gain of 34% in relation to anhydrous ethanol (Martinez-Frias et al., 2007). In this regard, several studies in the literature have explored the direct use of wet ethanol in CI engines (Rosa, Altafini, et al., 2019; Rosa, Lorenzini, et al., 2019). The authors in Saxena et al. (2012, 2014) evaluated an HCCI engine using ethanol with up to 30% water per volume, with equivalence ratios of 0.5–0.55 and intake pressures of 1.8 and 2 bar. The authors used a heat exchanger system as a heat source to achieve the wet ethanol autoignition conditions. The increase in equivalence ratio and the intake pressure was reported to result in higher gross IMEP values. In addition, as the concentration of water in ethanol increases, IMEP tends to decrease. The maximum IMEP values achieved were 8.15 bar for pure ethanol, 7.92 bar for 90% ethanol mixture, 7.67 bar for 80% ethanol mixture, and 7.26 bar for 70% ethanol mixture. The authors reported that these maximum values were reached with delayed combustion phases.

According to Gohn et al. (2020), the combustion duration increases and the thermal and combustion efficiencies tend to decrease with wet ethanol blends with more than 20% water by mass. Furthermore, blends containing more than 20% water required a significantly higher intake temperature, which was not viable for commercial usage. However, wet ethanol with 20% water exhibited acceptable combustion and thermal efficiencies comparable to that of the mixture with 7.7% water (by mass).

It was also reported that as the concentration of water in ethanol increased, the CO and HC emissions increased. According to the authors, the presence of water in ethanol acts as a heat sink as a result of its high specific and latent heat. In this way, the presence of water in ethanol results in lower temperatures inside the cylinder, which leads to less oxidation of the fuel and carbon monoxide, thus increasing the levels of HC and CO emissions. Similar results were found by Telli et al. (2023) as presented in Figure 9.7 – indicated efficiency (a), peak in-cylinder temperature (b), CO emissions (c), and organic gas emissions (d) in relation to the combustion phasing for ethanol-water mixtures at 4 bar IMEP. In this study, the ethanol-water mixtures have 5, 20, 30, and 40% water by vol. in ethanol. The water injection strategy was employed to control the combustion phasing, and the autoignition conditions were reached by recirculating all exhaust gases from another conventional combustion diesel cylinder (Telli et al., 2023).

9.4 PREMIXED CHARGE COMPRESSION IGNITION WITH ALCOHOL FUELS

PCCI combustion uses a homogeneous mixture for the premixed combustion, as in HCCI combustion. The difference is when the fuel is injected into the cylinder. While HCCI combustion is carried out during admission time (early injection) in PCCI combustion, a part of the fuel is introduced into the combustion chamber during the compression stroke, from 30 to 160 CAD BTDC, while another part can be supplied with

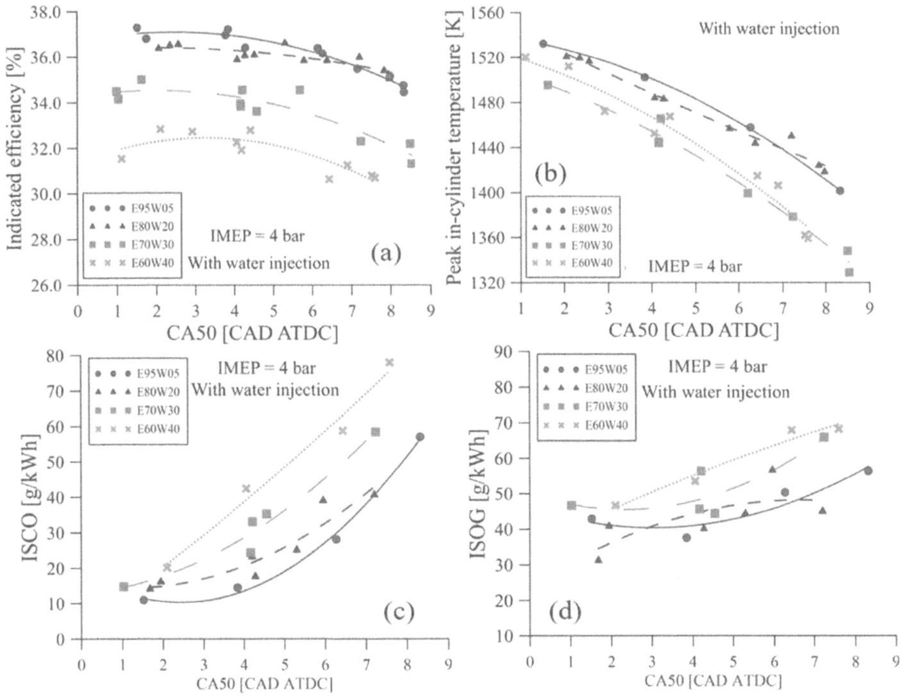

FIGURE 9.7 Indicated efficiency (a), peak in-cylinder temperature (b), CO (c) and organic gases (d) emissions for different ethanol-water mixtures in relation to the CA50 at 4.0 bar IMEP and with water injection. (Adapted and reprinted from Telli et al. (2023) with permission of Elsevier.)

the intake manifold. Furthermore, PCCI combustion also allows for optimizing the combustion phases through multi-split injection strategies (Shim et al., 2020).

In general, PCCI combustion employs two fuels with distinct reactivity, with the LRF being supplied by a PFI into the engine intake manifold, and the HRF being directly injected. However, according to the literature, there is no impediment to using a single fuel that is delivered through the intake manifold and directly injected into the cylinder or simply injected inside the cylinder. Shim et al. (2020) employed a PCCI-premixed charge combustion strategy with only one HRF. They advanced the start of injection (SOI) to 90 CAD BTDC and split the injection into three short periods. In this case, the fuel injection was carried out during the compression time, allowing time for the formation of a premixed charge (even if not homogeneous) with only a single fuel.

Alcohols are considered LRFs, and they can only be used in a PCCI engine in a dual-fuel strategy. In this regard, the alcohols are usually injected with a PFI in the intake manifold. Nibin et al. (2021) carried out experiments with a single-cylinder air-cooled engine in PCCI mode, injecting bioethanol as LRF and wheat germ oil (WGO) as HRF without any adjustments in the SOI. They evaluated different bioethanol fractions to optimize PCCI operations. The engine was less efficient when operating with WGO compared to operating with diesel fuel (original condition). However, when operating with WGO and bioethanol, the brake thermal efficiency increased, reaching 29.1% of efficiency with 30% of bioethanol energy fraction, which is similar to the efficiency of 29.8% obtained for the baseline condition. At maximum load, with a 30% bioethanol energy fraction, the NO concentration in the exhaust gases was reduced by 7% in relation to the diesel fuel condition. The smoke opacity increases when using pure WGO in relation to the baseline diesel condition but reduces when using bioethanol. Figure 9.8 illustrates the results of brake thermal efficiency (a), CO concentration (b), and HC concentration (c) under different conditions.

It is possible to observe in Figure 9.8, the behavior of the parameters under the maximum brake power condition. Figure 9.8(a) shows that the operation with 100% WGO is less efficient than operating with 100% diesel fuel. On the other hand, Figure 9.8(b) and (c) shows that CO and HC emissions increasing when using pure WGO fuel, which is expected since these gases represent unburned portions of fuel emitted in the combustion products. In addition, it can be noticed that when the share of bioethanol increases, the brake thermal efficiency increases progressively, while HC and CO emissions also progressively reduce.

Singh et al. (2020) carried out experimental tests using a single-cylinder engine at four different loads and operating at a constant speed of 1500 rpm. The authors investigated a diesel fuel engine operating in CDC mode and PCCI combustion

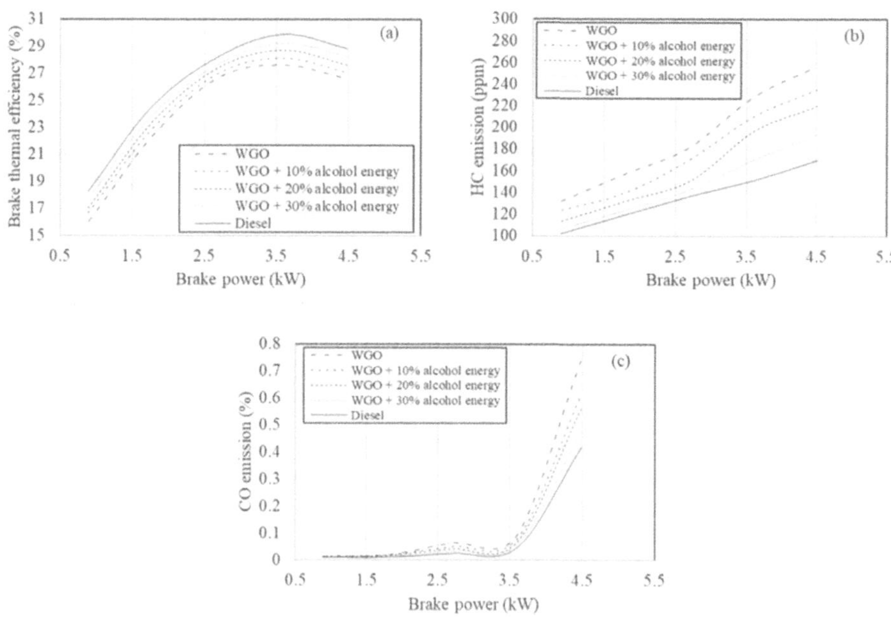

FIGURE 9.8 Brake thermal efficiency (a), HC concentration (b), CO concentration for different test conditions. (Data extracted from Nibin et al. (2021).)

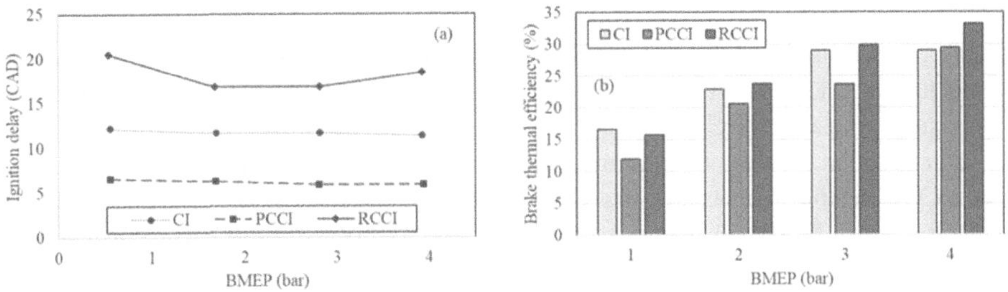

FIGURE 9.9 Ignition delay (a), Brake thermal efficiency (b), of CDC (CI), PCCI, and RCCI engines. (Data extracted from Singh et al. (2020).)

mode. In the PCCI combustion case, the fuel injection was divided into an advanced injection at 35 CAD BTDC, followed by a pilot injection at 12 CAD BTDC, responsible for ignition. Note that the strategy in this work was a single-fuel PCCI engine. The authors also tested an RCCI combustion mode in which methanol was port fuel injected as LRF and diesel fuel was directly injected as HRF. The combustion of PCCI with a single fuel showed a shorter ignition delay compared to the CI and RCCI strategies, as presented in Figure 9.9(a). Reducing the ignition delay is a potential way to enhance the engine's thermal efficiency, but this does not occur in practice if the SOI is not optimized. In fact, Figure 9.9(b) shows that the combustion of PCCI of just one fuel, with a shorter ignition delay, results in lower efficiency compared to the combustion modes of CDC and RCCI. At higher loads, the PCCI combustion case achieved slightly higher efficiency than the CDC. The RCCI combustion mode demonstrated the highest efficiencies in this study, except at BMEP = 1.0 bar.

Finally, PCCI combustion basically consists of direct and early fuel injection with the aim of obtaining a more homogeneous charge. The ignition can be controlled, for example, with a pilot injection to start the combustion process. Thus, it is clear that PCCI combustion allows greater combustion control than HCCI combustion. As PCCI combustion still involves diffusive combustion, the formation of NO_x and PM is greater than in HCCI combustion, but the engine efficiency is improved. There is also a limitation on the engine load. Under high load conditions, strong knock events could occur, which is due to the high temperature inside the cylinder and the existence of a premixed load. Some research suggests that the engines' compression ratio could be reduced to compensate for this effect and possibly improve the operational map of PCCI combustion at higher load conditions.

9.5 REACTIVITY-CONTROLLED COMPRESSION IGNITION WITH ALCOHOL FUELS

Generally, RCCI combustion operates with two fuels that have distinct reactivity levels. Fuel with greater reactivity (HRF) is supplied to the engine by DI into the combustion chamber, and fuel with less reactivity (LRF) is provided to the engine in the air intake manifold using a PFI system (Nibin et al., 2021). The HRF is injected during the compression stroke at an advanced SOI compared to the common SOI of a

CDC engine. It results in more time for the HRF to diffuse into the LRF-air mixture that exists within the cylinder. The different reactivities cause a gradient that can be used to control the start of combustion (Rosa et al., 2020). Alcohols are excellent LRF fuels to use in RCCI combustion engines.

Zheng et al. (2018) investigated the effect of different biofuel properties on an RCCI engine. The LRFs used were ethanol, butanol, and 2,5-dimethylfuran (DMF), while biodiesel was used as HRF. The authors studied different levels of EGR. The findings indicated that the heat release rate curves change from two stages to just one stage when the EGR is increased. The biodiesel/ethanol combination presented a longer ignition delay in relation to the biodiesel/butanol and biodiesel/DMF blends. The authors also indicated that combustion behavior and emissions trends are similar for the three fuel combinations (LRF/HRF) under different injection times. However, the biodiesel/ethanol combination demonstrates a greater potential for the simultaneous reduction of soot and NO_x (soot <0.3 FSN and NO_x <1.5 g/kWh with an SOI of 8 CAD ATDC). The biodiesel/butanol combination demonstrates the best ITEs, reaching 47.5%. It is also noteworthy that the use of biofuels in RCCI combustion in a fraction of up to 80% allowed the reduction of soot emissions and the increase in HC and CO emissions, which are already widely presented and discussed in the literature. Finally, the authors report that the higher cetane number of butanol compared to ethanol, which was previously shown in Table 9.1, limits its application to RCCI at high loads or substitution rates, which can be seen mainly in Figure 9.10(a), with 80% FPI and 60 mg/cyc of load. However, note in Figure 9.10(b) that this reduction in ignition delay does not imply a reduction in the ITE.

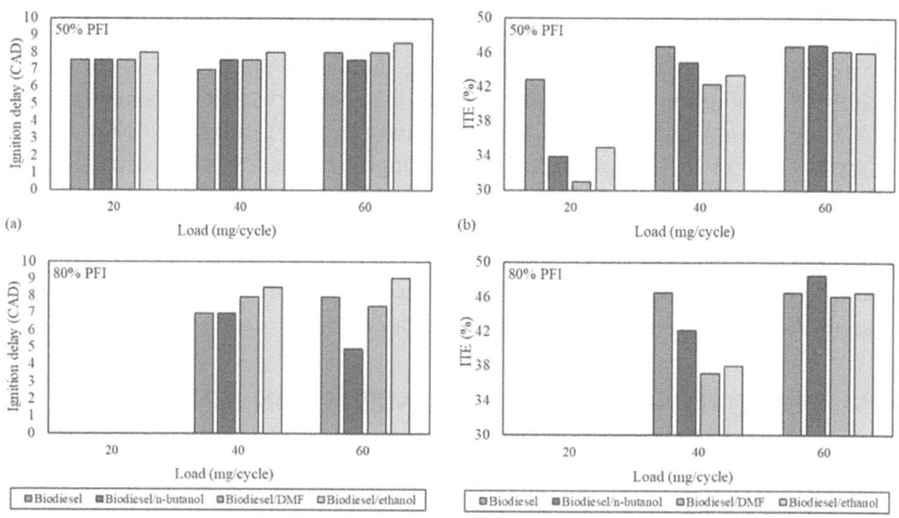

FIGURE 9.10 Ignition delay (a), indicated thermal efficiency (b) of the RCCI engine for different alcohols as LRF. (Data extracted from Zheng et al. (2018).)

Pan et al. (2021) studied the cyclic variations (COV) in a RCCI engine fueled with ethanol and diesel. They investigated the influence of diesel SOI and the premixed ratio (PR) on engine operation. The PR is described as the energy supplied by the premixed fuel in relation to the total energy supplied. The authors detected that the COV of different analyzed parameters, such as maximum cycle pressure, maximum heat release rate (HRR), IMEP, and maximum pressure rise rate (PRR), were affected similarly by SOI diesel. The highest cyclic variability was achieved for the SOI at 35 CAD BTDC, and since the SOI was delayed (closer to the TDC), with 24 and 20 CAD BTDC, the cyclic variability reduced, increasing again at the SOI equal to 15 CAD BTDC. The current engine literature recommends a COV of IMEP values not greater than 3%. If it is greater than 3%, it might affect the drivability of the vehicle. The highest COV of the IMEP value obtained by the authors was 2.77%, still below the threshold limit. Thus, this work demonstrated that RCCI combustion can be promising in terms of cyclic variation. In this research, a multi-cylinder engine equipped with common rail direct fuel injection was used, a fact that possibly contributed to the good results in terms of cyclic variability. A single-cylinder engine would tend to present less promising results.

Rosa et al. (2020) also investigated the effects of SOI on an RCCI engine, using ethanol as LRF and diesel as HRF. Furthermore, they investigated the effects caused by the concentration of water in ethanol and the ethanol energy fraction (EEF) on the combustion process. They performed the experiments in a single-cylinder, water-cooled engine at 1800 rpm with a constant equivalence ratio of 1.9. Among the 34 different cases tested, the highest COV of IMEP obtained was 3.1%, which occurred at the most advanced SOI diesel angle (34 CAD BTDC), with the highest concentration of water in ethanol (36% v/v) and the highest fraction of energy in ethanol (0.82). As in the results obtained by Pan et al. (Pan et al., 2021), this study shows that there are promising prospects in terms of combustion stability when using ethanol as LRF in an RCCI engine. However, this fact does not mean better engine thermal efficiency or even a reduction in emissions, as also evidenced by the study by Rosa et al. (Rosa et al., 2020). The best indicated efficiency was obtained when the engine operated with SOI = 14 CAD BTDC. Increasing the SOI to 27 or 34 CAD BTDC reduced the engine's ability to operate stably, restricting good operation only at high EEF values. This occurs because ethanol has a high capacity to delay combustion. In particular, it was only possible to operate with SOI at 34 CAD BTDC (which can be considered to be a significantly advanced SOI) when ethanol was supplied with high concentrations of water (24 and 36% v/v). With respect to CO emissions, they increase with the increase in EEF and decrease with the advancement of SOI diesel. NOx emissions show exactly the opposite behavior. It can be noted that there are promising aspects that indicate some feasibility of using ethanol as LRF in RCCI engines. Other alcohols that may be viable in this application are butanol and methanol.

Ganesan et al. (2020) investigated the use of n-butanol as an LRF in an RCCI engine and mahua oil as an HRF in a single-cylinder engine. A significant reduction in NO emissions and smoke opacity of 33.1% and 45.4% was reported, respectively, compared to monofuel operation with only mahua oil at maximum load. However,

HC and CO emissions increased. With up to a 20% share of butanol, the brake thermal efficiency increased by around 6.2%. Note that the results in terms of emissions are similar to those reported in the literature for other types of alcohol. The reduction in PM emissions and smoke opacity in RCCI engines fueled with alcohol as LRF could be attributed to different reasons. Firstly, the reduction in the amount of HRF supplied to the engine reduces the amount of fuel in the diffusive burning phase. As a consequence, it reduced the rich combustion regions, where the formation of these types of materials typically occurs. In the same way, the longer ignition delay results in more fuel in the premixing phase (uncontrolled) and less fuel in the diffusive combustion phase. It is also possible to mention the high flame propagation speed due to the presence of the LRF in the RCCI mode as a potential contributor to reducing the smoke opacity.

The work of Singh et al. (2020), cited in the previous section, also involved RCCI combustion with diesel and methanol. It can be seen in Figure 9.9(a) that despite longer ignition delays, RCCI combustion demonstrates higher values of thermal efficiency. Additionally, the findings revealed that in this combustion mode, the fuel consumption and the temperature of the exhaust gases are lower than in the CDC and PCCI combustion modes, which indicates a better use of fuel energy in the RCCI mode. The RCCI mode with diesel and methanol also resulted in lower combustion noise at all loads. Finally, these results are in agreement with the results already widely presented in the literature, where the RCCI combustion mode with diesel and methanol resulted in higher CO and HC emissions and lower NO_X emissions compared to the CDC and PCCI modes.

9.6 CONCLUSIONS

Several studies have explored the use of alcohols in LTC combustion modes. The literature presented in this chapter has shown the potential of using alcohol in LTC techniques to enhance the performance and emissions characteristics of internal combustion engines by increasing thermal efficiency and reducing pollutant emissions. The main conclusions of this chapter are as follows:

- The use of alcohols in HCCI engines is a good alternative since they are easy to vaporize and mix with air. However, it is still difficult to control the start of combustion, and some techniques need to be used.
- In general, the alcohol HCCI combustion decreased NO_X emissions due to the low combustion temperature in relation to conventional fuels (gasoline). On the other hand, CO emissions are usually higher for alcohol than for gasoline. HC emissions are very similar or slightly higher for gasoline than for alcohol fuels.
- Regarding the use of alcohols in PCCI and RCCI engines, these fuels are a good alternative to reduce the consumption of fossil fuels that are generally used in CI engines. PCCI or RCCI engines fueled with alcohol have shown the potential to at least maintain the energy efficiency performance of CDC combustion, in many cases surpassing the performance of the CDC engine.

- Since alcohol fuels have a high latent heat of vaporization, more heat is needed to vaporize these fuels, which reduces the charge temperature and directly affects the ignition delay, the start of the combustion, and combustion temperature.
- In general, it is also observed that the use of alcohols in PCCI and RCCI engines increases CO and HC emissions because of the incomplete combustion of the "cold" premixed charge, but there are lower levels of NO_X and particulate emissions since it is formed in high combustion temperatures.

Finally, it can be seen that several studies published have demonstrated the potential of using alcohol in LTC modes, but the utilization of LTC concept in commercial engines has been limited to date. However, due to environmental concerns, the transportation sector urgently needs more efficient and sustainable engine technologies to reduce greenhouse gas and pollutant emissions. Thus, the use of alcohols in LTC engines seems like an interesting pathway to substitute conventional fossil fuels and reduce emissions, especially in those countries where alcohols are produced on a large scale, such as Brazil, India, and the USA.

REFERENCES

Agarwal, A. K., Singh, A. P., García, A., & Monsalve-Serrano, J. (2022). Challenges and opportunities for application of reactivity-controlled compression ignition combustion in commercially viable transport engines. *Progress in Energy and Combustion Science*, *93*(August 2021), 101028. https://doi.org/10.1016/j.pecs.2022.101028

Agarwal, A. K., Singh, A. P., & Maurya, R. K. (2017). Evolution, challenges and path forward for low temperature combustion engines. *Progress in Energy and Combustion Science*, *61*, 1–56. https://doi.org/10.1016/j.pecs.2017.02.001

Akkoli, K. M., Kamate, S. C., Topannavar, S. N., Bhavimani, A. R., Banapurmath, N. R., Veza, I., Soudagar, M. E. M., Khan, T. M. Y., El-Shafay, A. S., Kalam, M. A., Shivashimpi, M. M., & Gulli, A. M. (2022). Influence of injection pressure and aluminium oxide nano particle-added fish oil methyl ester on the performance and emission of compression ignition engine. *Energies*, *15*(24), 9491. https://doi.org/10.3390/en15249491

Bendu, H., Deepak, B. B. V. L., & Murugan, S. (2016). Application of GRNN for the prediction of performance and exhaust emissions in HCCI engine using ethanol. *Energy Conversion and Management*, *122*, 165–173. https://doi.org/10.1016/j.enconman.2016.05.061

Bendu, H., & Sivalingam, M. (2016). Experimental investigation on the effect of charge temperature on ethanol fueled HCCI combustion engine. *Journal of Mechanical Science and Technology*, *30*(10), 4791–4799. https://doi.org/10.1007/s12206-016-0951-6

Bharath, A. N., Yang, Y., Reitz, R. D., & Rutland, C. (2015, April 14). *Comparison of Variable Valve Actuation, Cylinder Deactivation and Injection Strategies for Low-Load RCCI Operation of a Light Duty Engine*. https://doi.org/10.4271/2015-01-0843

Bissoli, M., Frassoldati, A., Cuoci, A., Ranzi, E., & Faravelli, T. (2018). A model investigation of fuel and operating regime impact on homogeneous charge compression ignition engine performance. *Energy & Fuels*, *32*(2), 2282–2298. https://doi.org/10.1021/acs.energyfuels.7b00893

Çelebi, Y., & Aydın, H. (2019). An overview on the light alcohol fuels in diesel engines. *Fuel*, *236*, 890–911. https://doi.org/10.1016/j.fuel.2018.08.138

Chen, H., Zhang, P., & Liu, Y. (2018). Investigation on combustion and emission performance of a common rail diesel engine fueled with diesel-ethylene glycol dual fuel. *Applied Thermal Engineering, 142*, 43–55. https://doi.org/10.1016/j.applthermaleng.2018.06.073

Dempsey, A. B., Das Adhikary, B., Viswanathan, S., & Reitz, R. D. (2012). Reactivity controlled compression ignition using premixed hydrated ethanol and direct injection diesel. *Journal of Engineering for Gas Turbines and Power, 134*(8). https://doi.org/10.1115/1.4006703

Duan, X., Lai, M. C., Jansons, M., Guo, G., & Liu, J. (2021). A review of controlling strategies of the ignition timing and combustion phase in homogeneous charge compression ignition (HCCI) engine. *Fuel, 285*(August 2020), 119142. https://doi.org/10.1016/j.fuel.2020.119142

Gainey, B., Gohn, J., Hariharan, D., Rahimi-Boldaji, M., & Lawler, B. (2020). Assessing the impact of injector included angle and piston geometry on thermally stratified compression ignition with wet ethanol. *Applied Energy, 262*, 114528. https://doi.org/10.1016/j.apenergy.2020.114528

Gainey, B., & Lawler, B. (2021). The role of alcohol biofuels in advanced combustion: An analysis. *Fuel, 283*(August 2020), 118915. https://doi.org/10.1016/j.fuel.2020.118915

Gainey, B., O'Donnell, P., Yan, Z., Moser, S., & Lawler, B. (2021). LTC performance of C1–C4 water-alcohol blends with the same cooling potential. *Fuel, 293*(December 2020), 120480. https://doi.org/10.1016/j.fuel.2021.120480

Ganesan, N., Masimalai, S., Ekambaram, P., & Selvaraju, K. (2020). Experimental assessment of effects of n-butanol on performance, emission, and combustion characteristics of Mahua oil fueled reactivity controlled compression ignition (RCCI) engine. *Emission Control Science and Technology, 6*(3), 345–357. https://doi.org/10.1007/s40825-020-00163-1

Gharehghani, A. (2019). Load limits of an HCCI engine fueled with natural gas, ethanol, and methanol. *Fuel, 239*(April 2018), 1001–1014. https://doi.org/10.1016/j.fuel.2018.11.066

Gharehghani, A., Abbasi, H. R., & Alizadeh, P. (2021). Application of machine learning tools for constrained multi-objective optimization of an HCCI engine. *Energy, 233*, 121106. https://doi.org/10.1016/j.energy.2021.121106

Gohn, J., Gainey, B., Zainul, S., & Lawler, B. (2020). Wet ethanol in LTC: How water fraction and DTBP affect combustion and intake temperature at naturally aspirated and boosted conditions. *Fuel, 267*(September 2019), 117094. https://doi.org/10.1016/j.fuel.2020.117094

Gong, C., Li, Z., Yi, L., & Liu, F. (2019). Comparative study on combustion and emissions between methanol port-injection engine and methanol direct-injection engine with H2-enriched port-injection under lean-burn conditions. *Energy Conversion and Management, 200*, 112096. https://doi.org/10.1016/j.enconman.2019.112096

Gong, C., Li, Z., Yi, L., Huang, K., & Liu, F. (2020). Research on the performance of a hydrogen/methanol dual-injection assisted spark-ignition engine using late-injection strategy for methanol. *Fuel, 260*, 116403. https://doi.org/10.1016/j.fuel.2019.116403

Harari, P. (2018). Comprehensive review on enabling reactivity controlled compression ignition (RCCI) in diesel engines. *Integrated Research Advances Integr. Res. Adv, 5*(1), 5–19. http://pubs.iscience.in/ira

He, B.-Q., Liu, M.-B., & Zhao, H. (2015). Comparison of combustion characteristics of n-butanol/ethanol–gasoline blends in a HCCI engine. *Energy Conversion and Management, 95*, 101–109. https://doi.org/10.1016/j.enconman.2015.02.019

Hossein Fakhari, A., Gharehghani, A., Mahdi Salahi, M., & Mahmoudzadeh Andwari, A. (2024). RCCI combustion of ammonia in dual fuel engine with early injection of diesel fuel. *Fuel, 365*, 131182. https://doi.org/10.1016/j.fuel.2024.131182

IEA. (2023). CO_2 emissions in 2022. *International Energy Agency.* https://doi.org/10.1007/978-3-031-25984-5_300288

Kakoee, A., Bakhshan, Y., Aval, S. M., & Gharehghani, A. (2018). An improvement of a lean burning condition of natural gas/diesel RCCI engine with a pre-chamber by using hydrogen. *Energy Conversion and Management, 166*, 489–499. https://doi.org/10.1016/j.enconman.2018.04.063

Kale, A. V., & Krishnasamy, A. (2024). Experimental study on combustion, performance, and emission characteristics of a homogeneous charge compression ignition engine fuelled with multiple biofuel-gasoline blends. *Energy, 288*, 129621. https://doi.org/10.1016/j.energy.2023.129621

Kamimoto, T., & Bae, M. (1988). High combustion temperature for the reduction of particulate in diesel engines. *SAE Technical Paper*. https://doi.org/10.4271/880423

Krishnamoorthi, M., Malayalamurthi, R., He, Z., & Kandasamy, S. (2019). A review on low temperature combustion engines: Performance, combustion and emission characteristics. *Renewable and Sustainable Energy Reviews, 116*, 109404. https://doi.org/10.1016/j.rser.2019.109404

Krishnasamy, A., Gupta, S. K., & Reitz, R. D. (2021). Prospective fuels for diesel low temperature combustion engine applications: A critical review. *International Journal of Engine Research, 22*(7), 2071–2106. https://doi.org/10.1177/1468087420960857

Lanzanova, T. D. M., Dalla Nora, M., & Zhao, H. (2016). Performance and economic analysis of a direct injection spark ignition engine fueled with wet ethanol. *Applied Energy, 169*, 230–239. https://doi.org/10.1016/j.apenergy.2016.02.016

Mack, J. H., Schuler, D., Butt, R. H., & Dibble, R. W. (2016). Experimental investigation of butanol isomer combustion in homogeneous charge compression ignition (HCCI) engines. *Applied Energy, 165*, 612–626. https://doi.org/10.1016/j.apenergy.2015.12.105

Martinez-Frias, J., Aceves, S. M., & Flowers, D. L. (2007). Improving ethanol life cycle energy efficiency by direct utilization of wet ethanol in HCCI engines. *Journal of Energy Resources Technology, 129*(4), 332–337. https://doi.org/10.1115/1.2794768

Maurya, R. K., & Agarwal, A. K. (2014). Experimental investigations of performance, combustion and emission characteristics of ethanol and methanol fueled HCCI engine. *Fuel Processing Technology, 126*, 30–48. https://doi.org/10.1016/j.fuproc.2014.03.031

Morales Rojas, A. D., Quintana, S. H., & Caro, I. D. B. (2024). Experimental study of a homogeneous charge compression ignition engine using hydrogen at high-altitude conditions. *Sustainability, 16*(5), 2026. https://doi.org/10.3390/su16052026

Munuswamy, D. B., Devarajan, Y., Ramalingam, S., Subramani, S., & Munuswamy, N. B. (2022). Critical review on effects of alcohols and nanoadditives on performance and emission in low-temperature combustion engines: Advances and perspectives. *Energy and Fuels, 36*(14), 7245–7268. https://doi.org/10.1021/acs.energyfuels.2c00930

Nibin, M., Raj, J. B., & Geo, V. E. (2021). Experimental studies to improve the performance, emission and combustion characteristics of wheat germ oil fuelled CI engine using bioethanol injection in PCCI mode. *Fuel, 285*, 119196. https://doi.org/10.1016/j.fuel.2020.119196

No, S.-Y. (2016). Application of biobutanol in advanced CI engines – A review. *Fuel, 183*, 641–658. https://doi.org/10.1016/j.fuel.2016.06.121

Nour, M., Attia, A. M. A., & Nada, S. A. (2019). Combustion, performance and emission analysis of diesel engine fuelled by higher alcohols (butanol, octanol and heptanol)/diesel blends. *Energy Conversion and Management, 185*, 313–329. https://doi.org/10.1016/j.enconman.2019.01.105

Pan, S., Cai, K., Cai, M., Du, C., Li, X., Han, W., Wang, X., Liu, D., Wei, J., Fang, J., & Bao, X. (2021). Experimental study on the cyclic variations of ethanol/diesel reactivity controlled compression ignition (RCCI) combustion in a heavy-duty diesel engine. *Energy, 237*, 121614. https://doi.org/10.1016/j.energy.2021.121614

Panda, K., & Ramesh, A. (2024). Injection strategy of methanol for high loads and low NOx emissions in a neat methanol LTC engine. *International Journal of Engine Research, 25*(3), 573–588. https://doi.org/10.1177/14680874231200525

Paykani, A., Kakaee, A. H., Rahnama, P., & Reitz, R. D. (2016). Progress and recent trends in reactivity-controlled compression ignition engines. *International Journal of Engine Research*, *17*(5), 481–524. https://doi.org/10.1177/1468087415593013

Punasiya, M., & Sarangi, A. K. (2024). Machine learning-based modeling and predictive control of combustion phasing and load in a dual-fuel low-temperature combustion engine. *SAE International Journal of Engines*, *17*(4), 03–17-04–0030. https://doi.org/10.4271/03-17-04-0030

Rajesh Kumar, B., & Saravanan, S. (2016). Use of higher alcohol biofuels in diesel engines: A review. *Renewable and Sustainable Energy Reviews*, *60*, 84–115. https://doi.org/10.1016/j.rser.2016.01.085

Rameez, P. V., & Mohamed Ibrahim, M. (2024). A comprehensive review on the utilization of hydrogen in low temperature combustion strategies: Combustion, performance and emission attributes. *Journal of the Energy Institute*, *113*, 101511. https://doi.org/10.1016/j.joei.2023.101511

Reitz, R. D., & Duraisamy, G. (2015). Review of high efficiency and clean reactivity controlled compression ignition (RCCI) combustion in internal combustion engines. *Progress in Energy and Combustion Science*, *46*, 12–71. https://doi.org/10.1016/j.pecs.2014.05.003

Rosa, J. S., Altafini, C. R., Wander, P. R., Telli, G. D., & Rocha, L. A. O. (2019). Wet ethanol fumigation on a compression ignition engine: Effects of air intake throttled. *Journal of the Brazilian Society of Mechanical Sciences and Engineering*, *41*(11), 500. https://doi.org/10.1007/s40430-019-2023-1

Rosa, J. S., Lorenzini, G., Altafini, C. R., Wander, P. R., Telli, G. D., & Rocha, L. A. O. (2019). Performance effects and economic viability of high-hydrated ethanol fumigation and diesel direct injection in a small compression ignition engine. *Mathematical Modelling of Engineering Problems*, *6*(1), 1–9. https://doi.org/10.18280/mmep.060101

Rosa, J. S., Martins, M. E. S., Telli, G. D., Altafini, C. R., Wander, P. R., & Rocha, L. A. O. (2020). Exploring the effects of diesel start of injection and water-in-ethanol concentration on a reactivity controlled compression ignition engine. *Fuel*, *281*(July), 118751. https://doi.org/10.1016/j.fuel.2020.118751

Rosa, J. S., Telli, G. D., Altafini, C. R., Wander, P. R., & Oliveira Rocha, L. A. (2021). Dual fuel ethanol port injection in a compression ignition diesel engine: Technical analysis, environmental behavior, and economic viability. *Journal of Cleaner Production*, *308*, 127396. https://doi.org/10.1016/j.jclepro.2021.127396

Saxena, S., & Bedoya, I. D. (2013). Fundamental phenomena affecting low temperature combustion and HCCI engines, high load limits and strategies for extending these limits. *Progress in Energy and Combustion Science*, *39*(5), 457–488. https://doi.org/10.1016/j.pecs.2013.05.002

Saxena, S., Schneider, S., Aceves, S., & Dibble, R. (2012). Wet ethanol in HCCI engines with exhaust heat recovery to improve the energy balance of ethanol fuels. *Applied Energy*, *98*, 448–457. https://doi.org/10.1016/j.apenergy.2012.04.007

Saxena, S., Vuilleumier, D., Kozarac, D., Krieck, M., Dibble, R., & Aceves, S. (2014). Optimal operating conditions for wet ethanol in a HCCI engine using exhaust gas heat recovery. *Applied Energy*, *116*, 269–277. https://doi.org/10.1016/j.apenergy.2013.11.033

Shere, A., & Subramanian, K. A. (2024). Effects of hydrogen and EGR on energy efficiency improvement with ultra low emissions in a common rail direct injection compression ignition engine fueled with dimethyl ether (DME) under HCCI mode. *International Journal of Hydrogen Energy*, *52*, 1447–1474. https://doi.org/10.1016/j.ijhydene.2023.08.009

Shim, E., Park, H., & Bae, C. (2020). Comparisons of advanced combustion technologies (HCCI, PCCI, and dual-fuel PCCI) on engine performance and emission characteristics in a heavy-duty diesel engine. *Fuel*, *262*, 116436. https://doi.org/10.1016/j.fuel.2019.116436

Singh, A. P., Kumar, V., & Agarwal, A. K. (2020). Evaluation of comparative engine combustion, performance and emission characteristics of low temperature combustion (PCCI and RCCI) modes. *Applied Energy*, *278*, 115644. https://doi.org/10.1016/j.apenergy.2020.115644

Sjöberg, M., & Dec, J. E. (2006, April 3). Smoothing HCCI heat-release rates using partial fuel stratification with two-stage ignition fuels. *SAE 2006 World Congress & Exhibition*. https://doi.org/10.4271/2006-01-0629

Spanò, S., Savarese, M., Parente, A., Contino, F., & Jeanmart, H. (2024). Experimental and numerical investigation of methane combustion in a HCCI engine under enriched oxygen conditions. *Fuel*, *362*, 130772. https://doi.org/10.1016/j.fuel.2023.130772

Srivastava, D. K., Agarwal, A. K., Datta, A., & Maurya, R. K. (2018). *Advances in internal combustion engine research* (D. K. Srivastava, A. K. Agarwal, A. Datta, & R. K. Maurya (eds.)). Springer Singapore. https://doi.org/10.1007/978-981-10-7575-9

Taghavi, M., Gharehghani, A., Nejad, F. B., & Mirsalim, M. (2019). Developing a model to predict the start of combustion in HCCI engine using ANN-GA approach. *Energy Conversion and Management*, *195*, 57–69. https://doi.org/10.1016/j.enconman.2019.05.015

Telli, G. D., Altafini, C. R., Costa, C. A., Rosa, J. S., Rocha, L. A. O., & Lorenzini, G. (2020). Experimental study of a dual-fuel generator set operating on diesel fuel direct injected and hydrous ethanol fumigation at different loads. *International Journal of Design & Nature and Ecodynamics*, *15*(6), 777–784. https://doi.org/10.18280/ijdne.150602

Telli, G. D., Altafini, C. R., Costa, C. A., Rosa, J. S., Martins, M. E., & Oliveira Rocha, L. A. (2021). A comprehensive review of homogeneous charge compression ignition (HCCI) engines: Advantages, challenges and evolution. *SAE Technical Papers*, 1–12. https://doi.org/10.4271/2020-36-0042

Telli, G. D., Altafini, C. R., Rosa, J. S., & Costa, C. A. (2018a). Experimental analysis of a small engine operating on diesel–natural gas and soybean vegetable oil–natural gas. *Journal of the Brazilian Society of Mechanical Sciences and Engineering*, *40*(11), 547. https://doi.org/10.1007/s40430-018-1469-x

Telli, G. D., Altafini, C. R., Rosa, J. S., & Costa, C. A. (2018b). Experimental investigation of a compression ignition engine operating on B7 direct injected and hydrous ethanol fumigation. *Energy*, *165*, 106–117. https://doi.org/10.1016/j.energy.2018.09.171

Telli, G. D., Rocha, L. A. O., Zulian, G., Lanzanova, T. D. M., & Martins, M. E. S. (2022, March 29). Water Injection for Combustion Phasing Control and Load Expansion on an Ethanol HCCI Engine. *SAE Technical Paper*. https://doi.org/10.4271/2022-01-0451

Telli, G. D., Zulian, G. Y., Lanzanova, T. D. M., Martins, M. E. S., & Rocha, L. A. O. (2022). An experimental study of performance, combustion and emissions characteristics of an ethanol HCCI engine using water injection. *Applied Thermal Engineering*, *204*(November 2021), 118003. https://doi.org/10.1016/j.applthermaleng.2021.118003

Telli, G. D., Zulian, G. Y., Lanzanova, T. D. M., Martins, M. E. S., & Rocha, L. A. O. (2023). Investigation of the use of wet ethanol in an HCCI engine using water injection and direct exhaust heat recovery. *Energy Conversion and Management: X*, *18*(January), 100377. https://doi.org/10.1016/j.ecmx.2023.100377

Viggiano, A., & Magi, V. (2012). A comprehensive investigation on the emissions of ethanol HCCI engines. *Applied Energy*, *93*, 277–287. https://doi.org/10.1016/j.apenergy.2011.12.063

Wu, B., Zi, Z., Jin, S., Pei, Y., & Wang, D. (2024). Effect of diesel injection strategy and ammonia energy fraction on ammonia-diesel premixed-charge compression ignition combustion and emissions. *Fuel*, *357*, 129785. https://doi.org/10.1016/j.fuel.2023.129785

Xu, H., Wu, X., & Jia, H. (2024). Pilot injection impact on diesel PCCI combustion: An endoscopic study with a turbocharged engine. *Journal of the Brazilian Society of Mechanical Sciences and Engineering*, *46*(2), 76. https://doi.org/10.1007/s40430-023-04662-4

Yadav, N. K., Saxena, M. R., & Maurya, R. K. (2024). Inhalation toxicity characterization of nanoparticle and carbonyl emission from conventional diesel and methanol/gasoline-diesel RCCI engine. *Fuel*, *366*, 131353. https://doi.org/10.1016/j.fuel.2024.131353

Zheng, Z., Xia, M., Liu, H., Wang, X., & Yao, M. (2018). Experimental study on combustion and emissions of dual fuel RCCI mode fueled with biodiesel/n-butanol, biodiesel/2,5-dimethylfuran and biodiesel/ethanol. *Energy*, *148*, 824–838. https://doi.org/10.1016/j.energy.2018.02.015

Zulian, G. Y., do Prado F, L. M., Garlet, R. A., Martins, M. E. S., Lanzanova, T. D. M., & Telli, G. D. (2022). Exploring the use of water injection to control the combustion and expand the operation load of an ethanol HCCI engine. *SAE BRASIL 2021 Web Forum*, 1–11. https://doi.org/10.4271/2021-36-0053

10 Development of a Partially Premixed Combustion Technique for High Efficiency with Minimal Emissions

Satsimran Singh, Prem Kumar, and Sarbjot Singh Sandhu

10.1 INTRODUCTION

Internal combustion (IC) engines have been very instrumental in shaping modern society, ranging from powering vehicles such as cars and trucks to generating electricity with gensets. But the IC engines are currently facing some serious challenges. One of them is running out of fossil fuels and other is the emissions produced from them harming the environment as well as health (Wåhlin et al., 2004). Researchers are aiming for engines that are not only more efficient but also cleaner, with fewer harmful emissions. Low-temperature combustion (LTC) engine technologies are showing a great promise because of their ability to reduce emissions and achieve superior performance compared to the traditional combustion mode. LTC covers a number of techniques, each one distinguished with its individual approach, such as Homogeneous Charge Compression Ignition (HCCI), Premixed Charge CI (PCCI), Partially Premixed Charge Compression Ignition (PPCI), and Reactivity-Controlled CI (RCCI) (Noehre et al., 2006). Researchers have explored the LTC strategies such as PPCI and RCCI to minimize emissions.

These approaches have demonstrated the potential of reducing NO_x and PM emissions concurrently in CI engines, which can potentially eliminate the necessity for exhaust after-treatment technologies (Gwalwanshi et al., 2023). Multiple studies have established the effectiveness of these LTC methods in achieving cleaner engine performance. See Figure 10.1 for the variation of soot and NO_x formation with respect to local equivalence ratios and local temperature and it confirms this notion of reducing the NO_x simultaneously.

When it comes to LTC strategies, the distribution of equivalence ratio and temperature shifts to lower temperature regions within the soot formation zone (Neely et al., 2005). As for PCCI, while its φ-T distribution doesn't stray far from standard diesel combustion, the lower local equivalence ratio makes for cleaner, less smoky

DOI: 10.1201/9781003482277-10

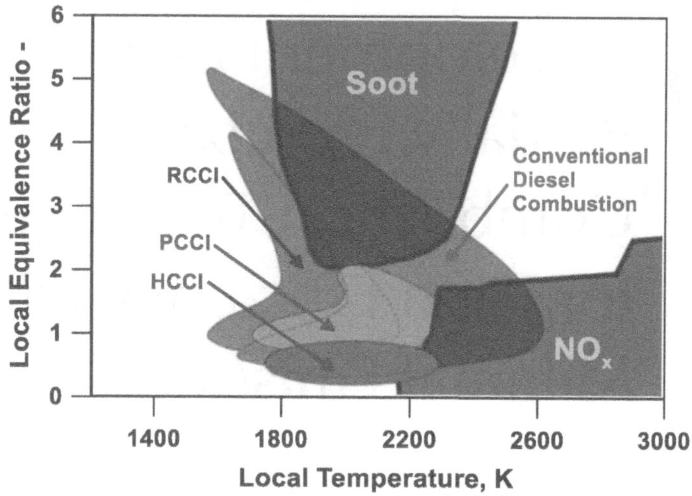

FIGURE 10.1 Mapping areas of NO_x and soot based on the local equivalence ratio (φ) and local temperature. (De Bellis et al., 2022.)

combustion, which shall be accomplished by better control over the injection timing and injection pressure (Bharadwaz & Kumari, 2023).

10.2 PARTIALLY PREMIXED CHARGE IGNITION

Homogenous charge ignition blends the elements of SI and CI engines by using a premixed charge like spark ignition engines while requiring autoignition like CI engines. HCCI has some challenges such as regulating combustion phasing, high noise levels, operating range, and cold start (Imtenan et al., 2014; Yao et al., 2009). PCCI has emerged as a progression, designed to address the limitations of the HCCI strategy, with key distinction between lying in the degree of air-fuel mixing. PCCI mode offers the versatility to utilize both diesel and gasoline fuels while substantially reducing NO_x and soot emissions to low levels (Manimaran et al., 2013). The combustion in PCCI approach can be accomplished by implementing early and late fuel injections, high injection pressures, and exhaust gas recirculation (EGR). PCCI provides significant benefits in managing emissions and maintaining the optimal thermal efficiency.

Partially premixed combustion ignition (PPCI) is basically an extension of PCCI. PPCI approach injects and premixes a fuel charge during the compression stroke of a CI engine. For this approach, the fuel can be introduced either in a single stage or in multiple stages injection before the start of combustion (SOC) during compression. This leads to the stratification of fuel concentration, which can provide more regulation over combustion (Zhang et al., 2012). The PPCI technique employs a longer ignition delay than the traditional diesel combustion, which then leads to enhanced mixing. The major difference between PPCI and HCCI is direct fuel injection, which enables better control over the combustion process. HCCI is dependent over the autoignition of a homogeneous fuel-air mixture, while PPCI combines premixed and direct injection (DI) for more precise combustion control. However, it's not as thoroughly mixed as HCCI.

PPCI combustion can be accomplished through the utilization of varying injection timings and different rates of EGR. While the PPCI can be achieved using single injection while using higher octane number fuels, such as alcohols, but this approach can suffer from some critical drawbacks, such as triggering a high heat release rate (HRR), while subsequently raising the pressure rise rate (PRR) and which can result in increased combustion noise (Lewander et al., 2011). The application of high research octane number fuel yields a shorter combustion duration, thereby promoting a rapid combustion.

To mitigate this challenge split injection (pilot/main) strategy can be used. This strategy uses separate injections sequentially. At first a pilot injection is introduced early in the cycle, which will lay down the premixed mixture. Then the main injection will take place, which will ignite this premixed mixture at the expected SOC timing (Rezaei et al., 2013). Research work done using the split injection has established that the pilot injection effectively restrains the extent of premixed combustion and thus provide simultaneous reduction in both combustion noise and PRR (Jain et al., 2017b; Rezaei et al., 2013). Moreover, dispersing the injected fuel volume inside the combustion chamber limits the cooling impact typically associated with alcoholic fuel, therefore facilitating complete combustion with successive autoignition of the charge (Huang et al., 2015). Consequently, combustion progresses more smoothly, contributing to a slower overall combustion process.

10.3 FACTORS EFFECTING PARTIALLY PREMIXED COMBUSTION STRATEGIES

PPCI can be achieved with help of different strategies, such as varying the injection timing, EGR rates, and injection pressure or by the combination of these (Deresso et al., 2023; Kumar & Sandhu, 2022; Liu et al., 2019). The combustion can be fine-tuned by sensibly adjusting the time for fuel to be injected inside the combustion chamber and how much of exhaust gas has to be recirculated. Implementing fuel injections at an early stage can result in a more uniform mixture of air and fuel, thereby reducing the production of nitrogen oxides (NO_x) and soot (Kumar & Sandhu, 2022). While on the other, if late fuel injections are employed, they can be useful to tailoring the combustion process for enhancing the efficiency and performance under specific conditions (Krishnamoorthi et al., 2019). In a similar fashion, varying the rate of EGR can provide a further control over combustion parameters, by reintroducing a portion of the exhaust gas back into intake air, the overall oxygen content inside the combustion chamber gets reduced, leading to lower combustion temperatures and reduced NO_x emissions (Manente et al., 2010). With the capability to modify both injection timings and EGR rates, the PPCI combustion process can be optimized with greater flexibility, leading to improved fuel efficiency, lower emissions, and enhanced performance under different operating conditions.

10.3.1 Fuel Injection Strategy

Fuel injection strategy is one of the most significant factors that affect the combustion, performance, and exhaust emissions of a CI engine. Most of the studies had applied the single injection in PPCI combustion whereas some of them used the two-stage injection.

During the single injection strategy in PPC, fuel gets injected inside the combustion chamber at the compression stroke, resulting in a stratified fuel-air mixture (Imtenan et al., 2014). Injecting fuel at a precise point during the compression stroke allows for a controlled delay between injection and ignition (see Figure 10.2). This delay is essential for achieving the desired fuel stratification and ensuring optimal preparation of the fuel-air mixture for combustion. Advanced single injection strategy achieves PPC without spray-piston interactions, improving soot emissions while maintaining NO_x levels by adjusting oxygen concentration (see Figure 10.3) (Benajes et al., 2008; De Ojeda et al., 2009). In spite of these benefits, it offers certain disadvantages such as rough combustion, increased noise, and reduced engine life and also limits its use during high loads due to high in-cylinder pressures (Benajes et al., 2008). The multi-injection strategy achieves PPC by enhancing fuel-air mixture preparation and controlling combustion.

Multiple injection strategy is implemented by injecting a small amount of fuel early during the compression stroke to create a premixed charge, followed by the main injection of remaining fuel later for autoignition and it helps in achieving regulated combustion and improved fuel-air mixing (Liang et al., 2019; Yao et al., 2010). The pilot injection at the start forms a preliminary fuel-air mixture, which allows partial vaporization. Following next, the main injection further mixes the fuel with air and create a more homogeneous charge while balancing the benefits of premixed and diffusion combustion (see Figure 10.2). See Figure 10.4, introducing a little bit quantity

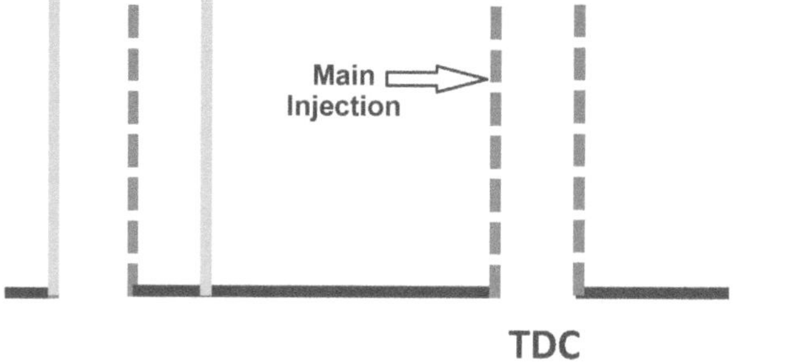

FIGURE 10.2 Single stage and split-stage injection. (Liang et al., 2019.)

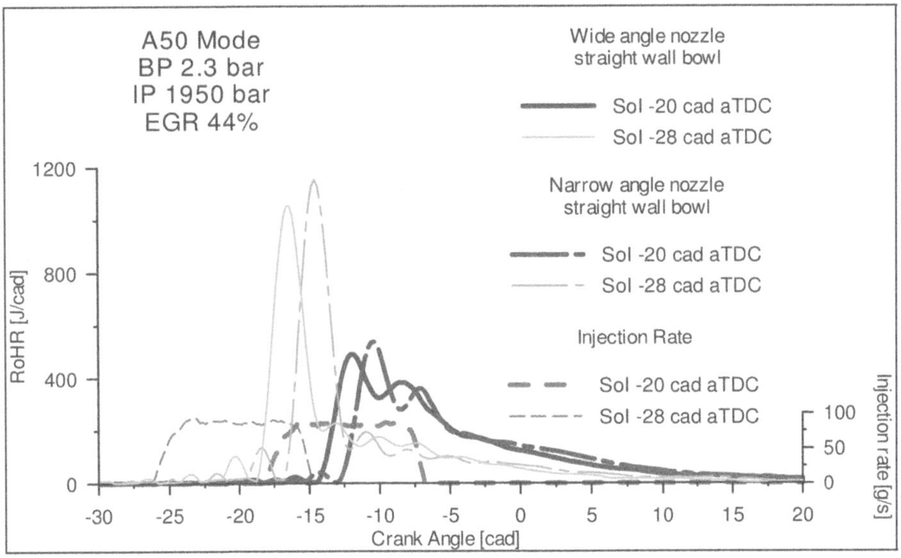

FIGURE 10.3 RoHR for the advanced single injection strategy in A50 mode. (Benajes et al., 2008.)

of fuel prior to main injection has shown to enhance fuel-air mixture homogeneity and improve combustion efficiency and emission parameters (Chen, 2000; Zheng et al., 2015). In some cases, a post-injection can also be implemented after the main injection for regulating the combustion duration and improve completeness, thereby reducing carbon monoxide (CO) as well as unburned hydrocarbons (UHC) emissions (Imtenan et al., 2014; Qiu et al., 2016a). See Figure 10.5, pilot injection facilitates the formation of a mixture during the ignition delay period and consequently slows down the oxidation reaction, resulting in an earlier rise and a reduced peak HRR (Khalid et al., 2013). The staged injections allow precise control over the combustion phasing and ignition delay, effectively managing rate of heat release, temperature, and pressure rise.

10.3.2 Combustion Control Techniques

Advanced combustion control is key to tuning the various engine parameters to control the combustion process for Premixed Charge CI (PPCI) engines. Advanced combustion control techniques such as dynamic modeling, real-time closed-loop combustion controllers and predictive combustion control model can be very useful for this purpose (Shim et al., 2020; Veera et al., 2012). Dynamic modeling provides a basis for understanding and predicting the combustion behavior in the engine. By developing an accurate dynamic model, the effects of different control strategies can be analyzed and the optimal operating condition can be found. This can be used for simulating the engine behavior and re-design the control strategies. Predictive combustion control models use historical combustion-related data and machine learning techniques to predict the future engine behavior (Norouzi et al., 2021).

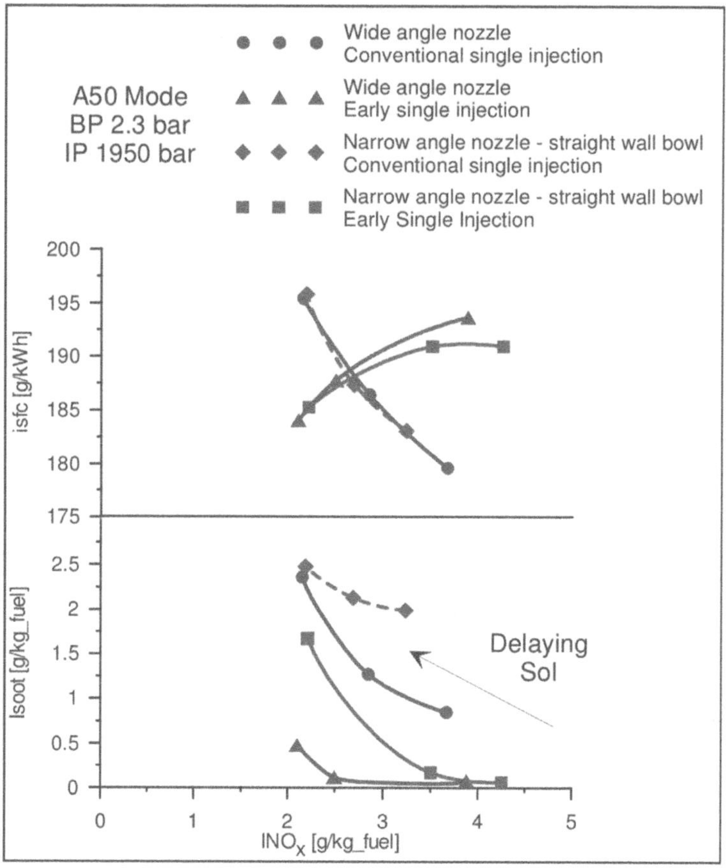

FIGURE 10.4 Emissions and ISFC for single injection strategy. (Benajes et al., 2008.)

See Figure 10.6, by anticipating the problems these models can adjust the control inputs to maintain the optimal PPCI combustion. Model Predictive Combustion Control (MPCC) can adjust the combustion in real-time, for optimal performance and reduced emissions (Yang et al., 2020). By integrating a predictive combustion model and feedback control, the multi-injection strategy was found to achieve the balance between combustion stability, efficiency, and emissions reduction (Korkmaz et al., 2018). Real-time closed-loop combustion controllers continuously monitor the parameters to maintain the desired performance. By controlling the combustion over a closed-loop approach by using actuators, sensors, and sophisticated control algorithms will be effective by continuously monitoring the variables such as injection timing, boost pressure, and EGR rate (Maurya, 2018). Based on the monitored results, the adjustments will be made to achieve the optimal combustion phasing and reduced emissions (Maurya, 2018). This will ensure the engine to operate at peak efficiency while minimizing the environmental impact.

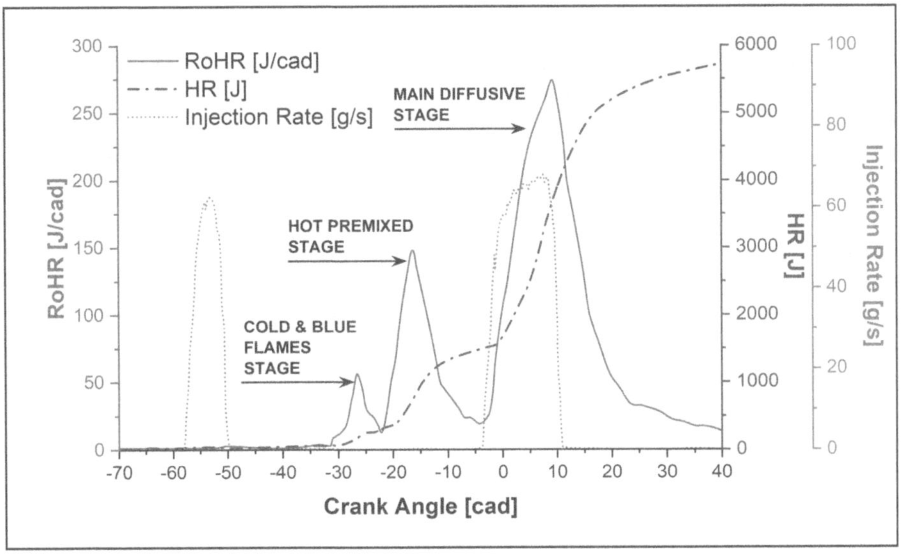

FIGURE 10.5 Injection rate, heat rise, and rate of heat rise for EPI strategy. (Benajes et al., 2008.)

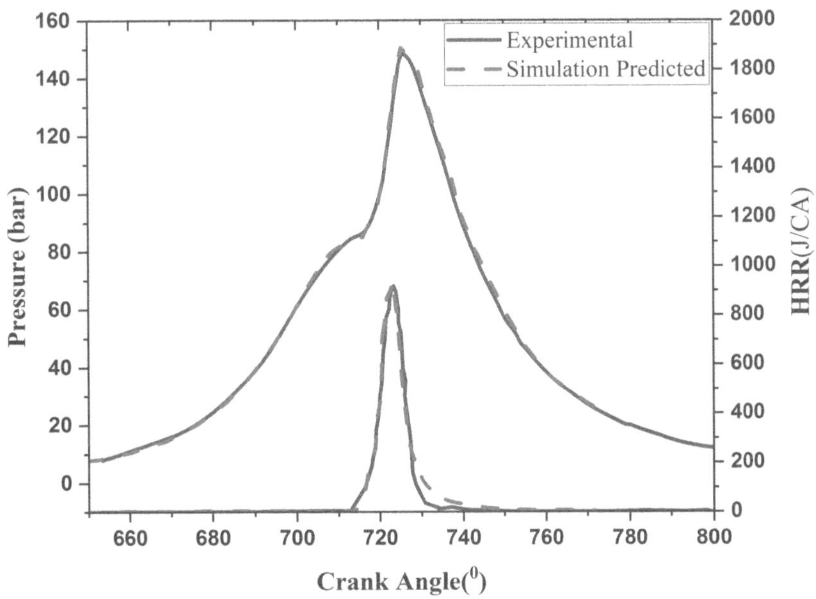

FIGURE 10.6 Simulation predicted and experimental data comparison. (Jin et al., 2021.)

10.3.3 Optimization of Injection Parameters

Fine-tuning and adjusting the injection parameters, such as injection pressure, timing of fuel injections, injection duration, can be used to have a precise control over the combustion phasing and ignition delay (Yang et al., 2020). See Figure 10.7, injection timing and injection pressure can be varied to change the ignition delay, which will then influence the mixing. Early start of injection (SOI) can promote better fuel-air mixing and reduce NO_x emissions, whereas a late SOI can help manage the combustion process under different engine loads by promoting the diffusion combustion while reducing the tendency to knock (Deep et al., 2017). The injection duration defines the time interval over which fuel is injected into the combustion chamber. Controlling the injection duration by making it either shorter or longer affects the air-fuel distribution inside the combustion chamber and the NO_x and soot emissions.

Increased injection pressure leads to increase in spray penetration length and a reduced combustion duration and an extended premixing period prior to the initiation of combustion. Due to the low in-cylinder temperature and pressure during the early injection period, this can result in significant spray-wall impingement. Either over-rich or over-lean zones will be formed due to incomplete fuel vaporization and oxidation caused by fuel impingement on the piston head or cylinder wall. It can, thus, sometimes create a contradictory result.

Higher injection pressure elevates the temperature inside the cylinder, which can cause the higher levels of NO_x emissions. Higher injection pressure enhances

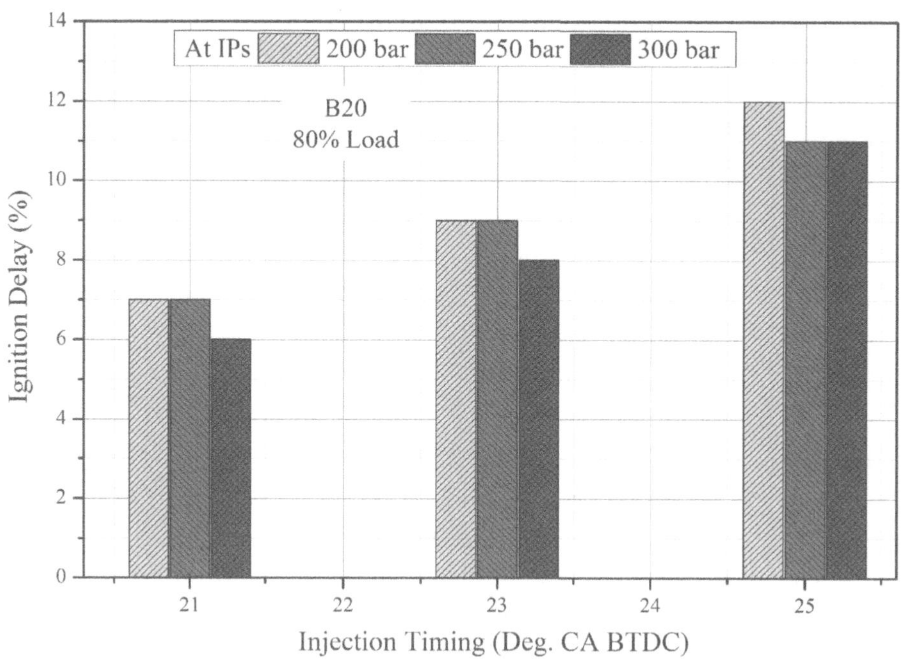

FIGURE 10.7 Effect on ignition delay at varied injection timings. (Deep et al., 2017.)

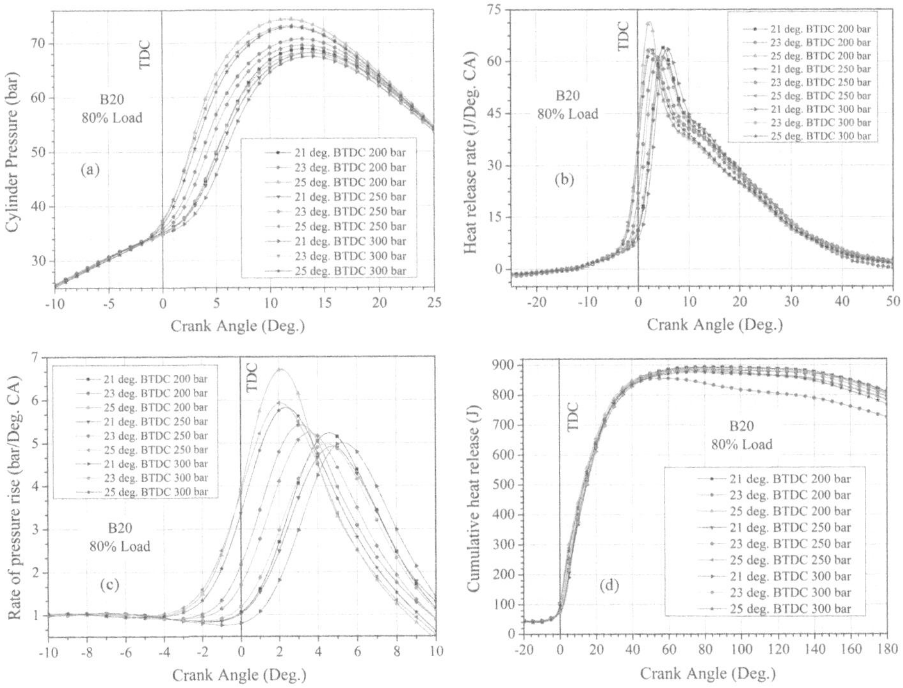

FIGURE 10.8 Effect of different injection timings and injection pressure on combustion parameters. (Deep et al., 2017.)

fuel atomization, resulting in improved air-fuel mixing creating more homogeneous mixture and reduced fuel-rich areas inside the combustion chamber (Natarajan et al., 2018). This is advantageous for achieving the LTC strategy, such as PCCI or PPCI and can help decreasing NO_x emissions. High injection pressure combined with more advanced injection timing is more effective for reducing NO_x concentrations. This phenomenon has shown to have steep peak in heat release and high in-cylinder temperature, accompanied by a fast rate of combustion and a reduced duration of combustion, which in turn improves the thermal efficiency of the engine (see Figure 10.8) (Jain et al., 2017a). This helps in achieving a more controlled and efficient combustion process while reducing both NO_x and soot emissions.

10.3.4 Engine Design Optimization

Designing the engine parts like piston by adjusting piston geometry varying compression ratio and intake/exhaust systems to improve combustion efficiency and performance in PPCI. Optimizing the engine components design like combustion chamber, including piston bowl shape and intake/exhaust valves, can significantly improve the air-fuel mixing (Kokjohn & Reitz, 2009; Noehre et al., 2006). This contributes to a more uniform charge and efficient combustion. The piston bowl shape has a big impact on airflow and fuel dispersion inside the cylinder. Well-optimized geometries can increase turbulence and mixing, key to combustion with minimal emissions. Besides the piston bowl shape, design of the injector nozzle, be it nozzle holes geometry, size, number, and shape, is another important factor that plays a big role in determining the fuel spray pattern. Larger diameters can increase turbulence kinetic energy (TKE) and equivalence ratio and hence more efficient combustion and less soot (Veera et al., 2012). The injector nozzle design affects spray formation, fuel penetration, and droplet size, all of which affects air-fuel mixture. More holes or smaller orifices give finer spray and better mixing. Optimizing the nozzle design can give more evenly distributed fuel charge which can help in cleaner combustion. Computational studies on the effect of hole size and distribution on spray characteristics and air-fuel mixing followed by experimental verification can help in this area. Based on that the design should be customized to meet the exact requirement of the engine air-fuel interaction. Computational Fluid Dynamics (CFD) simulations can be used to find the most optimum SOI, spray tip penetration length (STPL), piston bowl shape, and the exact location and topography of the stepped lip to minimize ISFC and emissions (Cengiz & Unverdi, 2023). For the selected piston bowl shape, the optimum spray inclusion angle can further improve combustion efficiency and emissions.

10.3.5 Fuel Selection

Choosing the right fuel and optimization of its properties, such as cetane number (CN) and volatility, can have a significant influence over ignition delay and combustion behavior. Knowing how fuel characteristics affect the requirements of the PPCI process can be helpful in achieving combustion that is consistent and controllable (Petersen et al., 2010). Fuels with high CN and varying fuel volatility can have some impact on combustion characteristics, while low viscosity can reduce injection pressure competence and increase brake-specific fuel consumption (BSFC) at high loads due to decreased mechanical efficiency (Mao et al., 2018a). Fuels with little aromatic amount and high CN can reduce particulate matter and NO_x emissions concurrently without trade-off. At higher NO_x levels, soot emissions may have a significant impact with respect to fuel being used in the engine. Ignition delay and fuel reactivity play critical roles in soot reduction at different loads, with low CN fuels being very sensitive to the speed variations (Mao et al., 2018a). At high loads molecular dilution effects become more important than CN in soot reduction. Moreover, optimal fuel properties like viscosity and oxygenated content are crucial for enhancing combustion efficiency and emissions reduction; higher injection pressures can further influence these outcomes (Du et al., 2015; Sun et al., 2016). These properties can also affect the mechanical efficiency and fuel/engine interactions under varying injection parameters in certain cases. The variations in the CO emissions while using different fuels like conventional ultralow sulfur diesel fuel (ULSD), low-temperature Fischer-Tropsch (LTFT) process from natural gas, and renewable diesel (RD) from camelina oil at different injection timings can be seen in Figure 10.9 (Sun et al., 2016). See Figure 10.10, the effect of fuel properties on pressures inside the cylinder and HRR at the engine-out while using diesel and its blend with gasoline (DG80), n-butanol (DB80), and n-heptane (DH80) (Mao et al., 2018b).

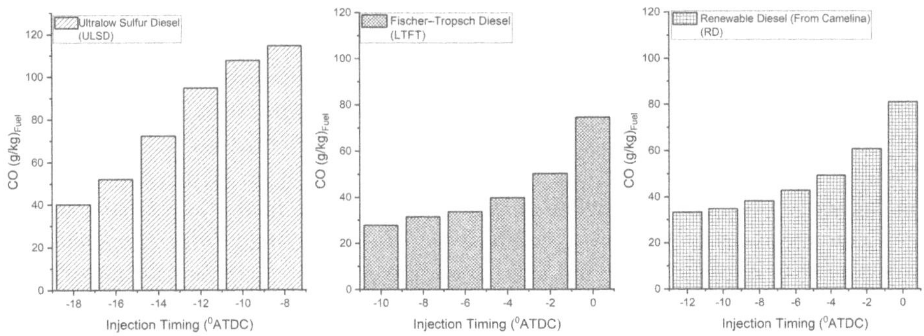

FIGURE 10.9 CO emissions for different fuels at changed injection timings. (Sun et al., 2016.)

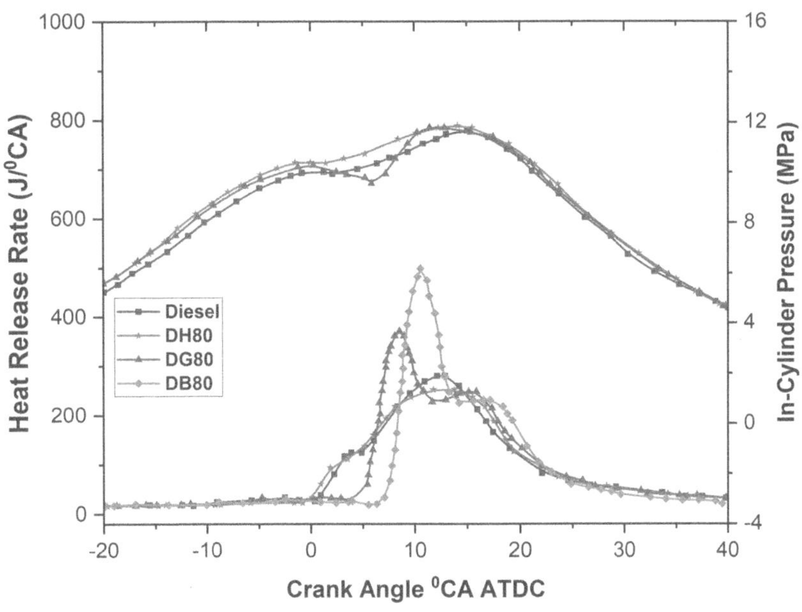

FIGURE 10.10 Influence of fuel characteristics on ICP and HRR at engine-out. (Mao et al., 2018b.)

10.3.6 Use of EGR

EGR plays a key role for achieving PPCI in diesel engines by recirculating exhaust gases inside the combustion chamber. EGR dilutes exhaust gases with the intake air, reducing the oxygen concentration and peak combustion temperatures, which can reduce formation of the nitrogen oxides (NO_x) (Zheng et al., 2004). EGR allows better fuel-air mixing before combustion by increasing the intake charge's specific heat capacity and extending the ignition delay. EGR can adjust the in-cylinder conditions to phase combustion optimally within the engine cycle and can enable the engines to run on different fuels and qualities (KIM et al., 2004).

By modulating EGR rates, engine combustion can run optimally with fuels of different CNs and volatilities for PPCI operation (Musculus et al., 2013). See Figure 10.2, Increasing the EGR rate effectively reduces NOx emissions across different fueling rates. While EGR reduces NO_x emissions, but higher EGR rates can result in increased soot and particulate matter emissions due to the lower oxygen concentration leading to incomplete combustion (see Figure 10.3). EGR can also lower the combustion temperature and slow down the process of combustion, which can reduce the thermal efficiency (Mao et al., 2018b). Controlling the levels of EGR is very important in achieving the PPCI combustion without the concerning disadvantages (Figures 10.11 and 10.12).

Use of high EGR while injecting fuel at different injection timing through the compression stroke can be beneficial for PPCI. However, using a high EGR rate can increase certain emissions and may result in reducing the engine performance. EGR may negatively impact fuel economy due to the reduced combustion efficiency because, to compensate the lower oxygen levels, engine may need to inject more fuel to maintain power output (Huang et al., 2016). High temperature of EGR can result in raising the temperature of the intake charge as well as increasing the

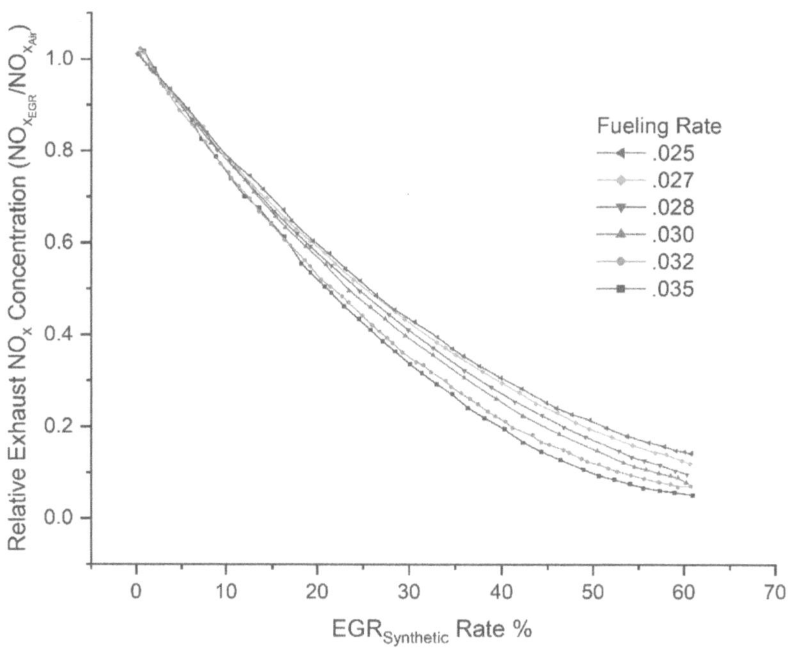

FIGURE 10.11 Reduction in NO_x compared with EGR rate in percentage. (Zheng et al., 2004.)

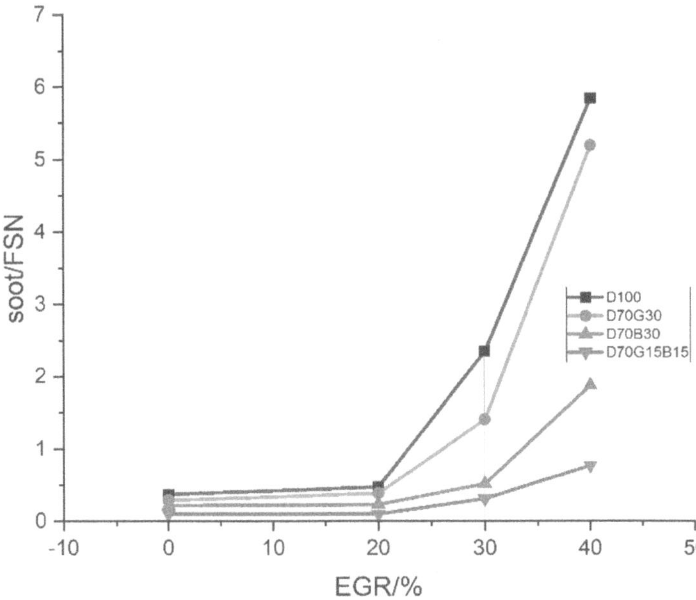

FIGURE 10.12 Soot emissions for blends at different EGR ratios. (Huang et al., 2016.)

cooling demand. During rapid changes in engine load, the EGR system may not respond quickly enough to optimize the EGR rate, leading to less efficient combustion and increased emissions during these transient conditions.

10.4 OCTANE NUMBER (ON)

The octane number is an important metric for evaluating resistance of a test fuel toward pinging or knocking during the combustion. The phenomenon of the knocking arises once the mixture air and fuel inside cylinder of an engine detonates early, which can lead for having an inefficient combustion and potential damage to the engine parts. The octane number can be ascertained through the comparison of knock intensity of a test fuel with one from blends of heptane and iso-octane. Heptane and iso-octane are used as the reference fuels because of high resistance to knocking offered by the iso-octane, while heptane knocks readily.

Based on the comparison with these reference fuels, the octane number can be defined with respect to the volume percentage of iso-octane present for a mixture of iso-octane and heptane matching corresponding to the knocking characteristics of test fuel for a standardized engine setting (Heywood, 2018). The rating assigned to the n-heptane is 0, while that assigned to pure iso-octane is 100. Thus, by comparison fuel having an octane number 90 will have same performance and knocking characteristics as 90% iso-octane and 10% n-heptane blend.

10.5 HIGH-OCTANE NUMBER (RON) FUELS

High research octane number fuels, shortened for RON, are formulated for resisting knock during combustion, which can show a crucial part in optimizing the efficiency and performance of IC engines (Lewander et al., 2011). Higher octane fuels can be beneficial for achieving the high-performance and high compression ratios in the engines designed to operate under higher pressures and temperatures, which can induce knocking (Cung et al., 2017). By using high-octane fuels, engines can achieve better efficiency and power output without the risk of knocking, thereby enhancing overall performance and longevity (Hildingsson et al., 2009). For meeting the stern environmental regulations, high-octane fuels can also result in cleaner combustion, due its resistance to premature ignition leading to the more controlled and complete burning of the fuel, resulting in reducing harmful emissions. See Table 10.1, These high-octane options, such as premium and super/ultra grades, are specifically developed for engines demanding higher compression, ultimately delivering enhanced performance and potentially contributing to lower emissions compared to regular and midgrade fuels designed for standard engine applications. While discussing high-octane fuels, it is important to distinguish between the two common octane ratings, which primarily measure the fuel's resistance to auto-ignition. See Table 10.2, RON and MON provide insights into a fuel's behavior under different engine operating conditions.

Engines designed with higher compression ratios or equipped with supercharging or turbocharging systems often necessitate the use of higher-octane fuels. This requirement or recommendation stems from the engine's capability to compress the air/fuel mixture to higher pressures, thereby extracting more mechanical energy from each combustion cycle. However, this increased compression also raises the risk of pre-detonation, or knocking, which can be mitigated by using higher octane fuels.

TABLE 10.1
Comparison of Different Grade of Octane Fuels

Octane Number	Fuel Type	Characteristics	Impact on Engine Performance	Environmental Benefits
87	Regular	Suitable for most engines without high compression ratios.	Lower performance compared to higher octane fuels.	Standard emissions for regular engines.
89–90	Midgrade	Better resistance to knocking than regular.	Slight improvement in performance and efficiency.	Slightly lower emissions than regular fuel.
91–94	Premium	Designed for high-performance engines with higher compression ratios.	Higher performance, allows for advanced engine tuning.	Potentially lower emissions due to better combustion efficiency.
95 or above	Super or ultra	Provides better fuel economy than lower octane fuels. Best choice for the latest high-performance engines.	Up to 3–4% lower fuel consumption than 91 octane fuel.	Reduced CO_2 emissions due to improved fuel economy.

TABLE 10.2
Comparison of RON and MON (Benavides et al., 2023; Rodríguez-Fernández et al., 2020)

Aspect	Research Octane Number	Motor Octane Number
Testing method	Tested for controlled settings in a laboratory at low speeds and mild conditions (600 RPM). Usually determined using standardized test methods like ASTM D2699.	Tested under more severe conditions (900 RPM), with higher temperatures and engine speeds using a different standardized test method, ASTM D2700.
Operating conditions	Simulates normal driving conditions.	Simulates high-load and high-speed driving conditions.
Temperature and humidity	Lower temperatures and controlled humidity.	Higher temperatures and less controlled humidity.
Relevance	More relevant to modern driving conditions and standard consumer vehicles.	More relevant for performance engines and high-stress driving conditions.
Typical values	Generally higher than MON for the same fuel.	Generally lower than RON for the same fuel.
Use in octane ratings	Part of Anti-Knock Index (AKI) (RON+MON)/2.	Part of the Anti-Knock Index (AKI) (RON+MON)/2.
Example values for regular fuel	Approximately 91–95.	Approximately 82–88.

10.6 HIGH-OCTANE NUMBER (ON) FUELS IN PPCI

Advanced combustion strategies like PPCI rely on precise control over ignition timing and fuel-air mixture preparation. High-octane fuels are particularly suited to these technologies due to their predictable and stable combustion characteristics, enabling more efficient and lower-emission engine operation. High-octane fuels are advantageous in PPCI engines because their ability to resist premature ignition as well as for increasing the delay in ignition, which are essential for accomplishing the PPC as they will able to withstand the higher combustion pressures and temperatures (Hildingsson et al., 2009). Higher octane fuels, such as methanol, ethanol, butanol, and pentanol, advance the cooling effect of charge, which can reduce nitrogen oxides emissions (NO_x). See Table 10.3, the alternative fuels with significantly higher RON than gasoline, which are advantageous for the controlled ignition and combustion phasing required in PPCI, also present a spectrum of energy densities that can impact overall engine efficiency and power output in these advanced combustion systems. Increasing the RON of any fuel can also improve engine power and acceleration.

Utilizing a high EGR rate, injecting fuel at varying injection timings during the compression stroke, and increasing fuel injection pressure can aid in achieving a desired outcome. Also using high RON fuel can serve the same purpose. Using higher RON fuels in such engines is must for having optimal performance and fuel economy. These fuels have a higher resistance to pre-detonation so that the engine

TABLE 10.3
Comparison of Popular Fuels with Their Energy Density and Octane No

Fuel Type	Octane Rating (RON)	Energy Content (MJ/kg)
Gasoline	87–98	44–47
Ethanol	108–113	26–30
Methanol	106–109	19–20
Butanol	96–100	29–36
Propane (LPG)	104–112	46–50
Methane (CNG)	120–130	50–55

can run more efficiently without knock-related issues. Therefore, in engines where higher compression ratios or forced induction systems are employed, the use of high-octane fuels is not just a recommendation but a requirement for smooth and efficient operation. Split injection or the pilot/main injection strategy can help to incorporate higher RON alcoholic fuel in contemporary CI engines. Utilization of alcoholic fuels with high-octane number, such as ethanol, methanol, and pentanol, is good option for Premixed Charge CI (PPCI) engines. These fuels can promote a homogeneous charge by extending the ignition delay and because of their high Lower Heating Value (LHV) and high research octane number (RON).

Higher oxygen content in alcoholic fuels, like butanol and methanol, can substantially lessen soot emissions when used in PPC strategy. The higher oxygen levels in these fuels promote leaner combustion, which decreases the formation of soot precursors (Uyumaz, 2015). Also fuels with higher LHV or greater latent heat of vaporization can cool the charge and lower the emissions of NO_x (Shamun et al., 2019). Combining the PPC strategy with higher alcoholic fuels such as butanol and pentanol, which have higher latent heat of vaporization than ethanol, methanol, and gasoline, seems to be a good approach for meeting stringent exhaust emission standards. When comparing the characteristics of n-butanol with those of methanol and ethanol, it is clear that higher alcoholic fuels have a higher density. So they can be mixed with diesel at any proportion beneath standard atmospheric conditions, unlike methanol and ethanol (Ning et al., 2020). Also, higher alcoholic fuels have better operational stability than ethanol because of its high energy density. To effectively execute the Premixed Charge CI (PPCI) approach, it is must to separate the ignition and fuel injection occurrences. This can be achieved by using advanced or retarded injection timings, employing hot EGR, and/or maintaining an elevated intake temperature using high RON fuel throughout the compression stroke (Rajesh Kumar et al., 2016).

10.6.1 Impact on Combustion and Performance Characteristics

Because of having high-octane ratings and oxygen content, alcoholic fuels can be beneficial for improving the performance and combustion characteristics in a CI engine while being used in either fumigated or blended form. Alcoholic fuels offer longer ignition delays in comparison to diesel fuels because of their lower CNs, which can be beneficial for attaining a more uniform charge during premixed combustion and CI modes, known for their minimal emissions. Alcohol blends significantly delay the SOC owing to the differences in fuel characteristics like CN, latent heat of evaporation, and viscosity. Incorporating high RON fuels can offer extended ignition delay into premixed CI combustion systems which, at low engine loads, can achieve remarkably low nitrogen oxide (NO_x) emissions and eliminate smoke production, even without the application of EGR (Hildingsson et al., 2009). Additionally, the greater volatility of petrol fuel enhances the flexibility of fuel injection strategies, which can further mitigate smoke emissions under specific conditions (Hildingsson et al., 2009). Researchers have indicated that using higher RON fuels in PPCI engines can significantly reduce soot and NOx emissions (Liu et al., 2017). High LHV helps draw thermal energy from the charge, thereby lowering the temperature inside cylinder and sharply reducing NOx emissions. Methanol, in particular, leads to a more substantial decrease in in-cylinder temperature compared to gasoline, due to its lower adiabatic flame temperature and greater charge cooling effect (Verhelst et al., 2019).

Brake-specific energy consumption (BSEC) can be lower in dual-fuel mode with alcohol injection compared to neat biodiesel, indicating that less fuel is required for generating the similar amount of power (El Shenawy et al., 2019). Study comparing ethanol, methanol, iso-butanol, and n-butanol detected that at middle and high loads, alcohol blends exhibited elevated peaks of HRR and in-cylinder pressure as result of increased premixed combustion (Li et al., 2019) (Pandey et al., 2019). Extended ignition delay allowed more fuel to combine with air before ignition, causing an elevated premixed HRR with the sequence of ignition delay ranging from shortest to longest is E13, NB20, and P25, correlating with their CNs (see Figure 10.13).

Gerardo Valentino et al. highlighted that diesel and butanol blends have longer ignition delays and higher volatility compared to diesel fuel, enhancing fuel-air mixing which can allow the partial premixed LTC achievable through the control of injection pressure, injection timing, and concentration of intake oxygen (Valentino et al., 2012). N-pentanol in blends can result in higher peak pressure and HRR at a 10% mass share, while methanol shows a significant improvement in HRR at a 30% mass share (Thiyagarajan et al., 2020). Alcoholic fuels can promote high thermal efficiency while reducing the emissions. High carbon alcohols such as pentanol and butanol can achieve higher brake thermal efficiency at lower mass basis in blends as compared to the lower alcohols such as methanol. A study showed that at 10% mass basis, n-pentanol shows higher efficiency than methanol, although at 30% mass basis, methanol leads to higher efficiency because of its higher heat release (Thiyagarajan et al., 2020). Ethanol showed similar characteristics when blended with diesel fuel (Qi et al., 2017). Qiang Fang et al. observed that ethanol's higher volatility and low CN resulted in an increased maximum HRR, decreased temperature during combustion, and delayed ignition, resulting in higher BTE with a BDE20 blend and advanced injection timing compared to diesel (Fang et al., 2013).

Haozhong Huang et al. reported that adding n-butanol to diesel enhanced combustion but increased fuel consumption (BSFC) due to an increased PRR (Huang et al., 2015). Advancing pilot injection in n-butanol blends additionally improved peak pressure and BSFC while reducing the main combustion pressure peak.

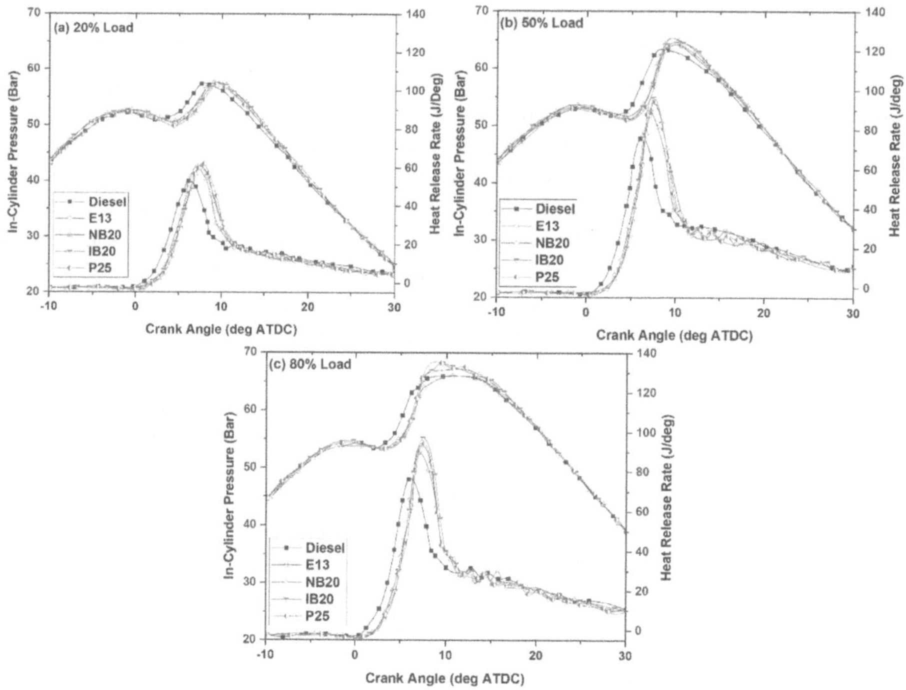

FIGURE 10.13 (a)–(c) Impact of alcoholic fuels on engine ICP and HRR. (Li et al., 2019.)

Research conducted with gasoline, n-heptane, and n-butanol blended with diesel showed that high cetane fuels (DH80, DG80) had minimal impact on combustion but exhibited lower injection pressure and higher fuel consumption at high load, likely due to low viscosity (Mao et al., 2018b). The CN was found to influence fuel mixing at part and intermediate loads, with low cetane fuels like DB80 having longer ignition delays, which are crucial for soot reduction but less effective at high load (Mao et al., 2018b). Study investigating the performance n-butanol/diesel blends during early and pre-injection partially premixed combustion found that these blends resulted in a reduced peak cylinder pressure and late SOC by more than 5 degrees crank angle (CA) equated to conventional diesel fuel (Liu et al., 2018). These findings specified that n-butanol/diesel blends have potential for mitigating engine noise and vibration by reducing peak in-cylinder pressure and delaying the onset of combustion. Research conducted for pentanol blends had confirmed that pentanol/diesel blends exhibit longer ignition delays, lower BSFC, and lower brake thermal efficiency (BTE) compared to pure diesel fuel (Rajesh Kumar & Saravanan, 2015). Research conducted on a DI, diesel engine employing a partially premixed LTC mode, iso-butanol and pentanol blends, when used with low EGR and delayed injection timing, achieved simultaneous deductions in NO_x and smoke emissions. Both blends exhibited a rapid premixed combustion phase compared to diesel. Although the B40 blend showed elevated peak pressures because of ignition delay, both blends demonstrated higher efficiency and increased HRRs (Rajesh Kumar & Saravanan, 2016). Compared with low-carbon alcohols (ethanol and methanol), high alcohols such as pentanol has better miscibility with diesel, reducing the risk of phase separation and ensuring consistent fuel blend properties. Overall, alcohol blends enhance fuel/air mixing and atomization, resulting in faster combustion and reduced combustion duration in comparison with neat diesel and also be beneficial in achieving the PPC combustion.

10.6.2 Impact on Emissions Characteristics

Partially premixed charge low-temperature combustion (PPC) can concurrently decrease the both NO_x and soot emissions in engines by operating outside the high-temperature zone. While a single early injection approach reduces particle size and quantity, but excessively early timing can backfire. Split injection can be more effective for soot control. Increasing the pilot fuel ratio in split injection further reduces particle formation but it can also affect the particle size. Optimizing injection strategies is key to balancing particle control, combustion efficiency, and minimizing other emissions in PPC engines. Alcoholic fuels, due to their increased oxygen content, can facilitate more thorough combustion, causing the reduction in the production of soot precursors and particulate matter and with low CN can achieve the LTC, further reducing the amounts of NO_x. Fuels with low CN, are more resistant to self-ignition, can extend the range of PPC load and are particularly suitable for high compression ratios and high engine loads (Zhen & Wang, 2015). This suitability arises from its prolonged ignition delay, which allows for an extended operating range of the PPCI engine, thereby optimizing performance and efficiency (Ning et al., 2020).

Late injection timing significantly reduced NO_x emissions to less than 200 ppm and smoke emissions to less than 0.5 l/m, although it can increase the CO and HC emissions at a brake mean effective pressure (BMEP) of 0.86 MPa (Qiu et al., 2016b). Qiang Fang et al. reported that medium EGR with retarded injection timing reduced NO_x and smoke releases by lowering oxygen concentration and prolonging ignition delay, aided by ethanol's high latent heat of vaporization (Fang et al., 2013). However, this can lead to in slightly increased HC and CO emissions. Haozhong Huang et al. observed that a specific n-butanol blend with 4 mg pilot injection significantly reduced both soot and NO_x emissions though minimally impacting BSFC and efficiency, suggesting a potential optimal balance for emissions and performance (Huang et al., 2015). The late injection timing and high injection pressure will also decrease smoke emissions but can result in increased NO_x emissions for diesel fuel. EGR improved NO_x emissions by increasing ignition delay and promoting partial premixed combustion, which benefits smoke and NO_x emissions but comes at the cost of higher fuel consumption and reduced engine thermal efficiency (Valentino et al., 2012). The butanol-diesel blends found to be exhibiting prolonged ignition delay, elevated peak pressure, and faster heat release during premix, which provided better tolerance for higher EGR rates, leading to lower smoke emissions (Rajesh Kumar & Saravanan, 2016). Both methanol and n-pentanol injections reduced the smoke emissions with further reductions observed at higher mass percentages of alcohol and it also reduced the smoke opacity and emissions of nitrogen oxides while causing high HC and CO emissions (Li et al., 2019). The application of EGR will additionally reduce NO_x emissions, although it also leads to decreased BTE and increased levels of smoke opacity, HC, and CO emissions. High RON fuel is particularly suitable for achieving superior partially premixed combustion.

10.7 CHALLENGES AND OPTIMIZATION

Using high-octane fuels for partially premixed combustion (PPC) in diesel engines has been very exciting and beneficial but there are some challenges in doing so. High-octane fuels can help in longer ignition delay but too much of delay can lead to misfires and knocking if not managed properly (Wang et al., 2021). Ignition delay and reactivity of high-octane number fuels need to be precise as per the loading conditions to ensure proper air-fuel mixing for an efficient combustion. Misfires and knocking can be regulated by optimizing fuel injection timing to reach the desired combustion phasing (Kolhe et al., 2023). Higher octane number fuels although reduce the NO_x, but in some cases at high mass ratio and engine conditions can produce higher release of NO_x because due to high temperature caused by elevated HRR and in-cylinder pressure, which requires additional emission control strategies (Vargün et al., 2023). Studies have observed increased cyclic variability for the combustion parameters, especially for the maximum mass-averaged temperature (T_{max}) and maximum in-cylinder pressure (P_{max}) and these fluctuations can lead to decreased combustion stability and increased engine roughness (Wang et al., 2016). To mitigate these effects, the fuel injection timing, injection pressure, intake temperature and other engine parameters are required to be adjusted for attaining a stable combustion process.

10.8 CONCLUSION

Studies have shown that LTC can reduce NO_x and particulate matter emissions so that there may not need to have exhaust after-treatment systems. High-octane fuels are great for optimizing partially premixed combustion in IC engines. Promising attributes of higher alcohols like butanol and pentanol (higher ignition temperature, better combustion efficiency, lower emissions, and knocking) make them more suitable for PPC than ethanol or methanol. High-octane fuels can provide stable combustion, better fuel-air mixing and reduce NO_x and soot emissions without compromising cleaner and more efficient engine technologies. But we need to overcome the hurdles like proper fuel injection parameters and intake pressure and temperature adjustments to fully utilize the benefits of high-octane fuels for PPC and reduce cyclic variability. More research and technology development are needed to overcome these hurdles and widespread acceptance of n-butanol and other high-octane fuels in PPC engines and have a more environmentally friendly and sustainable future for IC engines.

REFERENCES

Benajes, J., Molina, S., Novella, R., & Arthozoul, S. (2008). Advanced injection strategies to attain partially premixed combustion process in a heavy duty diesel engine. *SAE Technical Papers, 2008*(724), 776–790. https://doi.org/10.4271/2008-01-0642

Benavides, A., Zapata, C., Benjumea, P., Franco, C. A., Cortés, F. B., & Ruiz, M. A. (2023). Predicting octane number of petroleum-derived gasoline fuels from MIR spectra, GC-MS, and routine test data. *Processes, 11*(5), 1437. https://doi.org/10.3390/pr11051437

Bharadwaz, Y. D., & Kumari, A. S. (2023). PCCI combustion of low-carbon alternative fuels: A review. *Journal of Thermal Analysis and Calorimetry, 148*(12), 5179–5207. https://doi.org/10.1007/s10973-023-12108-1

Cengiz, C., & Unverdi, S. O. (2023). A CFD study on the effects of injection timing and spray inclusion angle on performance and emission characteristics of a DI diesel engine operating in diffusion-controlled and PCCI modes of combustion. *Energies, 16*(6). https://doi.org/10.3390/en16062861

Chen, S. K. (2000, August 21). *Simultaneous Reduction of NOx and Particulate Emissions by Using Multiple Injections in a Small Diesel Engine.* https://doi.org/10.4271/2000-01-3084

Cung, K. D., Ciatti, S. A., Tanov, S., & Andersson, Ö (2017). Low-temperature combustion of high octane fuels in a gasoline compression ignition engine. *Frontiers in Mechanical Engineering, 3.* https://doi.org/10.3389/fmech.2017.00022

De Bellis, V., Malfi, E., Lanotte, A., De Felice, M., Teodosio, L., & Bozza, F. (2022). A tabulated chemistry multi-zone combustion model of HCCI engines supplied with pure fuel and fuel blends. *Energies, 16*(1), 265. https://doi.org/10.3390/en16010265

Deep, A., Sandhu, S. S., & Chander, S. (2017). Experimental investigations on the influence of fuel injection timing and pressure on single cylinder C.I. engine fueled with 20% blend of castor biodiesel in diesel. *Fuel, 210*(August), 15–22. https://doi.org/10.1016/j.fuel.2017.08.023

De Ojeda, W., Zoldak, P., Espinosa, R., & Kumar, R. (2009). Development of a fuel injection strategy for partially premixed compression ignition combustion. *SAE Technical Papers, January.* https://doi.org/10.4271/2009-01-1527

Deresso, H., Ancha, V. R., Nallamothu, R. B., Singh, B., & Yoseph, B. (2023). Review: Investigation of partially pre-mixed charge ignitions (PPCI) engine mode. *Journal of Harbin Institute of Technology (New Series), 30*(3), 85–94. https://doi.org/10.11916/j.issn.1005-9113.2022001

Du, J., Sun, W., Guo, L., Xiao, S., Tan, M., Li, G., & Fan, L. (2015). Experimental study on fuel economies and emissions of direct-injection premixed combustion engine fueled with gasoline/diesel blends. *Energy Conversion and Management, 100*(x), 300–309. https://doi.org/10.1016/j.enconman.2015.04.076

El Shenawy, E. A., Elkelawy, M., Bastawissi, H. A. E., Panchal, H., & Shams, M. M. (2019). Comparative study of the combustion, performance, and emission characteristics of a direct injection diesel engine with a partially premixed lean charge compression ignition diesel engines. *Fuel, 249*(February), 277–285. https://doi.org/10.1016/j.fuel.2019.03.073

Fang, Q., Fang, J., Zhuang, J., & Huang, Z. (2013). Effects of ethanol-diesel-biodiesel blends on combustion and emissions in premixed low temperature combustion. *Applied Thermal Engineering, 54*(2), 541–548. https://doi.org/10.1016/j.applthermaleng.2013.01.042

Gwalwanshi, M., Chauhan, M. K., & Kumar, R. (2023). A review of low temperature combustion techniques and the evolution of combustion strategies. *Lecture notes in mechanical engineering*, 75–83. Springer. https://doi.org/10.1007/978-981-19-6945-4_6

Heywood, J. B. (2018). *Internal combustion engine fundamentals.* McGraw-Hill Education.

Hildingsson, L., Kalghatgi, G., Tait, N., Johansson, B., & Harrison, A. (2009). Fuel octane effects in the partially premixed combustion regime in compression ignition engines. *SAE Technical Papers*, *4970*. https://doi.org/10.4271/2009-01-2648

Huang, H., Liu, Q., Wang, Q., Zhou, C., Mo, C., & Wang, X. (2016). Experimental investigation of particle emissions under different EGR ratios on a diesel engine fueled by blends of diesel/gasoline/n-butanol. *Energy Conversion and Management*, *121*, 212–223. https://doi.org/10.1016/j.enconman.2016.05.034

Huang, H., Liu, Q., Yang, R., Zhu, T., Zhao, R., & Wang, Y. (2015). Investigation on the effects of pilot injection on low temperature combustion in high-speed diesel engine fueled with n-butanol-diesel blends. *Energy Conversion and Management*, *106*, 748–758. https://doi.org/10.1016/j.enconman.2015.10.031

Imtenan, S., Varman, M., Masjuki, H. H., Kalam, M. A., Sajjad, H., Arbab, M. I., & Rizwanul Fattah, I. M. (2014). Impact of low temperature combustion attaining strategies on diesel engine emissions for diesel and biodiesels: A review. *Energy Conversion and Management*, *80*(x), 329–356. https://doi.org/10.1016/j.enconman.2014.01.020

Jain, A., Singh, A. P., & Agarwal, A. K. (2017a). Effect of fuel injection parameters on combustion stability and emissions of a mineral diesel fueled partially premixed charge compression ignition (PCCI) engine. *Applied Energy*, *190*, 658–669. https://doi.org/10.1016/j.apenergy.2016.12.164

Jain, A., Singh, A. P., & Agarwal, A. K. (2017b). Effect of split fuel injection and EGR on NOx and PM emission reduction in a low temperature combustion (LTC) mode diesel engine. *Energy*, *122*, 249–264. https://doi.org/10.1016/j.energy.2017.01.050

Jin, S., Li, J., Deng, L., & Wu, B. (2021). Effect of the HPDI and PPCI combustion modes of direct-injection natural gas engine on combustion and emissions. *Energies*, *14*(7). https://doi.org/10.3390/en14071957

Khalid, A., Jaat, M., Zaman, I., Manshoor, B., & Ali, M. F. M. (2013). Effect of pilot injection on mixture formation, ignition process and flame development in diesel combustion. *Applied Mechanics and Materials*, *390*, 327–332. https://doi.org/10.4028/www.scientific.net/AMM.390.327

Kim, D. S., Kim, M. Y., & Lee, C. S. (2004). Combustion and emission characteristics of partial homogeneous charge compression ignition engine. *Combustion Science and Technology*, *177*(1), 107–125. https://doi.org/10.1080/00102200590883778

Kokjohn, S. L., & Reitz, R. D. (2009). Investigation of Design Parameters in Partially Premixed Compression Ignition Combustion Using Adaptive Injection Strategies. *ASME 2009 Internal Combustion Engine Division Spring Technical Conference*, 233–245. https://doi.org/10.1115/ICES2009-76030

Kolhe, A. V., Malwe, P. D., Chopkar, Y., Panchal, H., Ağbulut, Ü, Mubarak, N. M., Chowdhury, S., & Amesho, K. T. T. (2023). Performance analysis of biofuel–ethanol blends in diesel engine and its validation with computational fluid dynamics. *Environmental Science and Pollution Research*, *30*(60), 125117–125137. https://doi.org/10.1007/s11356-023-27086-y

Korkmaz, M., Zweigel, R., Jochim, B., Beeckmann, J., Abel, D., & Pitsch, H. (2018). Triple-injection strategy for model-based control of premixed charge compression ignition diesel engine combustion. *International Journal of Engine Research*, *19*(2), 230–240. https://doi.org/10.1177/1468087417730485

Krishnamoorthi, M., Malayalamurthi, R., He, Z., & Kandasamy, S. (2019). A review on low temperature combustion engines: Performance, combustion and emission characteristics. *Renewable and Sustainable Energy Reviews*, *116*(February), 109404. https://doi.org/10.1016/j.rser.2019.109404

Kumar, P., & Sandhu, S. S. (2022). An attempt to implement partially premixed combustion strategy in a multi-cylinder CRDI engine : A detailed experimental and wavelet transform analysis. *Fuel*, *323*(February), 124372. https://doi.org/10.1016/j.fuel.2022.124372

Lewander, C. M., Johansson, B., & Tunestal, P. (2011). Extending the operating region of multi-cylinder partially premixed combustion using high octane number fuel. *SAE 2011 World Congress and Exhibition*. https://doi.org/10.4271/2011-01-1394

Li, X., Guan, C., Yang, K., Cheung, C. S., & Huang, Z. (2019). Impact of lower and higher alcohol additions to diesel on the combustion and emissions of a direct-injection diesel engine. *Environmental Science and Pollution Research*, *26*(20), 21001–21012. https://doi.org/10.1007/s11356-019-05275-y

Liang, X., Zheng, Z., Zhang, H., Wang, Y., & Yu, H. (2019). A review of early injection strategy in premixed combustion engines. *Applied Sciences (Switzerland*, *9*(18), 1–34. https://doi.org/10.3390/app9183737

Liu, B., Cheng, X., Liu, J., Pu, H., & Yi, L. (2017). Combustion and particulate matter emission characteristics for diesel engine run by partially-premixed low-temperature combustion mode. *SAE Technical Papers, 2017-Octob.* https://doi.org/10.4271/2017-01-2398

Liu, B., Cheng, X., Liu, J., & Pu, H. (2018). Investigation into particle emission characteristics of partially premixed combustion fueled with high n-butanol-diesel ratio blends. *Fuel*, *223*(February), 1–11. https://doi.org/10.1016/j.fuel.2018.02.196

Liu, H., Tang, Q., Yang, Z., Ran, X., Geng, C., Chen, B., Feng, L., & Yao, M. (2019). A comparative study on partially premixed combustion (PPC) and reactivity controlled compression ignition (RCCI) in an optical engine. *Proceedings of the Combustion Institute*, *37*(4), 4759–4766. https://doi.org/10.1016/j.proci.2018.06.004

Manente, V., Zander, C.-G., Johansson, B., Tunestal, P., & Cannella, W. (2010). An Advanced Internal Combustion Engine Concept for Low Emissions and High Efficiency from Idle to Max Load Using Gasoline Partially Premixed Combustion. *SAE Technical Papers*.

Manimaran, R., Thundil Karuppa Raj, R., & Senthil Kumar, K. (2013). Premixed charge compression ignition in a direct injection diesel engine using computational fluid dynamics. *WSEAS Transactions on Heat and Mass Transfer*, *8*(1), 17–29.

Mao, B., Liu, H., Zheng, Z., & Yao, M. (2018a). Influence of fuel properties on multi-cylinder PPC operation over a wide range of EGR and operating conditions. *Fuel*, *215*(May 2017), 352–362. https://doi.org/10.1016/j.fuel.2017.08.099

Mao, B., Liu, H., Zheng, Z., & Yao, M. (2018b). Influence of fuel properties on multi-cylinder PPC operation over a wide range of EGR and operating conditions. *Fuel*, *215*(November 2017), 352–362. https://doi.org/10.1016/j.fuel.2017.08.099

Maurya, R. K. (2018). *Characteristics and control of low temperature combustion engines*. Springer International Publishing. https://doi.org/10.1007/978-3-319-68508-3

Musculus, M. P. B., Miles, P. C., & Pickett, L. M. (2013). Conceptual models for partially premixed low-temperature diesel combustion. *Progress in Energy and Combustion Science*, *39*(2–3). Elsevier Ltd. https://doi.org/10.1016/j.pecs.2012.09.001

Natarajan, S., Pitchandi, K., & Mahalakshmi, N. V. (2018). Optimization of performance and emission characteristics of PPCCI engine fuelled with ethanol and diesel blends using grey-Taguchi method. *Journal of Thermal Science*, *27*(1), 89–94. https://doi.org/10.1007/s11630-018-0988-1

Neely, G. D., Sasaki, S., Huang, Y., Leet, J. A., & Stewart, D. W. (2005). New diesel emission control strategy to meet US tier 2 emissions regulations. *SAE Technical Papers*, *114*, 512–524. https://doi.org/10.4271/2005-01-1091

Ning, L., Duan, Q., Chen, Z., Kou, H., Liu, B., Yang, B., & Zeng, K. (2020). A comparative study on the combustion and emissions of a non-road common rail diesel engine fueled with primary alcohol fuels (methanol, ethanol, and n-butanol) /diesel dual fuel. *Fuel*, *266*(September 2019), 117034. https://doi.org/10.1016/j.fuel.2020.117034

Noehre, C., Andersson, M., Johansson, B., & Hultqvist, A. (2006). Characterization of partially premixed combustion. *SAE Technical Papers, 724*, 776–790. https://doi.org/10.4271/2006-01-3412

Norouzi, A., Heidarifar, H., Shahbakhti, M., Koch, C. R., & Borhan, H. (2021). Model predictive control of internal combustion engines: A review and future directions. *Energies, 14*(19), 1–40. https://doi.org/10.3390/en14196251

Pandey, S., Bhurat, S., & Chintala, V. (2019). Combustion and emissions behaviour assessment of a partially premixed charge compression ignition (PCCI) engine with diesel and fumigated ethanol. *Energy Procedia, 160*(2018), 590–596. https://doi.org/10.1016/j.egypro.2019.02.210

Petersen, B. R., Ekoto, I. W., & Miles, P. C. (2010). An investigation into the effects of fuel properties and engine load on UHC and CO emissions from a light-duty optical diesel engine operating in a partially premixed combustion regime. *SAE Technical Papers, 3*(2), 38–55. https://doi.org/10.4271/2010-01-1470

Qi, D. H., Yang, K., Zhang, D., & Chen, B. (2017). Combustion and emission characteristics of diesel-tung oil-ethanol blended fuels used in a CRDI diesel engine with different injection strategies. *Applied Thermal Engineering, 111*, 927–935. https://doi.org/10.1016/j.applthermaleng.2016.09.157

Qiu, L., Cheng, X., Liu, B., Dong, S., & Bao, Z. (2016a). Partially premixed combustion based on different injection strategies in a light-duty diesel engine. *Energy, 96*, 155–165. https://doi.org/10.1016/j.energy.2015.12.052

Qiu, L., Cheng, X., Liu, B., Dong, S., & Bao, Z. (2016b). Partially premixed combustion based on different injection strategies in a light-duty diesel engine. *Energy, 96*(x), 155–165. https://doi.org/10.1016/j.energy.2015.12.052

Rajesh Kumar, B., & Saravanan, S. (2015). Effect of exhaust gas recirculation (EGR) on performance and emissions of a constant speed di diesel engine fueled with pentanol/diesel blends. *Fuel, 160*, 217–226. https://doi.org/10.1016/j.fuel.2015.07.089

Rajesh Kumar, B., & Saravanan, S. (2016). Effects of iso-butanol/diesel and n-pentanol/diesel blends on performance and emissions of a di diesel engine under premixed LTC (low temperature combustion) mode. *Fuel, 170*, 49–59. https://doi.org/10.1016/j.fuel.2015.12.029

Rajesh Kumar, B., Saravanan, S., Rana, D., & Nagendran, A. (2016). A comparative analysis on combustion and emissions of some next generation higher-alcohol/diesel blends in a direct-injection diesel engine. *Energy Conversion and Management, 119*, 246–256. https://doi.org/10.1016/j.enconman.2016.04.053

Rezaei, S. Z., Zhang, F., Xu, H., Ghafourian, A., Herreros, J. M., & Shuai, S. (2013). Investigation of two-stage split-injection strategies for a dieseline fuelled PPCI engine. *Fuel, 107*(x), 299–308. https://doi.org/10.1016/j.fuel.2012.11.048

Rodríguez-Fernández, J., Ramos, Á, Barba, J., Cárdenas, D., & Delgado, J. (2020). Improving fuel economy and engine performance through gasoline fuel octane rating. *Energies, 13*(13), 3499. https://doi.org/10.3390/en13133499

Shamun, S., Zincir, B., Shukla, P., Garcia Valladolid, P., Verhelst, S., & Tunér, M. (2018). Quantification and analysis of the charge cooling effect of methanol in a compression ignition engine utilizing PPC strategy. *Proceedings of the ASME 2018 Internal Combustion Engine Division Fall Technical Conference. Volume 1: Large Bore Engines; Fuels; Advanced Combustion*. San Diego, California, USA. ASME. https://doi.org/10.1115/ICEF2018-9657

Shim, E., Park, H., & Bae, C. (2020). Comparisons of advanced combustion technologies (HCCI, PCCI, and dual-fuel PCCI) on engine performance and emission characteristics in a heavy-duty diesel engine. *Fuel, 262*(November 2019), 116436. https://doi.org/10.1016/j.fuel.2019.116436

Sun, C., Kang, D., Bohac, S. V., & Boehman, A. L. (2016). Impact of fuel and injection timing on partially premixed charge compression ignition combustion. *Energy and Fuels, 30*(5), 4331–4345. https://doi.org/10.1021/acs.energyfuels.6b00257

Thiyagarajan, S., Sonthalia, A., Edwin Geo, V., Prakash, T., Karthickeyan, V., Ashok, B., Nanthagopal, K., & Dhinesh, B. (2020). Effect of manifold injection of methanol/n-pentanol in safflower biodiesel fuelled CI engine. *Fuel, 261*(September 2019). https://doi.org/10.1016/j.fuel.2019.116378

Uyumaz, A. (2015). An experimental investigation into combustion and performance characteristics of an HCCI gasoline engine fueled with n-heptane, isopropanol and n-butanol fuel blends at different inlet air temperatures. *Energy Conversion and Management, 98*, 199–207. https://doi.org/10.1016/j.enconman.2015.03.043

Valentino, G., Corcione, F. E., Iannuzzi, S. E., & Serra, S. (2012). Experimental study on performance and emissions of a high speed diesel engine fuelled with n-butanol diesel blends under premixed low temperature combustion. *Fuel, 92*(1), 295–307. https://doi.org/10.1016/j.fuel.2011.07.035

Valladolid, P. G. (2019). *Quantification and Analysis of the Charge Cooling Effect of.* 1–10.

Vargün, M., Özsezen, A. N., Botsalı, H., & Sayın, C. (2023). A study on the impact of fuel injection parameters and boost pressure on combustion characteristics in a diesel engine using alcohol/diesel blends. *Process Safety and Environmental Protection, 177*(July), 29–41. https://doi.org/10.1016/j.psep.2023.07.005

Veera, K. V., Masood, M., Ruan, S., Swaminathan, N., & Koll, H. (2012). Modelling of turbulent premixed and partially premixed combustion. *Advanced Fluid Dynamics.* https://doi.org/10.5772/26190

Verhelst, S., Turner, J. W., Sileghem, L., & Vancoillie, J. (2019). Methanol as a fuel for internal combustion engines. *Progress in Energy and Combustion Science, 70*, 43–88. https://doi.org/10.1016/j.pecs.2018.10.001

Wåhlin, F., Cronhjort, A., Olofsson, U., & Ångström, H. E. (2004). Effect of injection pressure and engine speed on air/fuel mixing and emissions in a pre-mixed compression ignited (PCI) engine using diesel fuel. *SAE Technical Papers, 724.* https://doi.org/10.4271/2004-01-2989

Wang, H., Feng, B., Zhang, L., Li, Y., Zheng, Z., & Yao, M. (2021). Experimental investigation on the effects of octane sensitivity on partially premixed low-temperature combustion. *Fuel, 287*, 119488. https://doi.org/10.1016/j.fuel.2020.119488

Wang, Q., Wang, B., Yao, C., Liu, M., Wu, T., Wei, H., & Dou, Z. (2016). Study on cyclic variability of dual fuel combustion in a methanol fumigated diesel engine. *Fuel, 164*, 99–109. https://doi.org/10.1016/j.fuel.2015.10.003

Yang, T., Yin, L., Meng, X., Tian, H., Long, W., & Tunestål, P. (2020). Partially premixed combustion optimization using double injection strategy in transient operation. *Applied Thermal Engineering, 169*(January), 114963. https://doi.org/10.1016/j.applthermaleng.2020.114963

Yao, M., Wang, H., Zheng, Z., & Yue, Y. (2010). Experimental study of n-butanol additive and multi-injection on HD diesel engine performance and emissions. *Fuel, 89*(9), 2191–2201. https://doi.org/10.1016/j.fuel.2010.04.008

Yao, M., Zheng, Z., & Liu, H. (2009). Progress and recent trends in homogeneous charge compression ignition (HCCI) engines. *Progress in Energy and Combustion Science, 35*(5), 398–437. https://doi.org/10.1016/j.pecs.2009.05.001

Zhang, F., Xu, H., Zeraati Rezaei, S., Kalghatgi, G., & Shuai, S. J. (2012). Combustion and emission characteristics of a PPCI engine fuelled with dieseline. *SAE Technical Papers.* https://doi.org/10.4271/2012-01-1138

Zhen, X., & Wang, Y. (2015). An overview of methanol as an internal combustion engine fuel. *Renewable and Sustainable Energy Reviews, 52*, 477–493. https://doi.org/10.1016/j.rser.2015.07.083

Zheng, M., Reader, G. T., & Hawley, J. G. (2004). Diesel engine exhaust gas recirculation––a review on advanced and novel concepts. *Energy Conversion and Management*, *45*(6), 883–900. https://doi.org/10.1016/S0196-8904(03)00194-8

Zheng, Z., Yue, L., Liu, H., Zhu, Y., Zhong, X., & Yao, M. (2015). Effect of two-stage injection on combustion and emissions under high EGR rate on a diesel engine by fueling blends of diesel/gasoline, diesel/n-butanol, diesel/gasoline/n-butanol and pure diesel. *Energy Conversion and Management*, *90*, 1–11. https://doi.org/10.1016/j.enconman.2014.11.011

11 Alcohol-Based CI Engines
Challenges, Solutions, and Future Directions

Sumit Kanchan, Sarbjot Singh Sandhu, and Prem Kumar

11.1 INTRODUCTION

Compression ignition (CI) engines, commonly known as diesel engines, have long been the backbone of the transportation industry, offering superior fuel efficiency and torque compared to their spark-ignition (SI) counterparts [1]. However, the emissions from conventional diesel engines have been a significant concern, with high levels of nitric oxides (NO_X) and particulate matter (PM) posing a threat to the environment and public health [2]. In recent years, there has been a growing interest in exploring the potential of alcohol-based fuels as a viable alternative to traditional diesel fuels. Alcohol-based fuels, such as ethanol and methanol, have the potential to address the emissions challenges associated with conventional diesel engines [3]. Dual-fuel gasoline-alcohol engines, for instance, have shown the ability to reduce greenhouse gas emissions while maintaining the performance characteristics of diesel engines [4]. Moreover, the use of alcohol fuels can lead to improved thermal efficiency and reduced aromatic hydrocarbon, sulfur, and lead content, further contributing to the reduction of air pollution [5]. One of the significant advantages of alcohol-based CI engines is their ability to operate at higher compression ratios, which can lead to enhanced efficiency [6]. Additionally, the increased oxygen content of alcohol-based fuels can improve the combustion process, resulting in more complete and efficient burning. An emerging option of interest involves oxygenated biofuels derived from sustainable sources [7, 8]. Among these, alcohols stand out as highly promising due to their favorable physicochemical properties, abundant raw material availability, established production processes, and inherent oxygen content. Figure 11.1 shows the advantages of using alcohol fuels in CI engines. They demonstrate the ability to mitigate soot formation and decrease PM emissions from internal combustion engines, leveraging their intrinsic oxygenated characteristics.

However, the use of alcohol fuels in CI engines presents challenges. The higher cost and limited availability of alcohol fuels compared to traditional diesel fuel may pose barriers to widespread adoption [9, 10]. Additionally, the lower energy density of alcohol fuels compared to diesel can result in reduced driving range, which may be a concern for some applications. Addressing these challenges through further research and development will be crucial for the successful implementation of alcohol-based CI engines in the transportation sector [11]. However, a variety of

DOI: 10.1201/9781003482277-11

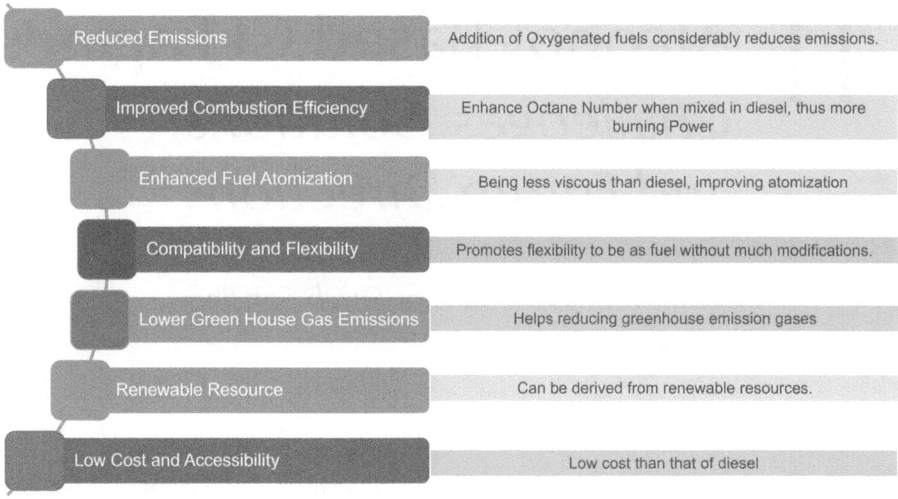

FIGURE 11.1 Advantages of using alcohol fuels in CI engines.

alcohol fuels can be utilized in CI engines, either in raw or blended form; ethanol and methanol are two of the most explored alcohol-based fuels for diesel applications. Ethanol, which can be produced from renewable biomass sources, has been used in blends with conventional diesel fuel to create "E-diesel" mixtures. Methanol, on the other hand, can be derived from natural gas, coal, or biomass, and has also been investigated for direct use in CI engines or as a blend component. The choice of alcohol fuel type and blend ratio can impact engine. Using pure alcohol fuels directly in unmodified diesel engines can lead to compatibility issues and poor cold start performance. However, blending alcohol fuels such as ethanol and methanol with conventional diesel can offer advantages like reduced emissions and improved combustion efficiency, while mitigating the challenges of lower energy density and cold start problems [12].

The optimal alcohol-diesel blend ratios and engine modifications required warrant further investigation to enable the widespread adoption of alcohol-based CI engines. Several studies have highlighted the key challenges in using alcohol fuels in CI engines. Ethanol, for example, has a lower energy density compared to diesel, which can result in reduced driving range and power output [13]. Methanol has a high latent heat of vaporization, posing difficulties in cold start and combustion stability [14, 15]. Butanol, while having properties more like diesel, can suffer from lower lubricity and higher viscosity, leading to fuel injection and engine wear issues [16]. Pentanol has shown promise, but its higher cost and limited availability may hinder widespread adoption [17]. This study aims to augment thoughtful understanding of the factors influencing combustion, with a focus on how engine performance and exhaust emissions are formed by optimized combustion processes. Considering the necessity for comprehensive study on the effects of alcohol and engine parameters in CI engines, this work critically examines combustion characteristics and addresses challenges posed by alcohol fuels. Later in the study a comparative analysis is done, summarizing the effectiveness of each considered alcohol fuel at parameters like combustion characteristics, material compatibility, lubricity issues, and fuel stability to name a few. It also explores potential solutions and future research directions to improve engine performance when using alcohol-based fuels. Finally, this review identifies key research gaps and suggests paths for further study.

11.2 STUDY OF VARIOUS LIGHT ALCOHOL FUELS FOR ITS CHALLENGES AND POTENTIAL SOLUTIONS IN CI ENGINES

Each alcohol fuel possesses unique physical and chemical properties that greatly affect engine-out characteristics such as performance, combustion, and emissions. Figure 11.1 summarizes the advantages of using alcohol fuels in CI engines. The physical properties of a fuel impact its vaporization and atomization processes, while its chemical properties affect ignition and combustion. Table 11.1 provides a comparison of the physicochemical properties of diesel and other alcohol fuels used in the CI engine. This section aims to outline the advantages, challenges faced, and

TABLE 11.1
Comparison of the Physicochemical Properties of Diesel and Other Alcohol Fuels Used in the CI Engine

Property	Units	Diesel	Methanol	Ethanol	Propanol	n-Butanol	1-Pentanol
Molecular formula	–		CH_3OH	C_2H_6O	C_3H_8O	C_4H_9OH	$C_5H_{11}OH$
Density	kg/m^3	840	729	792	804	810	814
Boiling point	°C	180–360	64	78	97	117.1	138
Viscosity @ 40°C (Taking 1cst = 1 mm^2/s)	mm^2/s	3.5	0.54	1.04	1.74	2.26	5.1
Research octane number	15–25	136	129	108	118	96	79
Motor octane number	–	104	102	91	103	78	72
Cetane number	–	51	5	7	12	17	20
Calorific value	MJ/kg	42.5	20	27	30.5	33	34.7
Latent heat of vaporization	kJ/kg	250	1178	840	728	584	308
Autoignition temperature	°C	256.85	470	434	350	385	300
Oxygen content	%	15	50	34	26.5	21.59	18.55
Specific gravity	kg/m^3	830–850	792	789	803	810	815
Freezing point	°C		–97	–114	–126	–89.3	–78.5
Flash point	°C	≥55	12	17	11.6	37	43

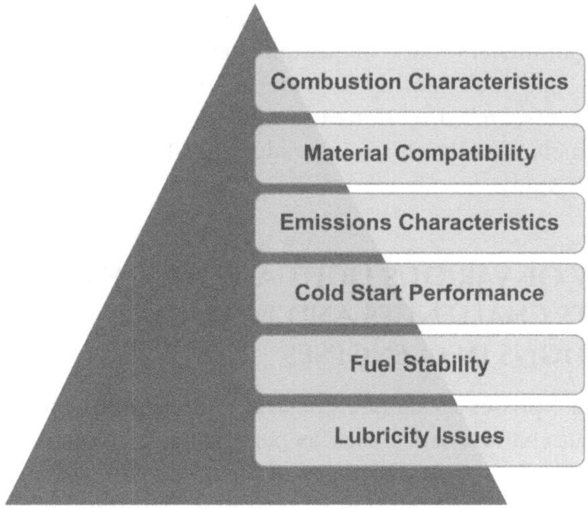

FIGURE 11.2 Key areas of study adopted during this work for highlighting the challenges and solutions.

potential solutions of methanol, ethanol, propanol, butanol and pentanol fuels' physical and chemical properties on combustion characteristics. Figure 11.2 shows the key areas of study considered for analysing and comparing the considered fuels.

11.2.1 Methanol

Methanol, known as methyl alcohol, is the simplest alcohol with one carbon atom having a chemical formula of CH_3OH. Produced from natural gas, biomass, or coal, methanol is known for its high-octane rating and clean-burning properties [18]. It is a colorless, tasteless liquid with a faint odor, polar solvent, and miscible with water and most organic solvents.

Methanol has a lower energy density compared to diesel, but it compensates with a high flame speed and the ability to produce less PM and nitrogen oxide (NO_X) emissions. Methanol has a flash point of 12°C, a melting point of −97.8°C, and a boiling point of 64°C, making it flammable and potentially explosive. Due to its higher oxygen content, it leads to more complete combustion and less stoichiometric air requirement. This results in lower emissions of carbon monoxide (CO) and hydrocarbons (HC). Methanol's lower viscosity compared to diesel also helps in achieving better atomization and mixing, enhancing combustion efficiency. Despite its lower calorific value, methanol's high latent heat of vaporization can reduce the intake air temperature, increasing the charge density and thereby improving engine performance. One significant benefit of methanol is its relatively high-octane number (this means a very low cetane number), which can prevent knocking in engines. Additionally, methanol has a lower boiling point, which aids in cold start conditions. Key advantages of using methanol as fuel are shown in Figure 11.3.

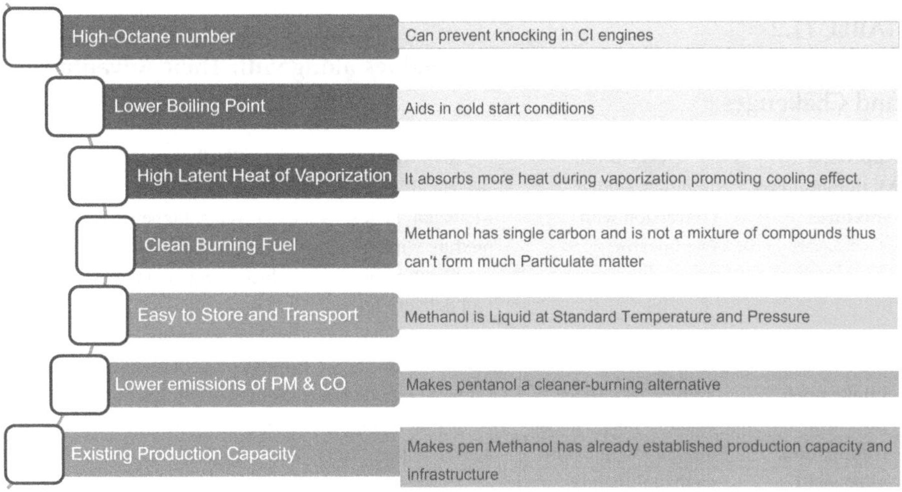

FIGURE 11.3 Advantages of using methanol as fuel in CI engines.

The dual-fuel approach, with separate introduction of methanol and diesel, is more common. Various methods to introduce methanol in CI engines are summarized in Table 11.2.

Methanol's lower flash point and higher vapor pressure pose safety risks, including vapor lock and handling issues. Its low cetane number means it cannot be directly used in CI engines due to poor autoignition characteristics. To use methanol in CI engines, it must be combined with a fuel that has good autoignition properties. The Detroit Diesel Company (DDC) [25] and MAN developed modified engines using improved methanol with ignition improvers, but this solution requires a suitable fuel system. Mixing methanol with diesel simplifies engine conversion with one injector but needs emulsifiers or co-solvents due to poor mixing. Higher methanol fractions reduce energy content, affecting power, and require lubricating additives due to low lubricity [26].

Moreover, due to its hygroscopic nature, methanol can absorb water vapor from the atmosphere, leading to phase separation and dilution of fuel value in methanol-diesel blends. Methanol's major issue is its high toxicity, which poses risks through ingestion, skin or eye contact, and inhalation. Studies [19–21, 23] have demonstrated that methanol can be used in CI engines with certain modifications, such as adjusting the fuel injection system and using ignition improvers to address its lower cetane number [27].

11.2.1.1 Key Challenges and Associated Solutions

11.2.1.1.1 Combustion Characteristics

Using methanol in CI engines presents several combustion challenges. First, the maximum substitution ratio, i.e., the proportion of diesel replaced by methanol, varies significantly among engines, influenced by factors such as compression ratio and turbocharging pressures. This variability affects end-gas temperature history and knock occurrence. Additionally, using the fumigation method can lead to a drop in

TABLE 11.2
Methods to Introduce Methanol in CI Engines along with Their Advantages and Challenges

Approach	Description	Advantage	Challenge	Ref.
Methanol-diesel mixtures	Simplifies engine conversion with one injector.	Easy Handling of Methanol – Diesel mixture without modification in fuel delivery line.	Requires emulsifiers or co-solvents; higher methanol fractions reduce energy content and necessitate lubricating additives.	[19]
Methanol intake port injection (fumigation approach)	Methanol is introduced into the engine's intake ports using a low-pressure methanol fueling system and port fuel injectors.	Easier retrofit solution; reported widely in engine conversions.	Limits power output due to end-gas autoignition; methanol fraction limited at higher and lower loads; potential issues with engine oil life and emissions.	[20]
Dual-fuel approach	Methanol and diesel are introduced into the engine separately.	More common approach; allows flexible fuel use.	Requires modifications to fuel injection system; potential for uneven combustion stability.	[21]
Glow plug ignition	Uses glow plugs to ignite methanol, eliminating the need for a pilot fuel.	Enables CI engine to run solely on methanol; straightforward ignition.	Needs careful timing of methanol injection; potential issues with methanol ending up in lube oil.	[22]
HCCI and PPC modes	Research engines run on methanol in homogeneous charge compression ignition (HCCI) and partially premixed combustion (PPC) modes.	Higher efficiency; reduced soot formation.	Requires precise control of ignition timing; long ignition delays due to high heat of vaporization and autoignition temperature.	[23, 24]

brake thermal efficiency (BTE), especially at low loads. The prominent challenges of using methanol on combustion performance are summarized in Table 11.3. At lower loads, the lower combustion efficiency can be attributed to the premixed methanol-air mixture approaching flammability limits, resulting in higher emissions of unburned fuel. Conversely, at higher loads, methanol's faster and cooler combustion can increase efficiency compared to the slower diffusion combustion of diesel [28].

11.2.1.1.2 Material Compatibility

Methanol poses several material compatibility challenges for CI engines, primarily due to its polarity and corrosiveness. Methanol can attack metals and elastomers used in fuel systems if they are not properly selected. This is aggravated by methanol's

TABLE 11.3
Key Challenges Observed by Researchers on Combustion Performance while Using Methanol/Methanol Blends in CI Engines

S. No.	Challenges Found in Published Literature	Possible Solution
1	Variability in maximum substitution ratio	Optimize engine parameters (compression ratio, turbocharging) to achieve consistent substitution ratios
2	Drop in brake thermal efficiency at low loads due to lower combustion efficiency	Improve fuel injection and air-fuel mixing techniques to enhance combustion efficiency at low loads
3	Increased ignition delay and flame lift-off length in methanol-air atmosphere	Adjust initial temperature and use advanced ignition timing controls to manage ignition delay and flame lift-off length
4	Difficulty in maintaining efficient combustion due to high latent heat of vaporization and lower energy content of methanol	Implement precise control of ignition and fuel mixture, and consider using additives to improve combustion characteristics

hygroscopic nature, which introduces water and contaminants that can trigger both dry and wet corrosion. Corrosion can be particularly severe in the presence of chloride ions, acetic acid, and ethyl acetate [29]. Methanol is also electrically conductive, increasing the risk of electrochemical and galvanic corrosion. Table 11.4 summarizes the ways these challenges can be mitigated. By using austenitic stainless steel,

TABLE 11.4
Key Challenges Observed by Researchers on Material Compatibility while Using Methanol/Methanol Blends in CI Engines

S. No.	Challenges Found in Published Literature	Possible Solution
1	Methanol's corrosiveness to metals and elastomers	Use austenitic stainless steel or metals coated with zinc or nickel alloy for components frequently in contact with methanol.
2	Hygroscopic nature leading to water-induced wet corrosion	Ensure the fuel system is designed to minimize water ingress and use hydrous methanol with caution.
3	Increased electrical conductivity leading to electrochemical and galvanic corrosion	Use materials resistant to electrochemical corrosion and manage fuel purity to limit contaminants.
4	Shrinkage, swelling, and hardening of elastomers	Employ alcohol-compatible elastomers like fluorocarbon elastomers or nitrile butadiene rubbers.
5	Sensitivity to temperature with most damage occurring during hot soak cycles	Utilize proper engine design and material selection to withstand temperature variations.

coated metals, alcohol-compatible elastomers, and appropriate additives can extend the life of a material [30].

11.2.1.1.3 Emission Characteristics

Coulier and Verhelst [19] reviewed the literature on dual-fuel diesel engines with methanol intake, noting several emission-related challenges. The maximum substitution ratio varies due to factors like compression ratio and turbocharging, affecting efficiency and emissions. Methanol fumigation generally reduces NO_X and PM emissions due to lower in-cylinder temperatures and faster combustion [7]. However, there is a significant increase in CO and HC emissions, especially at low loads. The NO_2 to NO ratio also increases due to HO_2 radical formation during methanol combustion. Table 11.5 summarizes the challenges observed by various investigators.

11.2.1.1.4 Cold Start Performance

Enough work has to be produced by the engine to cold start. Methanol, often considered as alternative fuel due to their renewable sources and lower emissions, present unique challenges and characteristics when used in internal combustion engines, particularly during cold starts. A few of the challenges are summarized below and in Table 11.6:

- *Lower Energy Density and Evaporation Requirements*: Methanol have lower energy densities compared to diesel, necessitating the evaporation of more mass to achieve a combustible air-fuel mixture.
- *Higher Heat of Vaporization*: Methanol have higher heats of vaporization than diesel, requiring more energy to evaporate the fuel. In cold conditions, this higher energy demand delays the formation of a vaporized fuel-air mixture suitable for combustion, potentially leading to increased cranking times before ignition.

TABLE 11.5
Key Challenges Observed by Researchers on Emission Characteristics while Using Methanol/Methanol Blends in CI Engines

S. No.	Challenges Found in Published Literature	Possible Solution
1	Increase in CO emissions due to lower temperatures	Enhance combustion temperatures through optimized engine tuning and improved fuel injection strategies
2	Increase in unburnt hydrocarbon (HC) emissions	Use advanced fuel injection systems and air-fuel mixing techniques to improve combustion efficiency and reduce HC emissions
3	Significant increase in NO_2 to NO ratio due to HO_2 radical formation	Employ catalytic converters and after-treatment systems to manage NO_x emissions effectively
4	Variability in particulate matter (PM) emissions, with a shift toward smaller particles	Implement advanced filtration and after-treatment systems to capture smaller particles and reduce overall PM emissions

TABLE 11.6
Key Challenges Observed by Researchers on Cold Start Performance while Using Methanol/Methanol Blends in CI Engines [31–34]

S. No.	Challenges Found in Published Literature	Possible Solution
1	Lower energy density, more mass to evaporate	Implement preheating systems such as glow plugs or intake air heaters to aid fuel vaporization.
2	Higher heat of vaporization	Utilize fuel heating systems to reduce viscosity and facilitate quicker vaporization.
3	Lower flammability limit, higher vapor requirement	Adjust engine control strategies for optimized injection timing and pressure.
4	High vapor pressure requirement for stoichiometric mixtures	Design engines with calibrated injection systems to achieve required vapor pressures.
5	Single-component fuel without volatile components	Develop engine calibration methods for precise fuel-air ratio adjustments during cold starts.

- *Lower Flammability Limit*: The lower flammability limit of methanol is higher than that of diesel. This implies that a higher concentration of vaporized fuel is needed in the air to achieve combustion, which can be challenging to achieve during cold starts when vaporization rates are reduced.
- *Vapor Pressure Requirements*: Despite having slightly higher saturated vapor pressures compared to diesel, methanol have low stoichiometric air-fuel ratios (AFR). Achieving the necessary vapor pressure for a stoichiometric mixture becomes more challenging in cold conditions due to reduced vaporization rates and increased demand for a precise fuel-air ratio.
- *Single-Component Fuels*: Methanol and ethanol are single-component fuels with defined boiling points, unlike diesel which contains volatile components that aid in cold startability. This characteristic makes achieving rapid and uniform vaporization more critical during cold starts to ensure proper combustion initiation.
- *Preheating Systems*: CI engines often utilize glow plugs or intake air heaters to preheat the combustion chamber and intake air, aiding in fuel vaporization. This is crucial for methanol due to its higher heat of vaporization.
- *Fuel Heating*: In extremely cold conditions, fuel heating systems may be employed to reduce the viscosity and improve vaporization of methanol before injection into the engine.

11.2.1.1.5 Lubricity Issue

Fuel additives are often used for improving fuel's lubricity, anti-corrosion, and anti-rust properties. Methanol has lower lubricity than diesel; therefore, adding lubricating additives is a must [35]. This property of methanol therefore imposes a necessity to use high-pressure methanol pumps and specially designed injectors to be used to inject fuel inside combustion chamber, which adds extra cost to the overall cost. Table 11.7 summarizes the key challenges and solutions as found in the literature published while using methanol/methanol blends on lubricity.

TABLE 11.7
Key Challenges Observed by Researchers on Lubricity while Using Methanol/Methanol Blends in CI Engines

S. No.	Challenge	Solution
1	Low lubricity of methanol leading to wear and tear	Add lubricity-enhancing additives to the methanol fuel to improve lubrication properties.
2	Increased wear of fuel injection system components	Use advanced coating materials on fuel system components to reduce wear and enhance durability.
3	Corrosion of fuel system components due to methanol	Incorporate corrosion inhibitors into the fuel to protect engine parts from methanol-induced corrosion.
4	Inadequate lubrication causing injector sticking	Implement regular maintenance schedules and use high-quality, methanol-compatible lubricants.
5	Compatibility issues with conventional lubricants	Develop and use specialized lubricants that are compatible with methanol to ensure proper lubrication.

11.2.1.1.6 Fuel Stability

Methanol's fuel stability is a critical concern for CI engines due to its tendency to absorb moisture and contaminants from the environment. This hygroscopic nature can lead to phase separation, where water-rich layers form, potentially causing corrosion and operational issues. Methanol's reactivity also makes it prone to oxidation, forming acidic by-products that can further degrade engine components. Additionally, methanol can degrade elastomers and metals if not properly managed. To ensure fuel stability, it is crucial to use appropriate additives, storage solutions, and material selections that can withstand methanol's chemical properties. Table 11.8 summarizes the key challenges and solutions as found in the literature published while using methanol/methanol blends on fuel stability.

TABLE 11.8
Key Challenges Observed by Researchers on Fuel Stability while Using Methanol/Methanol Blends in CI Engines

S. No.	Challenge	Solution
1	Hygroscopic nature leading to moisture absorption	Store methanol in sealed containers to limit exposure to moisture. Use desiccants and water-absorbing additives.
2	Phase separation caused by water absorption	Regularly monitor and remove any water layers. Use co-solvents to maintain homogeneity.
3	Oxidation forming acidic by-products	Add antioxidants and corrosion inhibitors to the methanol. Ensure proper storage conditions to minimize exposure to air.
4	Degradation of elastomers and metals	Use alcohol-compatible materials such as fluorocarbon elastomers and austenitic stainless steel. Apply protective coatings to vulnerable metals.
5	Contamination from environmental exposure	Implement rigorous fuel handling protocols to reduce contamination risk. Use high-quality methanol with minimal impurities.

11.2.2 Ethanol

Ethanol, also known as ethyl alcohol, is a two-carbon alcohol with the chemical formula C_2H_6O. It is a colorless, volatile, flammable liquid with a distinctive wine-like odor and taste. Ethanol is miscible with water and most organic solvents. It has a melting point of −112°C and a boiling point of 78°C. Ethanol is renewable, biodegradable, and can be produced from various biomass sources, making it an attractive alternative fuel for CI engines. Being an oxygenate, ethanol promotes more complete combustion, leading to reduced PM (soot) emissions, a significant environmental concern associated with diesel engines. This characteristic makes ethanol particularly attractive for reducing the environmental impact of CI engines, especially in blend forms where it can partially substitute for diesel fuel.

The use of ethanol in CI engines is gaining remarkable interest among researchers. The blend of ethanol-diesel is sometimes called diesohol [36]. Ethanol contains about 34% oxygen by weight, which can enhance combustion efficiency and reduce emissions. This attribute combined with other properties makes ethanol a fuel of interest for CI engines. As summarized by S.A. Shahir et al. [37], the most significant benefits are the reduction in PM emissions. Compared to low-sulfur diesel, diesohol can reduce PM emissions by approximately 15%, and a 10% ethanol addition to diesel can cut these emissions by 30–50%. Additionally, diesohol can achieve energy outputs like those of traditional fossil diesel. Ethanol also improves the cold flow properties of diesel fuel, enhancing performance in colder conditions. Furthermore, diesohol blends have a higher heat of vaporization, which contributes to more efficient combustion. Different strategies for using diesel-ethanol blends in CI engines are summarized in Table 11.9.

TABLE 11.9
Strategies Adopted by Various Investigators to Use Ethanol/Ethanol-Diesel Blend in CI Engines

S. No.	Strategy	Description	Advantage	Disadvantage	Ref.
1	Mixing of two fuels preceding injection	Blending diesel and ethanol before injecting into the engine.	Simple method, easy to implement.	Poor blend stability, dependent on chemical composition, temperature, and ethanol percentage.	[38]
2	Complete substitution by ethanol	Replacing diesel with approximately 95% ethanol.	Significant reduction in diesel usage.	Requires major engine modifications due to ethanol's weak autoignition properties.	[39, 40]
3	Fumigation of ethanol	Adding ethanol to the intake air charge.	Improves combustion efficiency.	Complexity in controlling precise ethanol-air mixture.	[41, 42]
4	Dual-fuel injection	Separate injection systems for diesel and ethanol.	Greater control over fuel mixture and combustion process.	Increased engine complexity and cost due to additional injection system.	[43, 44]

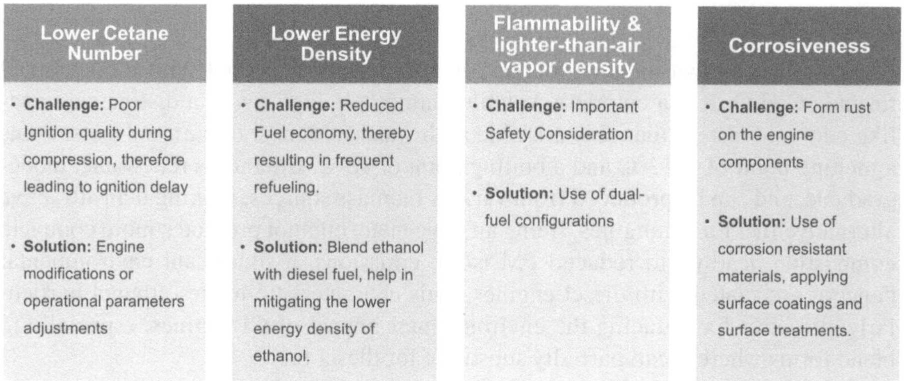

FIGURE 11.4 Crucial challenges and their potential solutions while using ethanol in CI engines.

11.2.2.1 Crucial Challenges and Associated Solutions

Figure 11.4 lists the crucial challenges and their potential solutions while using ethanol in CI engines.

Using diesohol (a blend of diesel and ethanol) in CI engines presents several challenges. These challenges primarily stem from the differences in the physical and chemical properties between diesel and ethanol. The cetane number of the diesohol blend is significantly lower than that of diesel fuel, which decreases ignition quality and increases ignition delay. Ethanol's incomplete miscibility with diesel also poses issues, leading to phase separation and inconsistent fuel mixture. Additionally, using diesohol affects various fuel properties, such as density, viscosity, lubricity, and energy content, which can impact fuel injection and combustion characteristics. A few key challenges as found in literature are highlighted in Table 11.10.

TABLE 11.10
Summary of Various Challenges as Encountered by Investigators while Using Ethanol/Ethanol-Diesel Blend in CI Engines [37–39]

S. No.	Disadvantage	Consequence	Potential Solution
1	Lower cetane number	Decreased ignition quality, longer ignition delay	Use cetane improvers or blend with higher cetane fuels.
2	Lower energy density	Lower fuel efficiency and reduced power output	Adjust engine calibration to optimize performance with lower energy density fuel.
3	High affinity for water	Fuel system corrosion, fuel filter plugging, microbial growth in the fuel system	Modify fuel system materials to resist corrosion and microbial growth; use water separation filters.

(Continued)

TABLE 11.10 *(Continued)*
Summary of Various Challenges as Encountered by Investigators while Using Ethanol/Ethanol-Diesel Blend in CI Engines [37–39]

S. No.	Disadvantage	Consequence	Potential Solution
4	Lower lubricity	Increased engine wear	Use lubricity enhancers or formulate the blend to maintain acceptable lubricity levels.
5	Incomplete miscibility of ethanol in diesel	Phase separation, inconsistent fuel mixture	Use emulsifiers or co-solvents to improve miscibility.
6	Variations in fuel delivery system	Minor modifications are required to ensure proper fuel delivery	Retrofit or adjust fuel delivery systems to handle the blend.
7	Altered fuel properties (density, viscosity, lubricity, energy content)	Affected fuel injection and combustion characteristics	Optimize engine calibration and fuel additives to maintain performance.
8	Lower viscosity	Potential wear and tear on engine components	Use lubricity enhancers or formulate the blend to maintain acceptable viscosity levels.
9	Lower calorific value	Reduced energy output per unit of fuel	Compensate by adjusting engine tuning or using higher energy density additives.
10	Increased soot formation	Higher particulate emissions	Implement advanced combustion strategies or after-treatment technologies to reduce soot emissions.

Additionally, fumigation of ethanol in diesel engines in low-temperature combustion modes is also catching researchers' attention. Table 11.11 summarizes the different low-temperature combustion strategies adopted in the literature along with its challenges.

Although ethanol's flammability and lighter-than-air vapor density are important safety considerations, they do not outweigh its inherent limitations as a direct diesel replacement in CI engines. Table 11.12 summarizes the effect of ethanol fumigation on engine performance and emission characteristics of diesel engine. Researchers have investigated the use of dual-fuel configurations as a potential solution to overcome these limitations. Studies [49–52] have shown that by combining ethanol with diesel in a dual-fuel system, the inherent drawbacks of using ethanol alone as a direct replacement in CI engines can be mitigated [50]. In this approach, a small amount of diesel fuel is used as the primary ignition source, while most of the energy is provided by ethanol, allowing the engine to operate with the benefits of alcohol-based fuels while maintaining safety and performance characteristics.

TABLE 11.11
Summary of Different Low-Temperature Combustion Strategies Adopted by Various Researchers Combined with Their Advantages, Disadvantages, and Potential Solutions

S. No.	Strategy	Description	Advantages	Disadvantages	Solution	Ref.
1	**Homogeneous charge compression ignition (HCCI)**	Premixed air-fuel mixture ignited by compression heat	Low NO_x and PM emissions, high efficiency	Difficult to control ignition timing and combustion phasing	Advanced control systems, real-time sensors	[45]
2	**Premixed charge compression ignition (PCCI)**	Partial premixing of fuel before compression ignition	Low NO_x emissions, good fuel efficiency	Limited by knock and misfire issues	Injection timing optimization, EGR strategies	[46, 47]
3	**Reactivity-controlled compression ignition (RCCI)**	Dual-fuel system with different reactivity fuels	Flexible control over combustion, low emissions	Complex fuel system, control challenges	Advanced fuel injection systems, optimized fuel blends	[48]
4	**Low-temperature combustion (LTC)**	General approach to reduce peak combustion temperatures	Lower NO_x and PM emissions, improved fuel efficiency	Higher CO and HC emissions, stability issues	Enhanced after-treatment systems, optimized operating conditions	[36]

TABLE 11.12
Effect of Ethanol Fumigation on Engine Performance and Emission Characteristics of Diesel Engine

Diesel Engine Specifications	Fuel Types and Blends	Engine Performance Parameters			Emissions	Optimal Range (%)	Reference
		BSFC	BTE	EGT			
Single-cylinder, 4-stroke, DI, VCR	Ethanol fumigated diesel	↓	↑	↓	Soot and smoke decrease and UHC and CO increase	20	[4]
Four-cylinder, 4-stroke, DI	Ethanol fumigated diesel at varying loads	↓	↓	↓	Smoke and NO_x emissions decrease but UHC and CO increase	15	[53]
Single-cylinder, 4-stroke, DI, Mitsuki, MIT-186FG model	Palm oil ethyl ester (POEE) and fumigated ethanol	↓	↓	↓	Smoke, NO_x, and PM emissions decrease but CO increases	30	[54]
Single-cylinder, 4-stroke, DI	Ethanol fumigated diesel	↓	↓	↓	Smoke, NO_x, and PM emissions decrease but CO increases	15	[55]

11.2.3 Propanol

Propanol, also known as propyl alcohol, is a three-carbon atom alcohol that exists in two isomeric forms: 1-propanol (n-propanol) and 2-propanol (isopropanol). Both isomers share the same chemical formula, C_3H_8O, and exhibit similar energy content, making them of interest for use as alternative fuels in CI engines. 1-Propanol is a linear aliphatic alcohol with the molecular structure $CH_3CH_2CH_2OH$. It is a colorless liquid with a mild, alcohol-like odor. 1-Propanol has a high heating value comparable to other alcohols, making it a potential candidate for CI engines. Its flammability, however, poses a significant risk of fire and explosion when exposed to heat, sparks, or open flames. Despite this, 1-propanol's complete miscibility with water and most common organic solvents enhances its versatility. In addition to its use as a solvent in the pharmaceutical and cosmetic industries, 1-propanol's energy density makes it a viable option for blending with diesel to improve combustion efficiency and reduce emissions in CI engines. However, safety precautions must be taken due to its potential to cause irritation to the eyes, skin, and respiratory tract, with high concentrations potentially leading to central nervous system depression.

2-Propanol, commonly known as isopropanol, has the molecular structure $CH_3CH(OH)CH_3$. This secondary alcohol also possesses a high energy content suitable for use in CI engines. Like 1-propanol, 2-propanol is highly flammable, with risks of fire and explosion under certain conditions. It is completely miscible with water and most organic solvents, making it adaptable for various applications. The characteristic alcohol-like odor of 2-propanol is generally stronger than that of 1-propanol. 2-Propanol's energy density makes it a strong candidate for use in CI engines, either as a pure fuel or blended with diesel, to enhance performance and reduce emissions. Nonetheless, exposure to 2-propanol can cause irritation to the eyes, skin, and mucous membranes, with high exposure levels potentially leading to symptoms such as headache, dizziness, and central nervous system depression. Figure 11.5 lists the crucial challenges and their potential solutions observed while

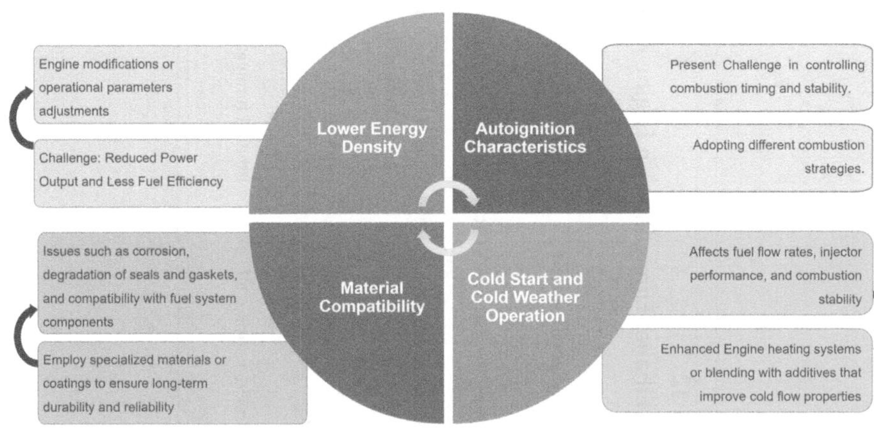

FIGURE 11.5 Crucial challenges and their potential solutions observed while using propanol/propanol blends in CI engines.

TABLE 11.13
Comparison of Important Properties of Different Propanol Isomers [56–58]

Property	Units	n-Propanol	Isopropanol
Chemical formula		C_3H_7OH	C_3H_8O
Molecular weight	g/mol	60.10	60.10
Boiling point	°C	97.2	82.5
Viscosity	mPa·s at 20°C	2.031	2.024
Flash point	°C	21	12
Density	g/cm³ at 20°C	0.8039	0.785

using propanol/propanol blends in CI engines. Table 11.13 compares the important properties of the different propanol isomers.

11.2.3.1 Crucial Challenges and Associated Solutions

The energy content and properties of both 1-propanol and 2-propanol make them promising alternatives to conventional fuels in CI engines. Their potential to improve combustion efficiency and reduce emissions positions them as valuable components in the future of sustainable fuel technologies. However, a few challenges are also imposed by these physicochemical properties.

- *Lower Energy Density*: Propanol exhibits a lower energy density compared to diesel fuel. This fundamental property translates into reduced power output and potentially lower fuel efficiency when used directly or in high concentrations in CI engines [59].
- *Viscosity Considerations*: The higher viscosity of propanol, relative to diesel, poses challenges in fuel atomization and spray characteristics during injection. Higher viscosity can lead to poor fuel-air mixing, incomplete combustion, and increased emissions of unburned hydrocarbons (UHC) and PM [60].
- *Autoignition Characteristics*: Propanol has different autoignition characteristics compared to diesel, including a lower autoignition temperature. While this property can facilitate easier ignition under certain conditions, it also presents challenges in controlling combustion timing and stability, particularly in CI engines designed and calibrated for diesel fuels.
- *Combustion Dynamics*: The combustion characteristics of propanol, such as flame speed and ignition delay, differ from those of diesel. These differences can influence engine performance parameters such as thermal efficiency and emissions.
- *Material Compatibility*: Propanol's chemical composition may interact differently with engine materials and components compared to diesel. Issues such as corrosion, degradation of seals and gaskets, and compatibility with fuel system components need careful consideration [37].

- *Cold Start and Cold Weather Operation*: Propanol-based fuels may pose challenges during cold starts and operations in colder climates. Their properties, such as viscosity and cold flow characteristics, can affect fuel flow rates, injector performance, and combustion stability at lower temperatures [37, 61].
- *Blending and Storage Stability*: Blending propanol with diesel or other fuels requires careful consideration of blend ratios and storage stability. Propanol's miscibility with water and other solvents can lead to phase separation issues and reduced fuel stability over time [62].

In a recent study by Bencheikh et al. [63], the performance, emission, and combustion characteristics of diesel engines using waste cooking oil and n-propanol/diesel blends were investigated. This research aimed to assess the feasibility of n-propanol as a potential alternative fuel or fuel additive in CI engines. They found that the inclusion of n-propanol improved combustion efficiency, leading to more complete fuel burning and consequently enhancing engine performance. Moreover, the blends exhibited significant reductions in carbon monoxide (CO) and PM emissions compared to pure diesel, indicating a potential for cleaner exhaust emissions. The study also highlighted an increase in BTE with n-propanol blends, suggesting improved fuel utilization within the engine. Zhao et al. [63] investigated the performance and emission characteristics of CI engines fueled with isopropanol-diesel blends. Their research revealed that these blends led to lower nitrogen oxides (NO_x) emissions and significantly reduced smoke opacity levels compared to pure diesel fuel. Despite a slight reduction in engine power output, which was deemed acceptable, the environmental benefits of reduced NO_x emissions and cleaner exhaust gases highlight the potential of isopropanol-diesel blends as a viable strategy for mitigating air pollution from CI engines.

11.2.4 Butanol

Butanol, a four-carbon alcohol, exists in four isomeric forms: n-butanol, 2-butanol, iso-butanol, and tert-butanol [58, 64]. Each isomer shares the chemical formula $C_4H_{10}O$ but differs in molecular structure, influencing its physical and chemical properties. These properties, along with their energy content, make butanol isomers viable candidates for use as alternative fuels in CI engines. Figure 11.6 shows the advantageous properties of butanol as fuel for CI engines.

n-Butanol (1-Butanol): n-Butanol, with the molecular structure $CH_3CH_2CH_2CH_2OH$, is a linear aliphatic alcohol. It is a colorless liquid with a mild, banana-like aroma and a strong alcoholic odor. n-Butanol's high heating value, comparable to other alcohols, makes it a potential fuel for CI engines. It is slightly hydrophobic and sparingly soluble in water but completely miscible with most organic solvents. This miscibility enhances its versatility in various applications, including as a solvent in the pharmaceutical and cosmetic industries. When used in CI engines, n-butanol can improve combustion efficiency and reduce emissions, although its flammability requires

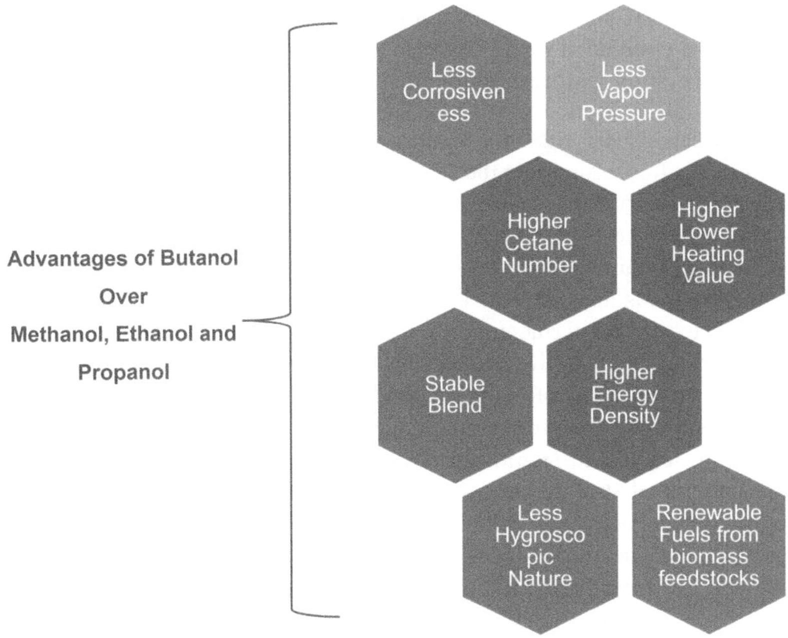

FIGURE 11.6 Advantages of using butanol over light alcohol fuels when used in CI engines.

careful handling. Direct contact may cause irritation to the eyes and skin, and inhalation of its vapors can have narcotic effects at high concentrations.

2-Butanol (Sec-Butanol): 2-Butanol, with the structure $CH_3CH_2CHOHCH_3$, is a secondary alcohol and a colorless, flammable liquid with a mild odor. It is slightly soluble in water and completely miscible with polar organic solvents. The energy density of 2-butanol makes it suitable for blending with diesel to enhance engine performance and reduce emissions. However, like n-butanol, its flammability poses risks of fire and explosion. Exposure to 2-butanol can cause irritation to the eyes and skin, and inhalation of high concentrations can lead to central nervous system depression.

Iso-Butanol (2-Methyl-1-Propanol): Iso-butanol, with the molecular structure $(CH_3)_2CHCH_2OH$, is a colorless liquid with a sweet, musty odor. It is miscible with all common organic solvents and slightly soluble in water. The energy content of iso-butanol makes it a strong candidate for CI engines. Its flammability and potential health effects are like those of other butanol isomers. Iso-butanol can be used as a fuel additive to improve combustion efficiency and reduce emissions, but safety precautions must be observed to prevent irritation and other health risks.

Tert-Butanol (Tert-Butyl Alcohol): Tert-butanol, with the structure $(CH_3)_3COH$, is a clear liquid with a camphor-like odor and a melting point slightly above room temperature, making it solid at room temperature. It is highly soluble in water and miscible with ethanol and diethyl ether. The high energy density of tert-butanol makes it a potential fuel for CI engines, either as a pure fuel or blended with diesel. However, its flammability and potential to cause irritation require careful handling and storage.

Since the late 1990s, there has been a growing interest in butanol as a sustainable automobile fuel, prompting the search for more cost-effective biobutanol production processes compared to petrochemical methods. Although commercial butanol is currently produced through petrochemical routes, microbial fermentation via the acetone-butanol-ethanol (ABE) process, particularly using *Clostridium acetobutylicum* bacteria [71], is an established method. This anaerobic fermentation technique, second only to ethanol fermentation by *Saccharomyces cerevisiae* in scale, was a major industrial process until the mid-20th century when it was replaced by oil refining due to the rising costs of biomass feedstock and the low sugar content in molasses. The oil crises and environmental concerns of the 1970s [72] reignited interest in biobutanol, leading to significant advancements in the ABE process, including improved microbial strains that enhance butanol production yields. In 1996, global butanol production was 2.49 billion kg, and today it is estimated at 10–15 billion pounds annually. The market for n-butanol is approximately 3 million tons per year, growing at 3.2% annually, with the overall biofuel market valued at $50 billion and expected to reach $250 billion [73]. Table 11.14 summarizes the important conclusions from experimental studies performed on CI engines while using butanol-diesel blends in CI engines. Due to its superior technical properties and sustainability, biobutanol is poised to become a leading biofuel alongside ethanol in the next decade.

11.2.4.1 Crucial Challenges and Associated Solutions

11.2.4.1.1 Combustion and Emission Characteristics

Butanol generally exhibits a lower cetane number compared to diesel fuel, which adversely affects ignition quality in CI engines. A lower cetane number can lead to poor combustion efficiency and increased ignition delay, potentially resulting in higher emissions of UHC and carbon monoxide (CO) [74, 75]. Table 11.15 highlights the primary challenges associated with using butanol-diesel blends and suggests potential solutions to mitigate these issues, ensuring better performance and emissions control.

11.2.4.1.2 Material Compatibility and Corrosiveness

Butanol fuels can be corrosive to certain materials commonly used in engine components and fuel systems. This corrosiveness can lead to the degradation of fuel lines, gaskets, and other engine parts, necessitating the use of corrosion-resistant materials or protective additives [76]. The compatibility of butanol with existing engine materials is a critical consideration for its widespread adoption. Modification like replacing the plastic and rubber parts with higher alcohol-tolerant materials could be the solution. To address the challenges posed by using butanol in CI engines, particularly its impact on plastic and rubber components, researchers are exploring a variety of materials that offer higher tolerance to alcohols. Table 11.16 summarizes the materials used in fuel systems, their properties, and their specific uses, along with the references.

TABLE 11.14
Summary of Various Experimental Studies Performed on CI Engines Using Butanol/Diesel Blends

S. No.	References	Engine Type Used	Butanol % in Diesel	Conditions	Important Conclusions	Ref.
1	Prem Kumar et al.	4-cylinder CRDI engine	10, 20, 30	Partially premixed combustion strategy, early, and late injection timings	• Coefficient of variance of indicated mean effective pressure decreased from 3.1 to 2.75% as n-butanol increased (0–20%) • CO and smoke emissions decreased	[65]
2	Ashish et al.	1-cylinder, four-stroke water-cooled diesel engine	10, 15, 20, 25	Varying compression ratio, start of injection, and injection pressure (n-butanol)	• BSFC increases • Lower soot and NO_x emissions • Higher CO_2 and unburned hydrocarbon emissions • 20% butanol blend showed better results	[66]
3	Sahin and Aksu	4-cylinder, 4-stroke turbocharged RENAULT K9K 700 water-cooled CRDI diesel engine	2, 4, 6	Varying speed and load (n-butanol)	• Marginal rise in peak pressure • Rise in BSFC • Lower soot emissions (observed for all blends)	[67]
4	Dogan et al.	Naturally aspirated, 1-cylinder 4-stroke DI diesel engine	5, 10, 15, 20	Varying engine loads (n-butanol)	• Rise in BTE and BSFC • Lower CO, NO_x, and smoke emissions	[68]
5	Karabektasans and Hosoz	Single-cylinder DI diesel engine	5, 10, 15, 20	Varying engine speed	• Lower BTE • Higher BSFC • Lower CO, NO_x, and higher UHC	[69]
6	Singh et al.	Mix case study	Varying conditions	Butanol diesel blend (Bu/B/D), biodiesel diesel blend, and diesel	Bu/B/D blends provide better fuel quality and significant results from the diesel engine characteristics	[70]

TABLE 11.15
Crucial Combustion and Emission Characteristics Challenges as Found in Literature of Butanol as Fuel in CI Engines along with Its Potential Solutions

S. No.	Challenge	Potential Solution
1	Increased HC emissions with higher butanol blends	Implement advanced combustion techniques and after-treatment systems to reduce HC emissions
2	Decreased vehicle drivability with 40% butanol blends	Fine-tune engine calibration and fuel injection strategies to improve drivability
3	Increased fuel consumption with higher butanol blends	Improve fuel injection timing and pressure to enhance combustion efficiency
4	Increased CO emissions in urban drive cycles	Use catalytic converters and optimized combustion control to reduce CO emissions
5	Slightly higher specific fuel consumption (SFC)	Improve engine design and fuel injection systems to maximize fuel efficiency
6	Slight decrease in brake thermal efficiency (BTE)	Enhance engine thermal management and fuel mixing processes
7	Engine performance deviations with 40% iso-butanol	Limit the concentration of iso-butanol to 30% or optimize engine calibration for higher blends

TABLE 11.16
Crucial Material Compatibility Challenges as Found in Literature of Butanol as Fuel in CI Engines along with Its Potential Solutions

Material	Properties and Uses	Ref.
Fluoroelastomers (FKM)	Exceptional resistance to heat, chemicals, and alcohols; ideal for fuel system seals	[76]
Polytetrafluoroethylene (PTFE) (Teflon)	Superior chemical resistance; used in gaskets and linings	[77]
Ethylene propylene diene monomer (EPDM)	Known for heat and oxidation resistance; being enhanced for improved alcohol tolerance for hoses and seals	[78]
Nitrile butadiene rubber (NBR)	Advanced to bolster resistance to alcohols; suitable for automotive fuel systems	[79]
Polyether ether ketone (PEEK)	Offers excellent chemical resistance and mechanical properties; suitable for components exposed to aggressive fuel blends	[80]
Silicone rubber	Maintains properties under extreme temperatures and exposure to alcohols; ideal for seals and gaskets	[81]
Polyvinylidene fluoride (PVDF)	Highly non-reactive thermoplastic fluoropolymer; used in fuel lines and storage tanks for solvent and alcohol resistance	[82]
Nylon (PA11 and PA12)	Being researched for improved alcohol resistance; suitable for fuel lines and other fuel-exposed components	[83]

These materials are critical in ensuring the durability and reliability of CI engines operating on butanol-blended fuels.

11.2.4.1.3 Cold Start and Low-Temperature Performance

Butanol fuels can present challenges in cold start and low-temperature conditions due to their higher viscosity and lower volatility compared to diesel. These properties can affect fuel atomization and mixing, leading to difficulties in starting the engine and achieving efficient combustion under cold conditions [5]. Higher lower heating value (LHV), and selfignition temperature (SIT), and lower CN of butanol as compared to diesel limit its straight usage in CI engines without any engine modifications.

11.2.5 Pentanol

Pentanol, specifically n-pentanol, is gaining attention as a viable fuel for CI engines due to its advantageous properties. Among higher alcohols, n-pentanol, with its five-carbon straight-chain structure, stands out as one of the most promising [84]. The production of pentanol commonly involves natural microbial fermentation using engineered microorganisms and biosynthesis from glucose. It has better fat-solubility than ethanol and therefore can be used for esterification reactions. Pentanol has a higher calorific value than diesel, which means it provides more energy per unit of fuel. This higher energy density makes it more efficient for use in CI engines [85]. The combustion characteristics of pentanol are favorable, as it burns more completely and with a higher flame speed, leading to improved engine performance and efficiency. Pentanol is less hygroscopic than lower alcohols like ethanol and methanol, meaning it absorbs less water from the atmosphere. This characteristic is crucial because water contamination in fuel can lead to phase separation, corrosion, and decreased combustion efficiency. Unlike methanol and ethanol, which are highly miscible with water, pentanol's lower water solubility reduces the risk of these issues. Pentanol has a lower vapor pressure compared to methanol and ethanol, which reduces the risk of vapor lock in fuel systems. Its higher boiling point also ensures better atomization and mixing with diesel, enhancing combustion efficiency. Additionally, pentanol's higher flash point makes it safer to handle and store. Studies [86, 87] have shown that pentanol can be blended with diesel up to 25% without any engine modifications, simplifying its adoption as a fuel. Furthermore, the use of pentanol-diesel blends results in lower emissions of PM and carbon monoxide (CO) compared to pure diesel. This makes pentanol a cleaner-burning alternative, aligning with stricter environmental regulations and sustainability goals. Figure 11.7 lists the various properties of pentanol which makes it a strong contender as a fuel for CI engines.

Despite the promising attributes of pentanol as a fuel for CI engines, recent studies have identified several challenges that need to be addressed. These challenges include issues related to combustion characteristics, emissions, material compatibility, and economic feasibility.

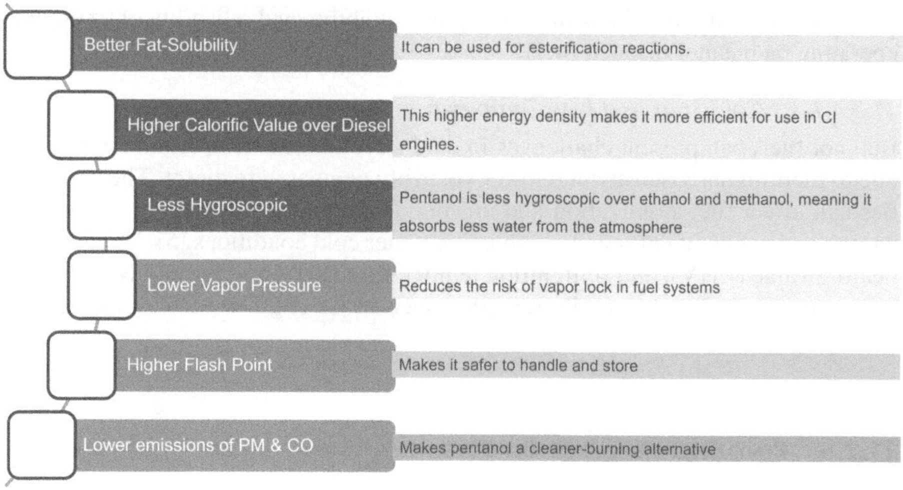

FIGURE 11.7 Advantages of using pentanol as fuel in CI engines.

11.2.5.1 Crucial Challenges and Associated Solutions

11.2.5.1.1 Combustion Characteristics

Table 11.17 summarizes the critical challenges observed while using pentanol as fuel for CI engines. One of the primary challenges highlighted is the variation in combustion characteristics when using pentanol blends. Studies have shown that while pentanol improves combustion efficiency, it can also lead to issues such as increased ignition delay and changes in the heat release rate. Wei et al. [88] investigated the effects of pentanol/diesel blends (up to 30% v/v) in a four-cylinder DI CI engine at various loads, comparing results to baseline procedure. The study found that all blended fuels exhibited similar pressure profiles and heat release curves, with increased ignition delay periods across low, medium, and high loads. A similar trend was observed by Rajesh Kumar B and Saravanan S [89] at higher loads. Their studies concluded that ignition delay period further increases with exhaust gas recirculation (EGR). Delayed ignition timings and peak pressures, along with peak heat release rates shifted away from top dead center (TDC), were observed with higher pentanol fractions due to its lower cetane number. The pentanol blend showed inferior combustion characteristics compared to the butanol blend due to its higher boiling point, lower oxygen content, and lower cetane number, regardless of EGR use and delayed injection timing. These variations can affect engine performance and require further optimization of engine parameters to fully utilize pentanol's benefits.

11.2.5.1.2 Material Compatibility

Pentanol's chemical properties can cause compatibility issues with certain engine materials, particularly elastomers and plastics used in fuel system components. Over time, exposure to pentanol can lead to material degradation, leaks, and failures in the fuel system. Research has highlighted the need for developing or

TABLE 11.17
Crucial Issues as Found in Literature on Usage of Pentanol as Fuel in CI Engines

S. No.	Key Issues	Studied Reason	Consequence	Possible Solution	Ref.
1	**Marginal combustion performance**	Lower cetane number	Increased ignition delay and changes in the heat release rate	Addition of cetane number improvers such as 2-ethylhexyl nitrate (EHN)	[108]
2	**Material incompatibility**	Exhibit corrosive properties	Degradation or swelling of materials in the fuel system Engine components such as pistons, piston rings, and valves Formation of oxide layers of Al_2O_3, FeO, and Fe_2O_3 formed on the surface of the engine components	Novel amino acid-based protic ionic liquids can be explored	[101]
3	**High UHC and NO_x emissions**	Lower cetane number	Increased emissions of unburned hydrocarbons (UHC) and nitrogen oxides (NO_x) Advanced injection timings result in decreased smoke emissions and extended ignition delays	Additional after-treatment devices Cetane number improvers such as EHN Research on the impact of varying EGR rates and injection strategies	[96, 108]
4	**Poor cold start performance**	Higher boiling point and high viscosity	Difficulties in starting engines at low temperatures Causes misfires	Need for preheating systems or blend optimization	[97, 98]
5	**Lubricity issues**	Low viscous-pressure coefficient and low viscosity	Struggles to form an efficient lubricating film Increased friction, resulting in accelerated wear and tear	Inclusion of additives Explore the applicability of novel amino acid-based protic ionic liquids Blending pentanol with diesel to ensure sufficient lubricity	[101]
6	**Fuel stability**	Oxidation between fuel molecules and oxygen	Darkening of the fuel, formation of acids and aldehydes	Application of fuel additives like BHA, DPA, BHT, and PA	[107]

selecting materials that are more resistant to pentanol's effects. Pentanol, being a higher-level alcohol, can exhibit corrosive properties that may impact the materials commonly used in fuel storage and delivery systems. Additionally, another concern is the potential for degradation or swelling of materials in the fuel system when exposed to pentanol. This can lead to leaks, malfunctions, and overall reduced efficiency of the fuel delivery system [90]. Furthermore, the compatibility of sealants and gaskets with pentanol must be carefully considered, as these components play a vital role in maintaining the integrity of the fuel system. Sealants and gaskets that are not designed to withstand the unique properties of pentanol may deteriorate over time, leading to leaks and safety hazards. Conducting thorough compatibility tests and using materials specifically designed for alcohol fuels can help mitigate these challenges and pave the way for the successful integration of pentanol as a viable alternative fuel option [91]. A study on material compatibility for key engine components such as pistons, piston rings, and valves was conducted by Sathish Kumar and Ashok Kumar [92] for light alcohol fuels like methanol. The findings revealed that these components are more prone to corrosion when using pure alcohol fuels compared to alcohol-diesel blends. Additionally, it was observed that oxide layers such as Al_2O_3, FeO, and Fe_2O_3 formed on the surface of the engine components. These oxide layers act as a barrier between the metal surfaces and the oxygen present in the fuel, thereby slowing down the oxidation process. Unlike methanol, pentanol is less hygroscopic, which means the deterioration of P100 (pure pentanol) is slower than that of M100 (pure methanol). As a result, pentanol undergoes less oxidation with engine components, producing fewer acids and consequently a lower total acid number (TAN).

11.2.5.1.3 Emission Characteristics

Although pentanol-diesel blends can reduce PM and carbon monoxide (CO) emissions, they may also lead to increased emissions of UHC and nitrogen oxides (NO_x). This trade-off between different types of emissions poses a challenge to meeting stringent emission regulations and necessitates additional after-treatment technologies. Liang Z et al. [93] conducted a study revealing that the n-pentanol/Fischer-Tropsch diesel blend shows lower PM emissions by nearly 85% compared to regular diesel. They observed that adding pentanol results in particles being arranged closer together, reduces graphitization, and increases surface functional groups like C–OH, C = O, and sp^3/sp^2, which can enhance post-treatment particle removal efficiency. However, due to its low cetane number, pentanol exhibits poor ignitability and cold start performance. To address these issues, researchers have explored the use of cetane number improvers such as 2-ethylhexyl nitrate (EHN). Additionally, Pan et al. [94] found that adding 50% n-pentanol to diesel reduces soot emissions, combustion speed, and BTE, while significantly increasing HC and CO emissions. The addition of EHN to the blend notably decreases HC and CO emissions. Several studies have stated the realism of using 50% pentanol-diesel blends without significant compromise in engine performance, although with a slight decrease in BTE. Research [95, 96] on the impact of varying EGR rates and injection strategies on CI engines using pentanol-diesel blends indicated that moderate EGR rates reduce NO_x and smoke emissions, but higher rates lead to further reductions in BTE. Similarly, advanced injection timings result in decreased smoke emissions and extended ignition delays.

11.2.5.1.4 *Cold Start Performance*

Cold start performance is another challenge, especially in colder climates. Pentanol has a higher boiling point compared to diesel, which can affect its volatility and cause difficulties in starting engines at low temperatures. Its high heat of vaporization, high viscosity, and low volatility properties contribute to difficulties such as prolonged ignition delay and increased likelihood of misfires during cold starts. The high heat of vaporization of pentanol requires more energy to vaporize and atomize the fuel, delaying the start of combustion. Additionally, its high viscosity makes it less likely to atomize properly at lower temperatures, leading to incomplete combustion and potentially causing misfires. Researchers [97, 98] have highlighted these challenges in various studies focused on alcohol fuels in CI engines. Addressing these issues is crucial for improving the usability of pentanol blends in cold climate regions and enhancing overall engine performance and reliability. This necessitates the need for preheating systems or blend optimization to ensure reliable cold start performance.

11.2.5.1.5 *Lubricity Issues*

Pentanol, with its low viscous-pressure coefficient and low viscosity, struggles to form an efficient lubricating film on frictional surfaces. Consequently, its lubricating performance and corrosion resistance are inferior to oil-based lubricants. However, the inclusion of additives can significantly enhance the physicochemical properties, film-forming capability, and lubricating performance of water-based lubricants [99, 100]. Diesel fuel naturally provides the necessary lubrication to these components, reducing friction and wear. Pentanol, while an attractive alternative fuel for CI engines, presents notable challenges due to its lower lubricity compared to diesel fuel. However, pentanol's lower lubricity can lead to increased friction, resulting in accelerated wear and tear, reduced engine life, and potentially higher maintenance costs. The increased wear and tear on fuel pumps and injectors due to pentanol's low lubricity is a significant challenge. Fuel pumps and injectors are precision components that rely on the lubricating properties of the fuel to operate efficiently. Without adequate lubrication, these parts can suffer from scoring, pitting, and other forms of mechanical wear. This not only affects the performance and efficiency of the engine but also leads to premature failure of critical components, necessitating costly repairs and replacements.

To address the lubricity issue, several solutions have been proposed and investigated. One effective approach is the use of lubricity additives in pentanol-diesel blends. In a recently published study [101], two novel amino acid-based protic ionic liquids were investigated for lubricating studies. These additives are designed to enhance the lubricating properties (anti-wear characteristics and extreme pressure loading capacities) of the base fluid, thereby reducing friction and wear on engine components. The study concluded that adding just 1% of the amino acid-based protic ionic liquid to the base lubricating fluid significantly improves its lubricating characteristics. Studies [102, 103] have shown that the addition of specific lubricity additives can significantly improve the performance of pentanol-diesel blends, making them more comparable to conventional diesel in terms of lubrication.

Another solution is blending pentanol with diesel in proportions that ensure sufficient lubricity while still taking advantage of pentanol's benefits. For example, blending up to 20–30% pentanol with diesel has been shown to maintain adequate lubricity and reduce the negative impacts on engine components [104]. This approach allows for a reduction in petroleum diesel use while mitigating the risks associated with low lubricity. Additionally, advancements in material science have led to the development of more wear-resistant materials for fuel system components. By using materials with higher resistance to wear and better compatibility with lower lubricity fuels, the adverse effects of pentanol can be further minimized.

11.2.5.1.6 Fuel Stability

Fuel deterioration includes sedimentation, darkening of the fuel, and formation of acids and aldehydes. This is due to oxidation process between fuel molecules and oxygen. Biodiesel is composed of unsaturated methyl esters like methyl linoleate, which is prone to oxidation, resulting in the formation of acids, aldehydes, esters, ketones, peroxides, and alcohols.

In a study [105], it was found that pentanol-diesel blends tend to separate when stored for extended periods, especially under varying temperature conditions. The phase separation was attributed to the different polarities and miscibility of the components, which affected the homogeneity of the blend. This inconsistency in the fuel mixture can cause irregular combustion, leading to inefficiencies and increased emissions. Another research by Yilmaz and Vigil [106] highlighted similar concerns. They observed that pentanol-diesel blends, when exposed to storage conditions typical of automotive environments, showed signs of phase separation after a few months. The study emphasized the importance of maintaining stable blending conditions and suggested that additives might be necessary to enhance the miscibility and stability of the blend. The phase-separated fuel exhibited varied combustion characteristics, causing fluctuations in engine performance and higher wear on engine components. The application of fuel additives can solve issues linked with phase separation. BHA (butylated hydroxyanisole), DPA (diphenylamine), BHT (butylated hydroxytoluene), PA (pyrogallol), TBHQ (tert-butyl hydroquinone), and PG (propyl gallate) are antioxidant additives used in diesel-biodiesel blends to avoid auto-oxidation [107]. The researchers found that the addition of pentanol to diesel increased the likelihood of water absorption, which further exacerbated the phase separation issue. This water absorption not only affected the blend's stability but also led to corrosion in fuel storage and delivery systems. The inconsistent fuel properties due to phase separation were linked to increased instances of incomplete combustion and engine knocking.

11.2.6 Conclusion and Future Directions

Future directions for CI engines emphasize the need to investigate the effects of methanol, ethanol, propanol, butanol, and pentanol as alternative fuels. Methanol-diesel blends should be studied for their impact on engine performance, emissions, and combustion behavior, with a focus on addressing phase separation issues by mixing methanol with propanol or isopropanol. Ethanol's potential as a co-solvent in diesel blends requires exploration to improve mixing stability, while also evaluating its impact on engine emissions such as NO_x, PM, and HC, and assessing its performance in advanced combustion modes like reactivity controlled compression ignition (RCCI), homogeneous charge compression ignition (HCCI), and pre-charge compression ignition (PCCI) [109]. Table 11.18 compares the effectiveness of various alcohol fuels under study at different studied parameters.

TABLE 11.18
Comparative Analysis Highlighting the Nature of Alcohol Fuels under Study at Different Considered Parameters

Parameters	Attributes	Methanol	Ethanol	Propanol	Butanol	Pentanol
Combustion characteristics	**Strength**	Faster and cooler combustion at high loads.	Rapid evaporative cooling ability at medium and high loads.	Improved combustion behavior at all loads.	Up to 20% blend showed better combustion behavior.	Burns more completely with a higher flame speed.
	Weakness	Poor combustion efficiency at low loads.	Combustion efficiency at low loads remains a challenge.	Lower energy density thereby offering reduced power output.	Rise in BSFC in most studies was observed.	Ignition delay period increases.
Material compatibility	**Strength**	Several material options which are methanol compatible.	Compatible with wide range of corrosion inhibitors.	Compatible with wide range of corrosion inhibitors.	Compatibility with various superior corrosion resistance materials.	Can be blended with diesel up to 25% thereby requiring no engine modification and minimal material modification.
	Weakness	Hygroscopic in nature.	Quality of ethanol controls the corrosion effects.	Degradation of seals, gaskets, and corrosion effects.	Leads to degradation of fuel lines, engine parts, and gaskets.	Less degradation of fuel lines, engine parts, and gaskets than propanol and butanol.
Emission characteristics	**Strength**	Compatibility with after-treatment devices like diesel.	Significantly reduce PM emissions.	Significant reduction in CO and PM.	Lower CO, NO_x, and smoke emissions.	Lower PM and CO emissions compared to diesel.
	Weakness	Higher NO_x and UHC emissions.	Increases soot formulation.	Increased soot formulation and NO_x.	Increased UHC emissions.	Increased UHC emissions.

(Continued)

TABLE 11.18 *(Continued)*
Comparative Analysis Highlighting the Nature of Alcohol Fuels under Study at Different Considered Parameters

Parameters	Attributes	Methanol	Ethanol	Propanol	Butanol	Pentanol
Cold start performance	**Strength**	Low viscosity at low temperatures.	Cold flow property is improved when blended with diesel.	Engine heating system or blending with additives could solve the cold start characteristics.	Cold flow property is improved when blended with diesel.	Cold start performance can be enhanced using cetane improvers like EHN.
	Weakness	Requirement of high concentration of vaporized fuel.	Limited miscibility which limits its usage at lower temperature.	Temperature dependency of viscosity is more significant than ethanol and methanol.	High viscosity and lower volatility compared to diesel resulting in delayed cold start.	Lower cetane number than other alcohol fuels.
Lubricity index	**Strength**	Less overly thickening of lubricating oil when used as fuel.	Minor variations in fuel delivery system are required.	Better lubricity than ethanol thus maintaining engine component longevity.	Better lubricity performance than gasoline and ethanol thereby less wear and tear.	High molecular weight and long HC chain provide natural lubricating characteristics than other lower alcohol fuel.
	Weakness	Low lubricity than diesel, therefore more requirement of additives.	Lubricity of a blended mixture is affected as its viscosity is lowered.	Lubricity characteristics are not good when compared with diesel.	Lubricity characteristics are not good when compared with diesel.	Lubricity efficiency is not a par with diesel despite being better than other alcohol fuels.
Fuel stability	**Strength**	Fuel stability can be increased when used as blends.	Minimal engine and fuel delivery modifications are required.	Less hygroscopicity than ethanol.	Lower hygroscopicity than ethanol thus maintaining better fuel stability.	Less hygroscopicity than methanol, ethanol, and propanol.
	Weakness	Highly reactive.	Poor fuel stability as a blend with diesel.	Oxidative degradation can occur over time.	Degrades more than diesel when exposed to light and air.	Prone to oxidation when exposed to light and air.

Propanol blends need examination for their stability, emission characteristics, and ability to enhance fuel injection processes and combustion efficiency [110]. Butanol stands out due to its higher CN, low vapor pressure, and high miscibility with diesel, making it a better additive than ethanol or methanol. Research should focus on butanol's effects on NO_x emissions, its role in dual-fuel injection strategies, and the use of branched-chain higher alcohols (e.g., iso-butanol, 1-butanol) to improve fuel properties and reduce emissions. The impact of butanol blends on engine performance metrics such as BTE, BSFC, and exhaust gas temperature also warrants study. Pentanol's potential in polyblends with higher and lower alcohols should be explored to improve mixing stability and reduce emissions, while investigating its effects on low-temperature ignition and main combustion processes. Overall, extensive research on alcohol-diesel blends is required to determine the optimum percentages for stability and performance, develop new fuel injection techniques to enhance engine performance and reduce emissions, and utilize fuel additives and CN improvers to improve the properties and stability of these blends [26, 111, 112]. Further experimentation at low temperatures is necessary to understand the behavior of alcohol-diesel blends and identify the best additive combinations for next-generation fuels. Conduct extensive research on alcohol-diesel blends to determine the optimum percentages for stability and performance. These can be summarized individually as below:

Methanol:

- Experimentation at low temperatures is necessary to understand the behavior of methanol-diesel blends and identify the best additive combinations for next-generation fuels.
- Study methanol's role in reducing engine emissions due to its higher oxygen content compared to fossil fuels.
- Address phase separation issues by mixing methanol with propanol/isopropanol in methanol/diesel blends.

Ethanol:

- Explore ethanol's potential as a co-solvent in diesel blends to improve mixing stability.
- Evaluate the impact of ethanol-diesel blends on engine emissions, particularly NO_x, PM, and HC [113, 114].
- Assess ethanol's performance in various advanced combustion modes such as RCCI, HCCI, and PCCI.

Propanol:

- Examine the stability and emission characteristics of propanol in methanol/diesel and ethanol/diesel blends.
- Study propanol's potential to enhance fuel injection processes and combustion efficiency.
- Evaluate the impact of propanol blends on NO_x, PM, and other exhaust emissions.

Butanol:

- Investigate butanol's advantages over ethanol and methanol due to its higher CN, low vapor pressure, and high miscibility with diesel.
- Study the effects of butanol on NO_x emissions and its role in dual-fuel injection strategies.
- Explore the use of branched-chain higher alcohols (e.g., iso-butanol, 1-butanol) in diesel engines to improve fuel properties and reduce emissions.
- Examine the impact of butanol blends on engine performance metrics such as BTE, BSFC, and exhaust gas temperature.

Pentanol:

- Explore pentanol use in polyblends with higher and lower alcohols to improve mixing stability and reduce emissions.
- Assess the effects of pentanol on low-temperature ignition and main combustion processes.
- Investigate the role of pentanol blends in enhancing engine performance and lowering NO_x and PM emissions in CI engines.
- Develop new fuel injection techniques to enhance diesel engine performance and reduce emissions.
- Utilize fuel additives and CN improvers to enhance the fuel properties and stability of all above-mentioned alcohol-diesel blends.
- Perform further experimentation at low temperatures to understand the behavior of alcohol-diesel blends and identify the best additive combinations for next-generation fuels.

REFERENCES

1. Leach, F., Kalghatgi, G., Stone, R., & Miles, P. (2020). The scope for improving the efficiency and environmental impact of internal combustion engines. *Transportation Engineering*, *1*, 100005.
2. Khan, M. I., & Amin, A. (2024). Exploring the impact of smog on healthy male rat physiology: A holistic approach integrating behavior, immunology, histopathology, and antibiotic-resistant gene.
3. Ghadikolaei, M. A., Wong, P. K., Cheung, C. S., Ning, Z., Yung, K. F., Zhao, J., …, & Berenjestanaki, A. V. (2021). Impact of lower and higher alcohols on the physicochemical properties of particulate matter from diesel engines: A review. *Renewable and Sustainable Energy Reviews*, *143*, 110970.
4. Abu-Qudais, M., Haddad, O., & Qudaisat, M. (2000). The effect of alcohol fumigation on diesel engine performance and emissions. *Energy Conversion and Management*, *41*(4), 389–399.
5. Kumar, B. R., & Saravanan, S. (2016). Use of higher alcohol biofuels in diesel engines: A review. *Renewable and Sustainable Energy Reviews*, *60*, 84–115.
6. Mahmudul, H. M., Hagos, F. Y., Mamat, R., Abdullah, A. A., & Awad, O. I.. (2016, November). Experimental investigation of the impact of using alcohol-biodiesel-diesel blending fuel on combustion of single cylinder CI engine. In IOP Conference Series: Materials Science and Engineering (Vol. 160, No. 1, p. 012038). IOP Publishing.
7. Wei, L., Yao, C., Wang, Q., Pan, W., & Han, G. (2015). Combustion and emission characteristics of a turbocharged diesel engine using high premixed ratio of methanol and diesel fuel. *Fuel*, *140*, 156–163.

8. Aydin, H., & Ilkılıc, C. (2010). Effect of ethanol blending with biodiesel on engine performance and exhaust emissions in a CI engine. *Applied Thermal Engineering*, *30*(10), 1199–1204.
9. Zhu, M., Ma, Y., & Zhang, D. (2012). Effect of a homogeneous combustion catalyst on the combustion characteristics and fuel efficiency in a diesel engine. *Applied Energy*, *91*(1), 166–172.
10. Işık, M. Z., & Aydın, H. (2016). Analysis of ethanol RCCI application with safflower biodiesel blends in a high load diesel power generator. *Fuel*, *184*, 248–260.
11. Zhao, P., Zeng, L., Li, P., Lu, H., Hu, H., Li, C., …, & Qi, Y. (2022). China's transportation sector carbon dioxide emissions efficiency and its influencing factors based on the EBM DEA model with undesirable outputs and spatial Durbin model. *Energy*, *238*, 121934.
12. Niculescu, R., Clenci, A., & Iorga-Siman, V. (2019). Review on the use of diesel–biodiesel–alcohol blends in compression ignition engines. *Energies*, *12*(7), 1194.
13. Mužíková, Z., Pospíšil, M., & Šebor, G. (2009). Volatility and phase stability of petrol blends with ethanol. *Fuel*, *88*(8), 1351–1356.
14. Milpied, J., Jeuland, N., Plassat, G., Guichaous, S., Dioc, N., Marchal, A., & Schmelzle, P. (2009). Impact of fuel properties on the performances and knock behaviour of a downsized turbocharged DI SI engine–focus on octane numbers and latent heat of vaporization. *SAE International Journal of Fuels and Lubricants*, *2*(1), 118–126.
15. Shamun, S., Shen, M., Johansson, B., Tuner, M., Pagels, J., Gudmundsson, A., & Tunestal, P. (2016). Exhaust PM emissions analysis of alcohol fueled heavy-duty engine utilizing PPC. *SAE International Journal of Engines*, *9*(4), 2142–2152.
16. Saravanan, S., Kumar, B. R., Varadharajan, A., Rana, D., & Sethuramasamyraja, B. (2017). Optimization of DI diesel engine parameters fueled with iso-butanol/diesel blends–response surface methodology approach. *Fuel*, *203*, 658–670.
17. Kumar, B. R., & Saravanan, S. (2016). Effects of iso-butanol/diesel and n-pentanol/diesel blends on performance and emissions of a DI diesel engine under premixed LTC (low temperature combustion) mode. *Fuel*, *170*, 49–59.
18. Goeppert, A., Czaun, M., Jones, J. P., Prakash, G. S., & Olah, G. A. (2014). Recycling of carbon dioxide to methanol and derived products–closing the loop. *Chemical Society Reviews*, *43*(23), 7995–8048.
19. Coulier, J., & Verhelst, S.. (2016). Using alcohol fuels in dual fuel operation of compression ignition engines: a review. In 28th CIMAC World Congress on Combustion Engine (pp. 1–12). CIMAC.
20. Yin, Z., Yao, C., Geng, P., & Hu, J. (2016). Visualization of combustion characteristic of diesel in premixed methanol–air mixture atmosphere of different ambient temperature in a constant volume chamber. *Fuel*, *174*, 242–250.
21. Cheung, C. S., Zhang, Z. H., Chan, T. L., & Yao, C. (2009). Investigation on the effect of port-injected methanol on the performance and emissions of a diesel engine at different engine speeds. *Energy & Fuels*, *23*(11), 5684–5694.
22. Richards, B. G. (1990). Methanol-fueled caterpillar 3406 engine experience in on-highway trucks. *SAE Transactions*, *99-4*, 1033–1045.
23. Fathi, M., Saray, R. K., & Checkel, M. D. (2011). The influence of exhaust gas recirculation (EGR) on combustion and emissions of n-heptane/natural gas fueled homogeneous charge compression ignition (HCCI) engines. *Applied Energy*, *88*(12), 4719–4724.
24. Valentino, G., Corcione, F. E., Iannuzzi, S. E., & Serra, S. (2012). Experimental study on performance and emissions of a high speed diesel engine fuelled with n-butanol diesel blends under premixed low temperature combustion. *Fuel*, *92*(1), 295–307.
25. Jackson, M. D., Unnasch, S., Sullivan, C., & Renner, R. A. (1985). Transit bus operation with methanol fuel. *SAE Transactions*, 50–61.
26. Verhelst, S., Turner, J. W., Sileghem, L., & Vancoillie, J. (2019). Methanol as a fuel for internal combustion engines. *Progress in Energy and Combustion Science*, *70*, 43–88.

27. Urban, C. M., Timbario, T. J., & Bechtold, R. L.. (1989). Performance and emissions of a DDC 8V-71 engine fueled with cetane improved methanol (No. 892064). SAE Technical Paper.
28. Liu, J., Li, Y., Li, G., Zhu, Z., He, H., & Liu, S. (2010). Effect of pilot diesel quantity and fuel delivery advance angle on the performance and emission characteristics of a methanol-fueled diesel engine. *Energy & Fuels*, *24*(3), 1611–1616.
29. Yuen, P. K. P. K., Beckett, J., & Villaire, W. (2010). Automotive materials engineering challenges and solutions for the use of ethanol and methanol blended fuels. *Green Technologies and the Mobility Industry, 81.*
30. Bromberg, L., & Cheng, W. K.. (2010). *Methanol as an alternative transportation fuel in the US: Options for sustainable and/or energy-secure transportation.* Sloan Automotive Laboratory, Massachusetts Institute of Technology.
31. Turner, J. W. G., Pearson, R. J., Dekker, E., Iosefa, B., Johansson, K., & Ac Bergström, K. (2013). Extending the role of alcohols as transport fuels using iso-stoichiometric ternary blends of gasoline, ethanol and methanol. *Applied Energy*, *102*, 72–86.
32. Nakata, K., Utsumi, S., Ota, A., Kawatake, K., Kawai, T., & Tsunooka, T.. (2006). The effect of ethanol fuel on a spark ignition engine (No. 2006-01-3380). SAE Technical Paper.
33. Marriott, C. D., Wiles, M. A., Gwidt, J. M., & Parrish, S. E. (2009). Development of a naturally aspirated spark ignition direct-injection flex-fuel engine. *SAE International Journal of Engines*, *1*(1), 267–295.
34. Aikawa, K., Sakurai, T., & Hayashi, A. (2009). Study of ethanol-blended fuel (E85) effects under cold-start conditions. *SAE International Journal of Fuels and Lubricants*, *2*(1), 196–203.
35. Xiang, H.. (2015). Introduction of jet sun methanol fuel additives and engine oils. In Tech. Rep. Guangzhou Jetsun Lubrication Technology Co., Ltd.
36. Pidol, L., Lecointe, B., Starck, L., & Jeuland, N. (2012). Ethanol–biodiesel–diesel fuel blends: Performances and emissions in conventional diesel and advanced low temperature combustions. *Fuel*, *93*, 329–338.
37. Shahir, S. A., Masjuki, H. H., Kalam, M. A., Imran, A., Fattah, I. R., & Sanjid, A. (2014). Feasibility of diesel–biodiesel–ethanol/bioethanol blend as existing CI engine fuel: An assessment of properties, material compatibility, safety and combustion. *Renewable and Sustainable Energy Reviews*, *32*, 379–395.
38. Azad, V. K., De, K., & Majumder, S. (2024). Ethanol blending and its environmental impacts: A case study of India. *Energy for Sustainable Development*, *79*, 101385.
39. Çakmak, A. (2024). Feasibility of complete substitution of petroleum diesel with biofuels in diesel engines: An experimental assessment of combustion, performance, and emissions characteristics. *Arabian Journal for Science and Engineering*, *49*(2), 2367–2387.
40. Domínguez, V. M., Hernández, J. J., Ramos, Á, Giménez, B., & Rodríguez-Fernández, J. (2024). Exploring the effect of methanol and ethanol on the overall performance and substitution window of a dual-fuel compression-ignition engine fueled with HVO. *Fuel*, *359*, 130529.
41. Suresh, P., Suja Mani Malard, R., Ram Kumarc, A., & Dhakshnamurthi, P. (2024). Effects of ethanol fumigation on the performance and emissions of diesel engines. *Physical Chemistry Research*, *12*(1), 135–144.
42. Lungu, J., Siwale, L., & Kashinga, R. J. (2024). Performance, combustion and emission characteristics of oxygenated diesel in DI engines: A critical review. *Journal of Power and Energy Engineering*, *12*(6), 16–49.
43. Nibin, M., Varuvel, E. G., JS, F. J., & Vikneswaran, M. (2024). Evaluation of wheat germ oil biofuel in diesel engine with hydrogen, bioethanol dual fuel and fuel ionization strategies. *International Journal of Hydrogen Energy*, *59*, 889–902.
44. Feng, S., Zhang, S., Zhang, H., & Shi, J. (2024). Effect of nozzle geometry on combustion of a diesel-methanol dual-fuel direct injection engine. *Fuel*, *357*, 129734.

45. Gawale, G. R., & Srinivasulu, G. N. (2020). Experimental investigation of ethanol/diesel and ethanol/biodiesel on dual fuel mode HCCI engine for different engine load conditions. *Fuel, 263*, 116725.
46. Mohammadi, A., Kee, S. S., Ishiyama, T., Kakuta, T., & Matsumoto, T. (2005). Implementation of ethanol diesel blend fuels in PCCI combustion (No. 2005-01-3712). SAE Technical Paper.
47. Kumar, P., & Sandhu, S. S. (2021). Impact analysis of partially premixed combustion strategy on the emissions of a compression ignition engine fueled with higher octane number fuels: A review. *Materials Today: Proceedings, 45*, 5772–5777.
48. Dempsey, A. B., Walker, N. R., & Reitz, R. (2013). Effect of cetane improvers on gasoline, ethanol, and methanol reactivity and the implications for RCCI combustion. *SAE International Journal of Fuels and Lubricants, 6*(1), 170–187.
49. Boretti, A. (2012). Advantages of converting diesel engines to run as dual fuel ethanol–diesel. *Applied Thermal Engineering, 47*, 1–9.
50. Lee, J., Lee, S., & Lee, S. (2018). Experimental investigation on the performance and emissions characteristics of ethanol/diesel dual-fuel combustion. *Fuel, 220*, 72–79.
51. Han, J., Somers, L. M. T., Cracknell, R., Joedicke, A., Wardle, R., & Mohan, V. R. R. (2020). Experimental investigation of ethanol/diesel dual-fuel combustion in a heavy-duty diesel engine. *Fuel, 275*, 117867.
52. Reitz, R. D., & Duraisamy, G. (2015). Review of high efficiency and clean reactivity-controlled compression ignition (RCCI) combustion in internal combustion engines. *Progress in Energy and Combustion Science, 46*, 12–71.
53. Tsang, K. S., Zhang, Z. H., Cheung, C. S., & Chan, T. L. (2010). Reducing emissions of a diesel engine using fumigation ethanol and a diesel oxidation catalyst. *Energy & Fuels, 24*(11), 6156–6165.
54. Chuepeng, S., Pirompugd, W., & Sutheerasak, E. (2023). Performance and emissions of a diesel engine fueled with palm oil ethyl ester combined with fumigated ethanol on a dual fuel mode. *Energy Reports, 9*, 470–477.
55. Chauhan, B. S., Kumar, N., Pal, S. S., & Du Jun, Y. (2011). Experimental studies on fumigation of ethanol in a small capacity diesel engine. *Energy, 36*(2), 1030–1038.
56. Şen, M. (2019). The effect of the injection pressure on single cylinder diesel engine fueled with propanol–diesel blend. *Fuel, 254*, 115617.
57. Atmanli, A. (2016). Comparative analyses of diesel–waste oil biodiesel and propanol, n-butanol or 1-pentanol blends in a diesel engine. *Fuel, 176*, 209–215.
58. Zhao, W., Yan, J., Gao, S., Lee, T. H., & Li, X. (2022). The combustion and emission characteristics of a common-rail diesel engine fueled with diesel, propanol, and pentanol blends under low intake pressures. *Fuel, 307*, 121692.
59. Moreau, A., Sobrino, M., Zambrano, J., Segovia, J. J., Villamañán, M. A., & Martín, M. C. (2021). Viscosities and densities of different alcohols (1-propanol, 2-propanol, 1-pentanol and 2-pentanol) at high pressures. *Journal of Molecular Liquids, 344*, 117744.
60. Konjević, L., Racar, M., Ilinčić, P., & Faraguna, F. (2023). A comprehensive study on application properties of diesel blends with propanol, butanol, isobutanol, pentanol, hexanol, octanol and dodecanol. *Energy, 262*, 125430.
61. Zare, A., Bodisco, T. A., Verma, P., Jafari, M., Babaie, M., Yang, L., …, & Stevanovic, S. (2022). Particulate number emissions during cold-start with diesel and biofuels: A special focus on particle size distribution. *Sustainable Energy Technologies and Assessments, 51*, 101953.
62. Wang, L., Zhang, J., Liu, L., Wang, Y., Chai, L., Yang, Z., …, & Tan, C. (2015). Stability and thermophysical properties of binary propanol–water mixtures-based microencapsulated phase change material suspensions. *Journal of Heat Transfer, 137*(9), 091019.
63. Bencheikh, K., Atabani, A. E., Shobana, S., Mohammed, M. N., Uğuz, G., Arpa, O., …, & Bokhari, A. (2019). Fuels properties, characterizations and engine and

emission performance analyses of ternary waste cooking oil biodiesel–diesel–propanol blends. *Sustainable Energy Technologies and Assessments, 35*, 321–334.
64. Sarathy, S. M., Vranckx, S., Yasunaga, K., Mehl, M., Oßwald, P., Metcalfe, W. K., …, & Curran, H. J. (2012). A comprehensive chemical kinetic combustion model for the four butanol isomers. *Combustion and Flame, 159*(6), 2028–2055.
65. Kumar, P., & Sandhu, S. S. (2023). An experimental investigation into the combustion stability and emissions of an n-butanol/diesel blended fueled partially premixed compression ignition (PPCI) engine. *International Journal of Engine Research, 24*(6), 2451–2464.
66. Nayyar, A., Sharma, D., Soni, S. L., & Mathur, A. (2017). Experimental investigation of performance and emissions of a VCR diesel engine fuelled with n-butanol diesel blends under varying engine parameters. *Environmental Science and Pollution Research, 24*, 20315–20329.
67. Şahin, Z., & Aksu, O. N. (2015). Experimental investigation of the effects of using low ratio n-butanol/diesel fuel blends on engine performance and exhaust emissions in a turbocharged DI diesel engine. *Renewable Energy, 77*, 279–290.
68. Doğan, O. (2011). The influence of n-butanol/diesel fuel blends utilization on a small diesel engine performance and emissions. *Fuel, 90*(7), 2467–2472.
69. Karabektas, M., & Hosoz, M. (2009). Performance and emission characteristics of a diesel engine using isobutanol–diesel fuel blends. *Renewable Energy, 34*(6), 1554–1559.
70. Singh, R., & Kumar, R. (2022). Insights into the influence of n-butanol with neat biodiesel and biodiesel-diesel blends on diesel engine characteristics. *International Journal of Energy Research, 46*(5), 5441–5466.
71. Cooksley, C. M., Zhang, Y., Wang, H., Redl, S., Winzer, K., & Minton, N. P. (2012). Targeted mutagenesis of the *Clostridium acetobutylicum* acetone–butanol–ethanol fermentation pathway. *Metabolic Engineering, 14*(6), 630–641.
72. Painter, D. S. (2014). Oil and geopolitics: The oil crises of the 1970s and the Cold War. Historical Social Research/Historische Sozialforschung, *39*(4), 186–208.
73. Agriculture Organization, & Agriculture Organization of the United Nations Staff (2000). *Commodity market review 1999-2000*. Food & Agriculture Org.
74. Rakopoulos, D. C., Rakopoulos, C. D., Giakoumis, E. G., Dimaratos, A. M., & Kyritsis, D. C. (2010). Effects of butanol–diesel fuel blends on the performance and emissions of a high-speed DI diesel engine. *Energy Conversion and Management, 51*(10), 1989–1997.
75. Rakopoulos, D. C., Rakopoulos, C. D., Hountalas, D. T., Kakaras, E. C., Giakoumis, E. G., & Papagiannakis, R. G. (2010). Investigation of the performance and emissions of bus engine operating on butanol/diesel fuel blends. *Fuel, 89*(10), 2781–2790.
76. Banerjee, S. S., & Bhowmick, A. K. (2015). Tailored nanostructured thermoplastic elastomers from polypropylene and fluoroelastomer: Morphology and functional properties. *Industrial & Engineering Chemistry Research, 54*(33), 8137–8146.
77. Dhanumalayan, E., & Joshi, G. M. (2018). Performance properties and applications of polytetrafluoroethylene (PTFE)—A review. *Advanced Composites and Hybrid Materials, 1*, 247–268.
78. Ismail, H., Pasbakhsh, P., Fauzi, M. A., & Bakar, A. A. (2008). Morphological, thermal and tensile properties of halloysite nanotubes filled ethylene propylene diene monomer (EPDM) nanocomposites. *Polymer Testing, 27*(7), 841–850.
79. Liu, J., Li, X., Xu, L., & Zhang, P. (2016). Investigation of aging behavior and mechanism of nitrile-butadiene rubber (NBR) in the accelerated thermal aging environment. *Polymer Testing, 54*, 59–66.
80. Zhang, G., Li, W. Y., Cherigui, M., Zhang, C., Liao, H., Bordes, J. M., & Coddet, C. (2007). Structures and tribological performances of PEEK (poly-ether-ether-ketone)-based coatings designed for tribological application. *Progress in Organic Coatings, 60*(1), 39–44.

81. Shit, S. C., & Shah, P. (2013). A review on silicone rubber. *National Academy Science Letters, 36*(4), 355–365.
82. McCollum, J., Morey, A. M., & Iacono, S. T. (2017). Morphological and combustion study of interface effects in aluminum-poly (vinylidene fluoride) composites. *Materials & Design, 134*, 64–70.
83. Fornes, T. D., & Paul, D. R. (2004). Structure and properties of nanocomposites based on nylon-11 and-12 compared with those based on nylon-6. *Macromolecules, 37*(20), 7698–7709.
84. Pinzi, S., Redel-Macías, M. D., Carmona-Cabello, M., Cubero, A., Herreros, J. M., & Dorado, M. P. (2021). Influence of 1-butanol and 1-pentanol addition to diesel fuel on exhaust and noise emissions under stationary and transient conditions. *Fuel, 301*, 121046.
85. Campos-Fernández, J., Arnal, J. M., Gómez, J., & Dorado, M. P. (2012). A comparison of performance of higher alcohols/diesel fuel blends in a diesel engine. *Applied Energy, 95*, 267–275.
86. Campos-Fernandez, J., Arnal, J. M., Gomez, J., Lacalle, N., & Dorado, M. P. (2013). Performance tests of a diesel engine fueled with pentanol/diesel fuel blends. *Fuel, 107*, 866–872.
87. Yaman, H., Yesilyurt, M. K., Shah, R. M. R. A., & Soyhan, H. S. (2024). Effects of compression ratio on thermodynamic and sustainability parameters of a diesel engine fueled with methanol/diesel fuel blends containing 1-pentanol as a co-solvent. *Fuel, 357*, 129929.
88. Wei, L., Cheung, C. S., & Huang, Z. (2014). Effect of n-pentanol addition on the combustion, performance and emission characteristics of a direct-injection diesel engine. *Energy, 70*, 172–180.
89. Saravanan, S. (2015). Effect of exhaust gas recirculation (EGR) on performance and emissions of a constant speed DI diesel engine fueled with pentanol/diesel blends. *Fuel, 160*, 217–226.
90. Muthuraman, V. S., Patel, A., Shreya, V., Vaidyanathan, A., Reshwanth, K. N. G. L., Karthick, C., …, & Nanthagopal, K. (2022). Progress on compatibility issues of alcohols on automotive materials: Kinetics, challenges and future prospects-a comprehensive review. *Process Safety and Environmental Protection, 162*, 463–493.
91. Primorac, T., Požar, M., Sokolić, F., Zoranić, L., & Urbic, T. (2018). A simple two dimensional model of methanol. *Journal of Molecular Liquids, 262*, 46–57.
92. Kumar, T. S., & Ashok, B. (2023). Material compatibility of SI engine components towards corrosive effects on methanol-gasoline blends for flex fuel applications. *Materials Chemistry and Physics, 296*, 127344.
93. Liang, Z., Liu, H., Fan, Y., Salehi, F., Zhang, Z., & Wang, C. (2024). Applying fischer tropsch and its pentanol blends into an aviation compression ignition engine for PM emissions control. *Journal of the Energy Institute, 116*, 101742.
94. Pan, M., Huang, R., Liao, J., Jia, C., Zhou, X., Huang, H., & Huang, X. (2019). Experimental study of the spray, combustion, and emission performance of a diesel engine with high n-pentanol blending ratios. *Energy Conversion and Management, 194*, 1–10.
95. Sabu, V. R., Thomas, J. J., & Nagarajan, G. (2020). Experimental investigation on the effects of multiple injections and EGR on n-pentanol–biodiesel fuelled RCCI engine. *RSC Advances, 10*(49), 29498–29509.
96. Liang, J., Zhang, Q., Chen, Z., & Zheng, Z. (2021). The effects of EGR rates and ternary blends of biodiesel/n-pentanol/diesel on the combustion and emission characteristics of a CRDI diesel engine. *Fuel, 286*, 119297.
97. No, S. Y. (2020). Utilization of pentanol as biofuels in compression ignition engines. *Frontiers in Mechanical Engineering, 6*, 15.
98. Yan, F., Zhong, W., Xiang, Q., Pachiannan, T., Wang, W., He, Z., & Wang, Q. (2023). Experimental investigation of the two-stage ignition delay and flame structure of pentanol/n-dodecane binary fuel. *Energy, 262*, 125546.

99. Dong, R., Yu, Q., Bai, Y., Wu, Y., Ma, Z., Zhang, J., …, & Cai, M. (2020). Towards superior lubricity and anticorrosion performances of proton-type ionic liquids additives for water-based lubricating fluids. *Chemical Engineering Journal*, *383*, 123201.
100. Dong, R., Bao, L., Yu, Q., Wu, Y., Ma, Z., Zhang, J., …. & Liu, W. (2020). Effect of electric potential and chain length on tribological performances of ionic liquids as additives for aqueous systems and molecular dynamics simulations. *ACS Applied Materials & Interfaces*, *12*(35), 39910–39919.
101. Sun, X., Liu, X., Chen, Q., Guo, P., Bai, Y., Zou, K., …, & Liu, W. (2024). Amino acid-based ionic liquids as water-ethylene glycol additives towards superior lubricity and corrosion resistance. *Journal of Molecular Liquids*, *401*, 124706.
102. Kreivaitis, R., Gumbytė, M., Kupčinskas, A., Kazancev, K., Ta, T. N., & Horng, J. H. (2020). Investigation of tribological properties of two protic ionic liquids as additives in water for steel–steel and alumina–steel contacts. *Wear*, *456*, 203390.
103. Avilés, M. D., Carrión, F. J., Sanes, J., & Bermúdez, M. D. (2018). Effects of protic ionic liquid crystal additives on the water-lubricated sliding wear and friction of sapphire against stainless steel. *Wear*, *408*, 56–64.
104. Santhosh, K., Kumar, G. N., & Sanjay, P. V. (2020). Experimental analysis of performance and emission characteristics of CRDI diesel engine fueled with 1-pentanol/diesel blends with EGR technique. *Fuel*, *267*, 117187.
105. Odziemkowska, M., Czarnocka, J., & Wawryniuk, K. (2018). Study of stability changes of model fuel blends. *Improvement Trends for Internal Combustion Engines*, *1*, 1–17.
106. Yilmaz, N., Vigil, F. M., & Donaldson, B. (2022). Fuel effects on PAH formation, toxicity and regulated pollutants: Detailed comparison of biodiesel blends with propanol, butanol and pentanol. *Science of the Total Environment*, *849*, 157839.
107. Lamba, B. Y., Joshi, G., Tiwari, A. K., Rawat, D. S., & Mallick, S. (2013). Effect of antioxidants on physico-chemical properties of EURO-III HSD (high speed diesel) and Jatropha biodiesel blends. *Energy*, *60*, 222–229.
108. Pan, M., Huang, R., Liao, J., Ouyang, T., Zheng, Z., Lv, D., & Huang, H. (2018). Effect of EGR dilution on combustion, performance and emission characteristics of a diesel engine fueled with n-pentanol and 2-ethylhexyl nitrate additive. *Energy Conversion and Management*, *176*, 246–255.
109. Hua, Y. (2024). Research progress of higher alcohols as alternative fuels for compression ignition engines. *Fuel*, *357*, 129749.
110. Çelebi, Y., & Aydın, H. (2019). An overview on the light alcohol fuels in diesel engines. *Fuel*, *236*, 890–911.
111. Jeyaseelan, T., Ekambaram, P., Subramanian, J., & Shamim, T. (2022). A comprehensive review on the current trends, challenges and future prospects for sustainable mobility. *Renewable and Sustainable Energy Reviews*, *157*, 112073.
112. Kumar, S., Cho, J. H., Park, J., & Moon, I. (2013). Advances in diesel–alcohol blends and their effects on the performance and emissions of diesel engines. *Renewable and Sustainable Energy Reviews*, *22*, 46–72.
113. Gupta, S., Kanchan, S., Kaur, R., & Sandhu, S. S. (2024). Investigation of performance and emission characteristics using ethanol-blended gasoline fuel as a flex-fuel in two-wheeler vehicle mounted on a chassis dynamometer. *Clean Energy*, *8*(3), 174–193.
114. Kanchan, S., Pradhan, S., Kumar, R., Sharma, S., Bhandari, O., Priyadarshini, M., …, & Dhiman, R. (2024). Developing a model for waste plastic biofuels in CRDi diesel engines using FTIR, GCMS, and WASPAS synchronisations for engine analysis. *Energy Exploration & Exploitation*, *42*(2), 648–684.

Index

For Product Safety Concerns and Information please contact our
EU representative GPSR@taylorandfrancis.com Taylor & Francis
Verlag GmbH, Kaufingerstraße 24, 80331 München, Germany